THUNDER OVER DAKOTA

The Complete History of Ellsworth Air Force Base, South Dakota

Lt. Col. George A. Larson, USAF (Ret.)

Schiffer Publishing Ltd®

4880 Lower Valley Road • Atglen, PA 19310

DEDICATION

When dealing with a book which covers 70 years of history, dedicating the book to one individual is not possible. Ellsworth Air Force Base was born during the build up for America's entry into WWII, and has played an important role in the defense of the United States through the decades and the nation's many conflicts until today. Therefore, I dedicate this book to all the officers, enlisted personnel, civil service personnel, spouses, and the citizens of western South Dakota that made this history possible.

Designed by Mark David Bowyer

Printed in China.
ISBN: 978-0-7643-4263-9

We are interested in hearing from authors with book ideas on related topics.

Published by Schiffer Publishing Ltd.
4880 Lower Valley Road
Atglen, PA 19310
Phone: (610) 593-1777
FAX: (610) 593-2002
E-mail: Info@schifferbooks.com.
Visit our web site at: **www.schifferbooks.com**
Please write for a free catalog.
This book may be purchased from the publisher.
Try your bookstore first.

In Europe, Schiffer books are distributed by:
Bushwood Books
6 Marksbury Avenue
Kew Gardens
Surrey TW9 4JF, England
Phone: 44 (0) 20 8392-8585
FAX: 44 (0) 20 8392-9876
E-mail: Info@bushwoodbooks.co.uk.
Visit our website at: www.bushwoodbooks.co.uk

CONTENTS

ACKNOWLEDGMENTS

I started pulling in information on Ellsworth Air Force Base while still on active duty with the Strategic Air Command, stationed at Offutt Air Force Base, 544th Technical Reconnaissance Wing, Deputy Commander for Resource Management. In my position I helped support Ellsworth Air Force Base during set up of a new intelligence facility. I moved to Rapid City in 1994, and with my military and aviation writing experience, since retiring from the Air Force in 1992, I started writing articles on Ellsworth Air Force Base. I was assisted in my writings by a series of 28th Bomb Wing historians, which culminated with the current historian, Paul Marcello. His grandfather, Al Marcello, trained on B-17s during WWII before shipping out with the 100th Bomb Wing to the Eighth Air Force, England. The span of seventy years covers many different people. I drew accounts of the B-29s from my book *The Superfortress and Its Final Glory: The Korean Air War: The Cold War's First Aerial Combat.* When doing the research on the Convair B-36s assigned to the base, I was fortunate to have the assistance of a friend and writer in Rapid City, John F. Welch, who allowed me to use sections of his book *RB-36 Days at Rapid City.* The naming of the base for Brigadier General Ellsworth after he crashed onboard an RB-36H on March 18, 1953, is a centerpiece of my book. Richard Stocker of St. John's, Newfoundland, climbed to the hill at Burgoyne's Cove, site of the RB-36H crash, and provided photographs. When discussing the building of the Intercontinental Ballistic Missiles and their deployment around the base, I was fortunate in gaining information on the Titan I ICBM from a fellow military historian, Milton R. Kuhl, Jr., currently living in Fort Bragg, CA. He was assigned to the contractor responsible for building the Titan I complexes around Ellsworth. The information on the installation of the Minuteman I and II ICBMs at the base was provided by the Association of Air Force Missileers of Breckenridge, Colorado. The Strategic Air Command Historian Office provided information on SAC B-52 and KC-135 alert on Ellsworth while I was writing a story on the Vietnam air war. The 28th Bomb Wing Command Section assisted in gaining approval from Air Command to research and write a book on Ellsworth Air Force Base, including the current B-1B mission. The last was Colonel Jeffrey B. Taliaferro, currently assigned to the Pentagon, Office of the Secretary of Defense. Many past and present airmen assigned to Ellsworth Air Force Base provided information and photographs for this book. Seventy years is a long time for a military installation to exist and remain an important contributor to the defense of the United States.

Lt. Col. George A. Larson, USAF (Ret.)

FOREWORD

When assigned to command Ellsworth Air Force Base in 2009, I looked extensively for a resource that captured the history of the 28th Bomb Wing, the base itself, and its namesake, Brig. Gen. Ellsworth. Unfortunately, not much was available in the public domain to capture the nearly 70 years of hard work, accomplishments, and sacrifices by its airmen and community alike. In this work, George Larson has compiled large amounts of raw data from a variety of hard to reach archives into one comprehensive resource, finally making this rich history available to the reader.

Here you'll find the stories of thousands of Ellsworth Air Force Base, 28th Bomb Wing, and Black Hills partners that form a powerful piece of Air Force heritage. The colorful stories of training B-17 crews for WWII, bombing the Japanese out of the Aleutian islands, interdiction and strategic attack in Vietnam, and continuing combat action worldwide in the B-1 are matched only by those of the Black Hills community donating land, pouring the concrete of the Pride Hangar roof, or gathering by the thousands to show their support of the Ellsworth mission.

In many ways, Ellsworth Air Force Base is the story of the United States Air Force growing up from early aviation, and then changing with the world's situation; through it all was a team of airmen and community partners dedicated to its particular mission and the broader defense of this nation – a story that most definitely deserves telling.

Colonel Jeffery Taliaferro, USAF
Principal Military Assistant to the Secretary of Defense

Chapter One
Construction of Rapid City Army Air Base

In the late 1930s, with the possibility of war rising in Europe due to the increasing militarism of Nazi Germany and its ally Italy, the United States government began to make the appropriate preparations. At the same time, threats to U.S. security increased from the Empire of Japan in the Pacific, which eventually joined Germany and Italy under the Tripartite Pact, signed on September 27, 1940. On January 8, 1941, looking at an inevitable future involvement in a European war, President Franklin D. Roosevelt requested a 17 billion dollar budget from the U.S. Congress to rearm, equip, and train the U.S. military for a world war (Europe and the Pacific Theaters of Operation). The political leadership of South Dakota recognized the economic impact its possible share in a major rearmament would bring to their state. U.S. Senators Chan Gurney and Karl Mundt, with U.S. Congressman Francis Case, were instrumental in helping to secure an Army air base in western South Dakota.[1]

Rock hangar at Rapid City's Municipal Airport. The twin-engine aircraft to the right of the photo is a Boeing 247 of Inland Airlines, which flew twice daily passenger flights from Cheyenne, Wyoming, through Rapid City to Huron, meeting other main airlines at these terminus airfields. The hangar was located on an approximately one square mile area, with unrestricted expansions in all directions to accommodate an Army air base. In 1939, at the time of this photograph, the large hangar also contained offices for the local Federal Weather Bureau. This site was determined to be a logical candidate for the construction of an Army air base. *Ellsworth Air Force Base, 28th Bomb Wing Historian.*

The Rapid City Army Air Base (as it eventually was officially designated) was conceived as an Air Corps Tactical Unit as part of the expansion program of the Air Corps of the United States Army, being delegated to the 7th Corps area. The location of the base, 12 miles northeast of the city of Rapid City, South Dakota, was investigated on October 1 and 2, 1941, by a Site Board consisting of five commissioned officers of the Army Air Force. The Site Board, accompanied by three officials from the city of Rapid City, after due investigation, reported the site desirable on the basis of excellent terrain and flying conditions, with year round training feasible. The undesirable points were the distance from Rapid City, remote railroad facilities, and inadequate water supply projects. Rapid City committed itself to lease land to the federal government for $1.00 per year, renewable for 25 years, to prevent obstruction of a 40-to-1 glide angle on prolongation of runways. It would reroute major county roads to furnish water in the quantity of 1,000,000 gallons per minute at 50 pounds pressure, to acquire necessary railroad right-of-way, and to cause to be constructed a hard-surfaced highway from Old Highway 14/16 to the reservation boundary, all at no cost to the federal government. However, because of the undesirable characteristics, the recommendation of the Site Board was that the Rapid City site not be acquired as a station for any type of aviation.

An Engineering Board report was made to the Chief of Engineers, U.S. Army, Washington, D.C., dated October 17, 1941, with the subject "Report of Engineering Feasibility of Proposed Air Base Site at Rapid City, South Dakota," with additional information and recommendations briefly as follows:

> The City of Rapid City is well served by railroads, highways, air lines, telephone and telegraph; adequate recreational facilities for both officers and enlisted men are afforded; skilled and unskilled labor are available in sufficient quantities; housing is available for officer personnel, but a Government housing project for non-commissioned officers will be mandatory; power, water, gas and telephone facilities will be provided at the reservation boundary at no cost to the Government; utility rates have been tested and appear reasonable; a sewage disposal plant for complete treatment will be required; total estimated cost of construction is $4,630,700. The proposed site is exceptionally feasible from an engineering standpoint for the construction of an Air Corps Tactical Unit with substantial savings possible compared to other sites favorably recommended.[2]

On December 7, 1941, military forces of the Empire of Japan attacked Pearl Harbor, Hawaii, damaging U.S. military forces. The U.S. entered the war against Japan, quickly followed by a declaration of war against Germany and Italy. This was a world war, and the U.S. needed to dramatically increase the size of its military forces. During the build-up to fight a global war, the United States Army Air Force (USAAF) built airfields across the U.S. to train pilots and aircrew. Airfields in South Dakota were under the administration and control of Second Air Force or Army Air Force Training Command (AAFTC).[3]

Advance planning was started immediately after the Engineering Board report was submitted. Three field parties were dispatched from the Omaha District Office around December 10, 1941, under the supervision of Mr. A. Wooley, Associate Engineer. The project was transferred to the Fort Peck District on December 21, 1941, with Mr. E.A. Fabian, Associate Engineer, being placed in charge. Two of the parties from Omaha, along with two survey parties from the Fort Peck District, continued the preliminary survey work, hampered somewhat by cold weather and intermittent snow. A complete topography and proposed layout map of the site was forwarded by messenger to the District Office on January 1, 1942, and approved by the Army Air Force Site Board January 2, 1942. The general layout was approved by the Army Air Force Site Board January 2, 1942. The first directive authorizing construction was issued immediately following this approval.

A factor probably influencing the ultimate approval of the site and designation as a Light and Medium Bomber Crew Training Unit was the existence of an uninhabited area suitable for a Bombing and Gunnery Range in the immediate flying vicinity of the site.

Offices were established in Rapid City, and a drafting section was organized to produce plot survey notes. Design work was delegated as follows: the Fort Peck offices were to do general layouts and plans, airfield layout plans, and profiles; the Kansas City office was to do cantonment roads, layout plans, and profiles; and A.E. Ellerbe and Company, architectural engineers, were given a contract to do water, sewer, and electrical utilities layout, plans, and profiles.

Elmer H. Oechsle, Major, Corp of Engineers, was named Area Engineer January 7, 1942, and continued in this capacity until August 11, 1942, at which time he was transferred. Beryl E. Howard, 1st Lt., Corps of Engineers, was named Area Engineer August 12, 1942, and continued in this capacity until completion of the job.[4]

The *Rapid City Daily Journal* ran the following articles concerning the air base and its construction:

> Contracts are expected to be let within a month for the construction of an $8,500,000 Army air base for the training of bomber pilots and bombardiers near Rapid City.
>
> Announcement that Rapid City had been selected for one of seven such schools in the United States, with personnel of 4,000 men or over, was made in Washington Wednesday morning.

7

The plant, with the local municipal airport as a nucleus, will cover 1,440 acres, in addition to a 750-square mile gunnery range and bombing field south of the Badlands National Monument. It will have an annual payroll of approximately $2,000,000. The base is expected to be permanent. The sites are within 20 miles of the site of Army Air Corps stratosphere flights of a few years ago, culminating in the world record flight on Armistice Day, 1935, from the Strato Bowl, 11 miles from Rapid City.

Consideration had also been given to Watertown and Sioux Falls as possible sites for the air base, which was a development of the Rapid City Airport. Runways will be paved to a width of 150-feet and a length of 6,000-feet. A huge concrete apron will also be laid out and numerous other improvements made. Over 200 buildings, for barracks, five hangars, mess halls, shops, and other facilities will be built.

Most of the expansion of the airport is expected to be north and west of the present site, necessitating the addition of about 800 acres.

The gunnery range where bombing pilots can *let go* with aerial cannons as they dive bomb over practice targets will be somewhere south of the Badlands National Monument, far enough removed from the recreation area so as not to interfere with the tourist business.

To what extent Rapid City will have to contribute to the development of the project is not yet definitely known. The City Commission has committed itself to assist in the procuring of the extra land need at the local airport. The Government also needs water, approximately 1,000,000 gallons per day.

The base is expected to be completed within six months after contracts are let.

Size of the gigantic project and what it means for South Dakota may be gleaned from the fact that $8,500,000 is as much as the Legislature appropriates for a year for all State government expenditures it controls. Paving the runways will require about 300,000 sacks of cement, or about 300 carloads of cement alone, and the apron will require a like amount.

Other facilities will also have to be provided. The Black Hills Power and Light Company and the Montana-Dakota Utilities Company have agreed to extend their power and gas systems, respectively, to the field.

The air base will not interfere with regularly scheduled, radio-controlled commercial aviation (Inland Airlines) in and out of Rapid City; in fact, it is expected to help expand facilities for that type of aviation. It will, however, exclude all private flying from the field.

Rapid City interests have been at work on this project for several months, and an Army Air Corps Inspection Board paid a visit here a few weeks ago and gathered data on a proposed site. The board also visited about 75 other sites in the United States, and since then the project, as far as Rapid City is concerned, has been *up in the air*.

All of South Dakota's Congressional delegation has been active in trying to bring the base to South Dakota. Congressman Francis Case (from Custer, South Dakota) has been particularly active on behalf of Rapid City. United States Senator Chan Gurney (from Yankton, South Dakota) and United States Senator William Bulow (from Beresford, South Dakota) also worked hard on the project and lent all assistance possible to all prospective, as did Congressman Karl Mundt (from Madison, South Dakota).[5]

Major Elmer Oechsle, Area Engineer, Army Engineering Corps in charge of the Rapid City Air Base project, discussed the local defense development in a talk about before the Cosmopolitan Club Monday noon.

Boyd Leedom had charge of the program and other guests included City Manager, A.S. Holm, Earl Brockelsby, M.F. Anderson, W.R. Milne and Earl Dake, Rapid City and George Cole, Milwaukee, Wisconsin.

Details of Major Oechsle's talk concerned mainly the type of base to be constructed and established here. Details of the letting of contracts are yet to be known.

It is hoped, he said, that the base will be ready for occupancy by August. It will be built to house about 400 officers and 3,600 men and will have its own cantonments, recreation halls, theater, chapel and other features. Present progress indicates three main paved runways, about 150-feet

8

wide by 6,500-feet long, with an additional width of 75-feet on each side of each runway of some type of improved construction. The field will be night lighted, with a railroad spur from somewhere near Box Elder, complete electric, gas, water and sewer systems, including a sewage disposal plant. Housing at the airport will be at the east side of the base, which at present includes the Municipal Airport field and extends about a mile and one-half east.

The base, he said, will be known as an all-purpose station with facilities for handling any type of plane up to the largest bomber now in use or contemplated. Facilities are being arranged for handling three bombing squadrons and a reconnaissance squadron, plus all necessary service and supply units. The site is a fortunate one, he added, with a minimum of grading required to put it in condition. Although nothing definite is known, it is hoped that construction may be underway in from 30-to-60 days.

An Army Engineering Corps Area Office has been set up in Room 202 of the Duhamel building and at the present time, Major Oechsle also has charge of the Provo Defense development.

As a final word, Major Oschsle discussed the *boom* angle of the development, pointing out that once the 3,000-to-4,000 construction workers have come and gone, most of the personnel of about 4,000 at the base will be housed at the base itself. He pleaded for a *fair standard of living* here during the *boom* period. He pointed to troubles that have occurred during some major developments, involving raising the cost of living *all out of proportion*.[6]

Contracts for construction of Rapid City Army Air Base will probably be let sometime within the next two or three weeks, Major Elmer H. Oschsle, Area Engineer, Army Engineering Corps, said here, Friday.

Major Oschsle returned late Thursday from the Twin Cities where he had been to attend a review of utility plans. He was also accompanied to Rapid City by Mrs. Oechsle and their two children, who have been in Omaha, Nebraska.

Utility plans, covering systems for distribution of water, gas and electricity, in addition to sewage and sewage disposal, were reviewed and some adjustments made. Detailed plans and specifications are now being made.

Detailed plans for barracks and hangars at the bombardier site will be standard Army plans and contracts for construction will be let at the same time that the utility construction work is let.

A total of 47 men is at work at the field now. Twenty-two men are engaged in consideration of an office building for the Corps of Army Engineers. The buildings will be standard barracks construction and when the field is completed will become one of the regular barracks.

As soon as the building is completed, probably sometime in February, depending upon weather conditions, the Engineering Corps will move into it. At present, the offices are in the Duhamel building.

Final details of employment and wage scales have not yet been announced, but are expected within a week or two, Major Oechsle said. Government coordinators have completed their local survey and have sent findings to Washington, where officials will establish minimum wage scale, based on prevailing local wages.

The field crew also now at work also includes 20 surveyors and other workers.

The general public will be excluded from the air base as soon as its final area is established (signs were posted that no photographs were allowed from an area near the air base) and a four-strand or five-strand barbed wire fence will be erected around the entire base. All workmen will be required to carry identification and others will be admitted only by special Army permission. The area will also be posted.[7]

United States Representative for South Dakota, Republican, Francis Case stated Air Corps officials…*anticipate no difficulty in permitting Inland Airlines to operate its two scheduled flights from Rapid City Airport on a lease basis…The airport site of a new Army air base allocated by the War Department would be available to all planes with radio facilities…After the Air Corps takes over radio equipment probably would be banned. Adjutant General E.S. Adams indicated that housing facilities would be needed for 150 officers and 100 non-commissioned officers with families in connection with the air base.*[8]

9

The United States Government will acquire actual ownership of most of the enlarged Rapid City Municipal Airport the City Commission learned in a postponed meeting Monday night. City Manager A.S. Holm and City Attorney Boyd Leedom, who returned Sunday from a week in Washington, D.C., reported to the Commission that the Federal government is starting condemnation proceedings to acquire the airport and its expansion areas with the exception of about 160 acres. The reason, they told the Commission is that the government wished at least a 25-year lease on the property in view of the better $8,000,000 being spent there, where under existing laws the City could give no better than a 10-year lease.

Under the plans, the government will pay the City for what lands it owns outright and will fulfill options or contracts applying to added acreages in the process of being acquired by the City. The City will also be paid for its original improvements at the field. Condemnation proceedings were chosen as the method of acquisition as it would cause delays in taking possession. The action may have been started already. It is expected that government purchase of the port will result in saving a portion of the proceeds of the $125,000 airport bond sale and possibly some of the bonds will be redeemed immediately. Such action would result in some taxation reduction.

Meanwhile, work at the air base site continues. Major Elmer H. Oechsle, Area Engineer, Army Corps of Engineers…*The Corps of Engineers expects to move into its office building, which eventually will become a barracks, at the air base during the coming weekend. A branch office will be maintained at the present location in the Duhamel building for interviewing applicants and as a general information bureau. Plans for further construction are also progressing. The letting of one contract was postponed, when the Department of Labor recalled an approved wage rate for further consideration and revision. Under existing laws, the engineers are not permitted to advertise for bids or approve contracts without a wage scale, established by the Department of Labor. As soon as a revised scale has been approved, bids will be taken, probably early in March.*

Holm and Leedom…*Not included in the acquisition plan is the site of the present hangar and the south end of the north-south runway, all located on the extreme west edge of the field. They were assured that under present plans and present conditions, Inland Airlines will be allowed to continue to use. The west side of the field and possibly other planes might, also, but there will be no private flying from the field. Private flying was barred when the Government first announced its construction plans.*

Also remaining in the possession of the City will be a small area surrounding the new well, in as much as the work under the jurisdiction of the City, is already well underway. The well reached 400-feet Saturday and the 16-inch casing was cemented to that depth. Drilling was to be resumed Tuesday from that depth in a hole 9 5/8-inches in diameter and drillers expect to hit the Dakota Sands within a week.

The City's commitments for which the citizens approved a $125,000 bond issue by over 99 percent remain much as they were except for the land acquisition. The City is to complete the well, which the government may purchase later, provide a right-of-way for the railroad spur, secure five acres for a radio range tower, gain a right-of-way, for sewage line below the sewage disposal plant and other items.

The present area of the port, outside of the 158 acres to be retained by the City, includes about 2,100 acres acquisition of the site by the Army Air Corps and the government was first broached to the City Commission several weeks ago, at which time the Board took the stand that it would rather retain possession, unless the government deemed it necessary to have the tile in its own name. That necessity has arisen, Rapid City's representatives were told in Washington, and they gave the informal opinion that such ownership by the government may make the local training plant, when completed, more permanent.

Although nothing definite has been announced in Washington, a large-scale aerial bombing range south of the Badlands is still very much in the picture, the Commission was told. The availability of such an area was originally responsible for Rapid City being selected as a bombing school site, the Commission has believed.[9]

10

The Armstrong Air-to-Air Gunnery and Bombing Range consisted of a total area of 404,439.41 acres six miles east of Eagle Butte, South Dakota. In 1943, the War Department acquired the land from the Department of Agriculture, Department of Interior, and private individuals in the counties of Armstrong and Sully. The U.S. Army Corps of Engineers constructed three flight markers on the range to guide aircraft to the target area, for discharge of .50 caliber machine guns at towed sleeve targets, and dropping of practice bombs by Boeing B-17 Flying Fortress aircrew from Rapid City Army Air Base. Fighters from bases in Nebraska also used the area for air-to-ground gunnery. The U.S. Army Corps of Engineers made no improvements to the range, most of the land for which was acquired from the Oglala Sioux Indian tribe on the Pine Ridge Reservation, located south of Rapid City and east of Hot Springs, South Dakota.[10] The area from the Pine Ridge Indian Reservation comprises 341,726 acres. In the mid-1960s, the South Dakota National Guard placed 100 car bodies on the retained area to be used as artillery targets until 1973. In 1977, the Air Force (Ellsworth Air Force Base) cleaned up 2,486 acres of land contaminated from bombing, machine gun, and artillery shelling, removing the soil to a depth of 18-inches. Even at that time it was known that additional cleanup would be required. This cleanup began in the summer of 2008 on contaminated land between South Dakota Highway 44 and between Scenic and Interior, referred to as the *impact area*. The cleanup was under the direction of the U.S. Army Corps of Engineers.[11]

Another article from the *Journal* stated:

City Commissioners also raised the question of the County fulfilling its promise to build an access road to the airport just west of Box Elder. Developments today indicated that the County is going ahead with acquisition of the right-of-way. It is estimated that with the City retaining part of one runway and the new hangar building that no further expansion of the City's airport will be necessary, but the possibilities of expanding to the west of the present building were mentioned briefly.

In commenting upon their Washington conversations in general, Holm…*There is not a chance that the bombing range will not be built. That…should quiet rumors that crop up once in a while, that the base has been abandoned. That surely, however, does not mean that we do not have to keep pushing and to see that nothing prevents the work from going forward. Rapid City has been praised for the progress already made on this port, in comparison with some in other areas, and we do not propose that anything here is going to spoil that record.*[12]

Field soil bearing tests were made by a special crew from the Division Office beginning January 4, 1942.

The first construction work was the erection of the Detention Barracks, Building No. T-14, by government plant and hired labor. This building, to be occupied by the U.S. Engineer Office staff throughout the construction period, was begun January 27, 1942, and completed March 1, 1942. The city started drilling a well in the approximate center of the air base February 4, 1942. On March 7, 1942, the first construction contract was let for the railroad and fence, and construction was started immediately. The balance of the major contracts was let March 30, 1942, with construction beginning immediately.

Railroad spur line from the current existing Dakota Minnesota Eastern Railroad, crossing Old Highway 14/16 into the south end of Ellsworth AFB. The spur is no longer active; it is a non-marked crossing on the highway, but has not been dug up and paved over. The track enters the air base, and sections inside have been removed where it crosses roads; these have been dug out and paved according to the surface of the road. It was used through the time when ICBMs were part of the base's operations. The bridge in the background is Interstate 90, east of Rapid City and north of Box Elder. *Lt. Col. George A. Larson, USAF (Ret.)*

11

The first buildings were completed and transferred to the using service July 31, 1942. Final transfer of the completed project was made to the Army Air Force on October 1, 1942. The total cost was $8,725,967, as taken from the final working estimate of the Field Progress Report:

I. General Description

1. Assigned Job No. Rapid City A-2. The Rapid City Army Air Base, assigned as Job Number Rapid City A-2, is located 12-miles northeast of Rapid City, South Dakota.

2. Located on Natural Bench. The base is situated on a natural bench approximately three miles long and two miles wide. The area is bounded on the north by a low range of hills. The east and west ends are clear, the south side breaks to a flat valley. The air base was laid out on this bench with the airfield proper on the west and the cantonment on the east.

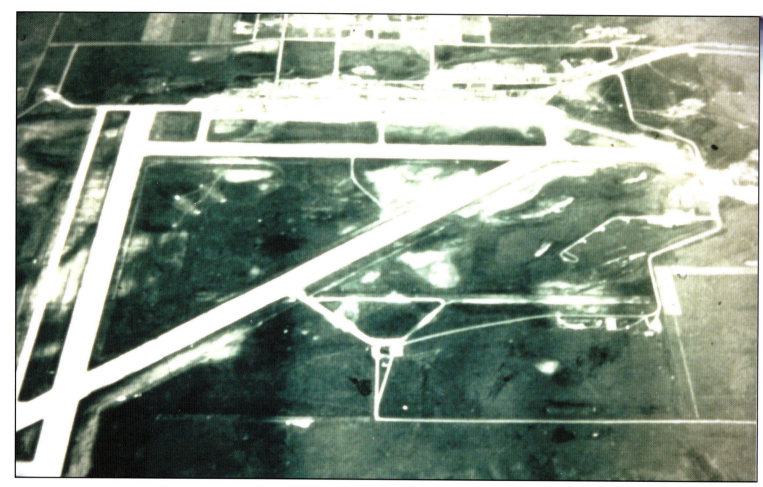

Aerial photograph of Rapid City Army Air Base, looking from the west with the top of the north end of the runway to the left and south end to the right. *Photograph: Ellsworth Air Force Base, 28th Bomb Wing Historian.*

3. Three 7,000-foot runways. The airfield consists of three runways: north-south, east-west and the third, northwest-southeast, in the form of a right triangle. The runways vary from 6,638 to 7,000-feet in length, and are bituminous surface, 300 feet wide with a 100-foot conditioned and seeded landing strip on each side and end.

4. Concrete apron 4,500-feet by 500-feet. Plane storage and servicing facilities are provided by a concrete apron 4,500-feet long and 500-feet wide parallel to the north-south runway, between the airfield and the cantonment area. Five hangars are provided along the east edge of the apron.

12

5. 221 buildings in the cantonment area. The cantonment proper consists of 221 buildings, including such groups as officers' quarters, enlisted men's quarters, hospital, warehouses, communications, service, headquarters and others.

6. Housing facilities provided. Housing facilities are provided for 460 officers and 3,790 enlisted men. An increase of 23 percent is possible by double bunking to a minimum of 500 cubic feet per man.

7. Temporary type buildings. The buildings are of the standard temporary type, the majority of which is wood construction. Footings, walls, and piers are mostly reinforced concrete. A few are tile. Floor construction consists of 1-inch sub-floor and 1-inch flooring carried by laminated wood joists. Kitchen, toilet, and heater room floors are concrete. Wall construction utilizes 2-by-4 and 2-by-6 studs with 1-inch sheathing, building paper and 1-inch siding on the exterior side, and 1/2-inch insulation board on the interior side. Roof construction consists of asphalt treated strip shingles over 1-inch sheathing supported by purloins and built up trusses 12-to-16 feet on center.

8. Coal used for fuel. Each building is equipped with a separate heating unit utilizing a hot air or steam system to best advantage. The use of coal exclusively for fuel was necessitated by the critical status of natural gas.

9. Adequate utilities provided. Water, sewage, and electrical services are provided for all buildings where they are required.

10. 1,000,000 gallons per day water from two 4,000-foot wells. Water is supplied in excess of the required rate of 1,000,000 gallons per day by two deep wells located on the base, each of which is approximately 4,500-feet deep.

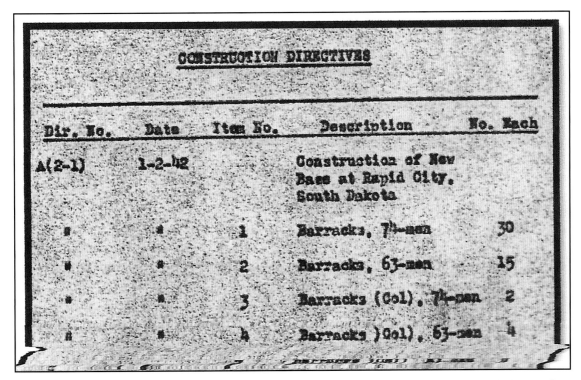

Buildings constructed in 1942 at Rapid City Army Air Force Base. "Completion Report. The Rapid City Army Air Base, Rapid City, South Dakota. Rapid City, South Dakota: United States Engineer Office. Report, October 30, 1942." (See pages 15 and 17)

13

These wells with pumping equipment were furnished by the City and are to be operated by the Government. The water is heated to 100 to 120 degrees Fahrenheit when pumped and is circulated through an evaporation type cooler before being put into the storage and distribution system.

11. 1,500,000 gallon water storage. Storage is prohibited by a 1,000,000 gallon concrete ground storage reservoir and a 500,000 gallon steel elevated tank. The system was designed for a normal population of 5,000 with the mains capable of handling three or fourfold expansion.

12. 720,000 gallons per day disposal plant. Sewage is collected by gravity flow water borne tank system and taken into a disposal plant of the sludge digestion type, located in the southeast corner of the air base. The main system consists of 32,730 lineal feet of vitrified clay pipe. The disposal plant is capable of handling 720,000 gallons per day, which is sufficient for 10,250 people, assuming 70 gallons per day per person. The effluent, after chlorination, is discharged into a natural drainage tributary of the area.

13. 26,000 volt electricity. Electrical service at 26,000 volts is provided by a local power company to a central sub-station with three 500 KVA transformers located in the approximate center of the area. Distribution is made by a trunk system using a 2,300/4,160 3-phase wye connected, grounded neutral system utilizing 50 miles of bare copper wire and 118 transformers provided at various high demand locations.

14. Hydraulic displacement gasoline system. Plane servicing is provided by a hydraulic displacement gasoline system with 275,000 gallon storage capacity. Truck loading facilities are provided at each end of the apron with servicing of planes to be accomplished by truck.

15. Reservation fenced with wire. Thirteen miles of five-strand barbed wire with wood posts is provided around the entire area. Critical areas that require extra protection are fenced with 8-foot Class A chain link fence on steel posts or 6-foot Class E barbed wire fence on wood posts.

16. 14 miles of blacktop roads. The area is serviced by 14 miles of bituminous surfaced roads, ranging from 24-to 12-feet in width. These roads are divided into streets and avenues, the streets being considered primary. Access to the area is provided by 1 1/4 miles of a 24-foot top bituminous surfaced road from U.S. Highway 16 to the south boundary of the area, where it joins the south end of sixth street.

17. 6,500-feet of gaveled walks. A total of six, 5,000 lineal feet of 4-foot graveled walks is incorporated with the system of cantonment roads. Graveled surface parking and service drives are provided for all buildings where necessary. The parking areas are limited to office buildings and service centers.

18. Railroad service. Railroad service is supplied by 2 1/2 miles of track from the mainline of the Chicago & Northwestern Railway Company [today the track to the south of Ellsworth AFB is operated by Dakota Minnesota Eastern Railroad] to a point in the approximate center of the air base. Two thousand eight hundred feet of side tracks with crossovers are provided, giving unloading space for seventy cars. The track is composed of 90 pound rerolled rail on 8 foot ties and 8-inches of gravel ballast.

II. Airfield

19. Area flat with 1% slope. The airfield proper involves an area of about 1,200 acres. The area is flat with a natural 1% slope to the south. Earth movement of 2,544,000 cubic yards was necessary for airfield construction and to provide glide angles on prolongation of runways and future expansion facilities including runways parallel to those constructed.

20. Bituminous surfaced runways. The runways are bituminous surfaced 300-feet wide with 100-foot conditioned and seeded landing strips on each side. A cut-off wall of bituminous material 12-inches wide and 24-inches deep is provided along the high side of each runway. The bituminous surface consists of 1 1/4-inch wearing course of 1-inch graded mineral aggregate and A.G. oil in the proportion of 95 percent aggregate to 5 percent oil by weight. An additional 2-inch course of mineral aggregate and oil as above was placed under the center 150-feet of the runways. The base course consists of a 6-inch layer of 1-inch graded aggregate stabilized with water and compacted to a density of 104 pounds per cubic foot. Oil used in the base course was approximately 4.5 percent of the total weight. Sealing of the bituminous was accomplished with two applications of RC-3 oil and mineral aggregate chips passing a 1/2-inch sieve. The priming application of oil was made at the rate of .20 gallons per square yard and the second application of oil was made at the rate of .25 gallons per square yard. The chips were applied at the rate of 17 pounds per square yard.

21. Luminous paint on the centerlines. Three 6-inch stripes of Chrome Yellow paint is provided on runway centerlines. Three 6-inch stripes of White are provided along each edge of the runways, and three 6-inch stripes of International Orange are provided along apron and taxi strip edges. Sixty-foot numerals corresponding to the first two numerals of magnetic azimuths are provided at the end of each runway. A luminous quality was obtained by an application of glass beads to the wet paint, which contains a special cement base for holding the beads in place.

Dir. No.	Date	Item No.	Description	No. Each
A(2-1)	1-2-42	24	Adm. Sq. 8, A-6	11
"	"	25	Finance Bldg. with Vault, A-10	1
"	"	26	Post Q.M., A-10	1
"	"	27	Fire Station, F-2	1
"	"	28	Guard House, GH-2	1
"	"	29	Detention Barracks, 63-men	1
"	"	30a	Ward, W-1	4
"	"	30b	Ward, W-2	2
"	"	30c	Ward, W-8	1
"	"	30d	Infirmary, I-2	1
"	"	30e	Administration, A-1	1
"	"	30g	Flight Surgeon's Unit, FS-1	1
"	"	30h	Storehouse, SH-5	1
"	"	30i	Storehouse, SH-7	2
"	"	30j	Mess, M-12	1
"	"	30k	Barracks, HB-54	3

Dir. No.	Date	Item No.	Description	No. Each
A(2-1)	1-2-42	31	Post Exchange, E-3	2
"	"	32	Post Office, PO-1	1
"	"	33	Flag Pole	1
"	"	34	Dispatcher's House, DH-1	1
"	"	35	QM Gasoline, GOS-2	2
"	"	36	Wash Rack (10 Car)	1
"	"	37	Repair Shop (Motor) SP-2	2
"	"	38	Grease Rack (Wood)	2
"	"	39	Light Maintenance Shop SP-14	1
"	"	40a	Carpenter Shop	1
"	"	40bc	Plumbing-Electric Shop	1
"	"	40d	Blacksmith Shop	1
"	"	40fe	Sheet Metal Paint Shop	1
"	"	40g	Warehouse, SH-18	1
"	"	40h	Utility Equipment Shed	1
"	"	40i	Adm. Bldg., A-5	1
"	"	41	TT-4, Tele. & Pole. Building	1
"	"	43	Recreation Bldg. HB-1	1
"	"	44	Theatre, TH-1038S	1
"	"	45	Warehouse, QM, SH-18	4
"	"	46	Commissary, SH-18 Mod. 1	2
"	"	47	Chapel, CH-1	

15

22. *North-south runway 6,750-feet.* The north-south runway, which is true north-south, is 6,750-feet long with a cut section on the north end extending 3,700-feet to secure a 40:1 glide angle. The south end is carried on a fill for 300-feet beyond the end of the runway.

23. *East-west runway 7,000-feet.* The east-west runway is normal to and west of the north end of the north-south runway. The runway is 6,950-feet long with a fill carried 300 feet beyond each end.

24. *Northwest-southeast runway 6,638-feet.* The northwest-southeast runway, which is 6,638-feet long, is at an angle of 42 degrees with the north-south runway. It forms the hypotenuse of a right angle triangle with the other two runways.

25. *Concrete apron 4,500-feet by 500-feet.* The concrete apron, which is parallel to the north-south runway, and 750-feet from its centerline, is 4,500-feet long by 500-feet wide with a total area of 250,000 square yards. The average section is 7-inches thick with all joints thickened to 9-inches. Joints are placed every 10-feet each way, with every 5th joint provided with slip joint dowels 12-inches on center. These dowels are only reinforcing steel used in this construction. Included in the apron are 4,500 tie down anchors, 15 flushing hydrants and 37 junction and electrical receptacle boxes. Plane servicing facilities are provided along the east edge of the apron, including hangars, water hydrants, gas and oil. The gas and oil are supplied to trucks, which are used to service the planes on the apron.

26. *5 hangars provided.* Hangar and repair space are provided by four squadron hangars, type OBH-1, with 11,520 square feet each and one repair hangar, type BEMI, with 24,000 square feet floor area.

27. *5 concrete taxi strips.* A total of five concrete taxi strips 50-feet wide are provided from the apron to the runways; two at the south end, one of which goes diagonally to and across the south end of the north-south; the other due west to the intersection of the north-south and northwest-southeast; a third to the middle of the north-south; a fourth from the north end of the apron due west to the north-south; and a fifth due north from the center of the end of the apron across the east end of the east-west. The last mentioned taxi strip is continued by bituminous surface for 1,200-feet to the target butt. The taxi strips are similar to the apron in construction details.

28. *2,544,000 cubic yards earth.* The grading involved 2,544,000 cubic yards of earth movement and finish grading of approximately 320 acres. All runways were graded 250-feet on one side and 750-feet on the other side of the runway centerline to accommodate future construction of dual runways. All earth movement areas were fertilized and seeded to eliminate erosion by both wind and water. The seeding was accomplished by single or double drilling areas according to need. The 100-feet strip adjacent to the runways received triple drilling to insure a heavy crop.

29. *Surface Drainage used for the airfield.* Drainage of the airfield is mostly handled on the surface with transverse grades being maintained on all runways and depressed sections adjacent to the landing strip on either or both sides to collect run-off water.

30. *12,850-feet reinforced used for the airfield.* The general slope of the airfield is to the southeast. That area southwest of the northwest-southeast runway is collected by a natural drainage tributary to the southeast where it eventually encounters the fill on the south end of the north-south. At this point, 897-feet of the 42-inch pipe was installed to carry water through the fill. Other pipe was installed under the apron and the north-south runway to remove the run-off water. Two lines of 60-inch pipe were installed under the north end of the north-south, one to maintain the ditch section along the north side of the east-west runway and the other to remove the water from the natural tributary cut-off by the taxi strips to the target butt.

31. *Apron drainage underground.* Drainage of the apron is provided by 19 inlets at regular intervals along the east edge of the apron. Carry-off is affected to the southeast edge of the apron. Carry-off is affected to the south by an underground pipe increasing in size from 15-inches at the upper end to 48-inches in diameter at the lower end. Drop inlets and pipes were used where necessary to carry the run-off water under the runways.

III. Buildings.

32. *241 structures on air base.* The Rapid City Army Air base embodies a total of 241 structures, the majority of these being temporary type, wood construction, buildings with the cantonment area. A few, such as the Ordnance Sewage Disposal, warehouse, target ranges and radio transmitter, are isolated due to their nature.

Dir. No.	Date	Item No.	Description	No. Each
A(2-1)	1-2-42	55	School Bldg. SB-12	1
"	"	58	Photo Lab, AC-1110	1
"	"	59a	Warehouse	1
"	"	59b	L-5 (Mod.) Heating Plant & Lavatory Bldg.	1
"	"	59c	Ammunition Assembly & Maintenance Shop	1
"	"	59d	SD-16 Trailer Shed 12 stalls	1
"	"	59e	6 Bays Magazine (Storage)	1
"	"	59g	SH-10 Magazine (Pyro)	1
"	"	59h	SH-11 Magazine (S.A.Am.)	1
"	"	59m	Bar. Igloo, Magazine (40'-4")	1
"	"	61	Paint & Oil Storage OSH-1	1
"	"	71	Sentry Boxes	2
"	"	75a	Air Corps Warehouse, Plan 1	1
"	"	75b	Air Corps Warehouse, Plan 4	1
"	"	19	Mess, Officers (Cadet) AC	1
"	"	48	Incinerator	1
"	"	51	Op. & Base Oper. Bldg. A-10	2
"	"	52	Bombsight Storage	1

Dir. No.	Date	Item No.	Description	No. Each
A(2-1)	1-2-42	53	Link Trainer (with heat(sealed)	1
"	"	54	Parachute Bldg.	1
"	"	56	Shop, Hangar	1
"	"	56a	Shop Hangar, SQD	4
"	"	57	Airdrome Traffic Control Tower	1
"	"	76	Base Engineering Shop, Mod.	1
"	"	77	Arm. Instrument & inspection & Adjustment Bldg., Mod.	1
A(2-2)	2-25-42		Day Room, RB-4	1
A(2-3)	2-27-42		Informal Approval of Ordnance Layout	
A(2-4)	4-6-42		Airways Transmitter Bldg., RS-1	1
A(2-5)	4-8-42		Mess Cafeteria, M-952 (deleted)	1
A(2-6)	5-13-42		Bomb Trainer Bldg.	1
A(2-7)	5-25-42 Ltr. - 5-30-42		Dental Clinic, DC-3	1
A(2-8)	6-20-42		Infirmary, Type 1-4	1
A(2-9)	7-7-42		Headquarters Bldg. Type T3A-6	1
A(2-10)	8-7-42		Supply Bldg. for Medical Detachment	1

Dir. No.	Date	Item No.	Description	No. Each
A(2-11)	8-11-42		Ordnance Oil & Paint Storage Warehouse	1

17

33. Cantonment divided in blocks. The cantonment is divided into blocks by a system of streets and avenues, each block being a complete service unit with assignment of purpose influenced by its respective location. The various blocks are designated as officers' quarters, enlisted men's quarters, hangars, training, hospital, recreation, operation, communication, service and detention area.

34. 65 blocks numbered consecutively. The blocks are numbered consecutively beginning with the hangar section along the east side of the concrete apron, numbering from south to north and continuing with the next row east numbering from north to south, etc., for a total of 60 blocks, blocks numbered 60 to 65 being assigned to the isolated areas as noted above.

35. Wood construction used. In general the buildings are constructed of wood with 1-inch sheathing building paper and 1-inch siding for the outside walls, 2-by-4 or 2-by-6 studs on two-foot centers and one-half inch insulation board on the inside walls. Inside partition, walls are finished on one side only to conserve materials.

36. Red roof asbestos strip shingle. The larger structures have roofs supported on built-up trusses twelve feet to sixteen feet on center with interior support by partitions where encountered. The roofs in general are constructed with 5:12 pitch; a few of the larger structures have flat roofs with roll roofing. They are one-inch sheathing covered with water proofed paper and Class C asphalt treated asbestos strip shingles.

37. Wood floors. Floors of buildings in general are one-inch sub-floor with one-inch tongue and grove flooring. Kitchens, heater rooms and toilets are provided with concrete floors.

38. 300# soil, 100 mph wind, 25# snow load. Chimneys are brick, clearing roof peak by five feet, and more when large heating units are accommodated. Footings with walls and pier supports are concrete, designed for 3,000-pound soil. The footings are minimum 4-foot 6-inches deep for frost protection. The buildings are designed to withstand one hundred mile per hour wind and 25-pound per square foot snow load.

39. Facilities for 440 officers. Officers' quarters established in block 43 will accommodate 440 officers in ten, type BOQ-44, wood frame two-story buildings. Each building is provided with a steam heating system, toilet facilities, hot and cold running water, electricity and double quarters. An Air Corps Cadet Officers' mess, seating 475 men, a recreation building, type ORBL-3 day room, and a type OM-1 club mess, are provided to complete this unit.

40. Facilities for 3,790 enlisted men. A total of 13 blocks is provided for enlisted men, two of which are isolated for colored troops. These blocks contain a total of thirty-four 74-man barracks and nineteen 63-man barracks. Each block is intended to accommodate a squadron. A typical block contains four 74-man two-story barracks complete with bunks, toilet facilities, hot and cold running water and a hot air heating unit, a type RB-4 day room, a type M-300 mess, a type A-6 squadron administration and a type SA-2 supply warehouse.

41. Headquarters type BH-1. Headquarters block is No. 14 and is located in the approximate center of the cantonment. This block utilizes one headquarters building, type BH-1, with 9,500 square feet for office space, an administration building, type A-10, and a finance building, type A-10. All three are equipped with hut air furnaces, toilet facilities, and electricity.

42. 24 buildings in hospital connected by closed corridor. The hospital group embodies 24 buildings, all connected by a closed corridor. The roof of the building nearest each of the four corners of the block carries a white border around a 25-foot red cross for identification from the air. The group consists of one officers' quarters, two nurses' quarters, one administration building, one dental clinic, one infirmary, one flight surgeons' unit, one mess, one recreation, seven wards, three barracks, four storehouses and a power and heating plant.

43. Accommodations for 20 officers, 31 nurses, 203 patients, 162 enlisted men. The hospital quarters will accommodate 20 officers, 31 nurses, 203 patients and 162 enlisted men. Facilities are provided for X-rays, major and minor operations, anesthetics, dental work, and isolation of acute cases with mortuary service. All buildings are steam heated and completely modern.

44. Automatic sprinkler alarm system, and central heating plant. An automatic sprinkler system and an automatic fire alarm system are provided for fire protection. The boiler house is of wood frame and corrugated metal construction with a concrete floor. A 60-inch radial brick chimney 95-feet high is provided for draft and smoke exhaust. The plant utilizes three stoker fired 174 horsepower boilers with necessary equipment. An auxiliary gasoline motor power plant rated at 30 KVA is located in the boiler house for operating equipment in the event of power failure.

18

45. Closed corridors connecting hospital. The closed corridors, connecting all buildings in the hospital group, are similar in construction to the buildings, with insulation and double flooring. They are 6-feet wide inside and 8-feet high, the peak of the roof houses sprinkler system pipes, steam mains and wiring. Double swinging doors are provided 15-feet from intersections as draft stops.

46. 2 chapels, 2 post ex's, 1 theater, 1 infirmary, 1 recreation. The recreation group includes two post exchanges, two chapels, one infirmary, one theater and one recreation hall. The post exchanges are complete with cafeteria service, soda fountain, barber shop, sales rooms and display counters. Steam heat utilizing unit heaters is provided. The chapels are located three blocks apart with seating capacity of 350 each complete with altar study and toiler facilities. The infirmary, similar to the one in the hospital group is used for examinations and has facilities for care of minor disorders and dispensing of medicine and drugs. The theater is complete with fire-proofed dual sound and projection equipment. Stream heating and air conditioning equipment are also provided. An elevated stage and sloping concrete floor, provide seating capacity for 1,038 persons. The recreation hall, type RB-1, has heating and ventilating equipment, elevated stage, and seating capacity for 357 persons. The seats can be cleared to provide space for dancing.

47. 4-corrugated squadron hangars. The hangar section located along the east edge of the concrete apron, embodies four type OBH-1, squadron hangars and one Base Engineering Maintenance and Inspection Building. The four OBH-1 hangars are structural steel frame covered with corrugated metal and insulated with 1-inch fiber board. The buildings are 120-feet long and 96-feet wide with panel sliding doors along the 120-foot front facing the apron. Heater room space and office space is provided a lean-to on the side opposite the doors. Heat for the hangar is provided a hot air furnace, stoker fed with automatic controls. Heat for the office space is provided by an upright stream boiler and a radiation system. These buildings were scheduled for Job A-2 but due to delayed shipment of steel, they were transferred and completed under Job A-3.

48. One Base Engineering Maintenance & Inspection wood on steel frame, 1 Base Engineering Shop. The Base Engineering and Inspection building is structural steel with one-inch siding over 1/2-inch gypsum board for fire protection, and 1/2-inch insulation board on the inside of the walls. This building was originally designed for Asbestos protected metal siding, but had to be substituted due to critical status of this material. The hangar space is 200-feet long and 120-feet wide parallel to the concrete apron with full opening panel sliding doors in each end. A 25-foot lean-to on each side provides two stories of office space on one side and office and repair space on the other. The offices adjacent to the apron accommodate the Dispatcher's office, Weather office, Engineer office, Map room and Pilot's locker room. The lean-to on the other side accommodates miscellaneous repair units and office space. A 6-inch concrete floor with drains and expansion joints is supplied for the entire building. A boiler house, housing two low-pressure boilers, is provided 25-feet to the south for heating the offices and hangars. A Base Engineers shop, type BES-2, is also provided across the street from the Base Engineering Maintenance and Inspection Building, with a total of 17,320 square feet floor space. This building is of wood construction with built-up wood trusses and full length monitors with sash on the roof. Facilities are provided for repair work of all kinds involving planes, including three 6-foot and one 24-foot spray-painted boots. Heating is accomplished by two stoker-fired boilers housed in a brick lean-to on the building. The floor is concrete. A total of five 21-foot by 10-foot exterior double doors are provided.

49. 55-foot control tower. A control tower, 15-foot by 15-foot free standing steel frame and 55-feet high is provided adjacent to the east side and near the middle of the apron. The tower house is metal covered with 4-inch insulation, wood floor, and glass windows on all four sides. It is equipped with radio, telephone and spot light for directing traffic on the apron and runways.

50. Target butt for machine guns. The target butt, located approximately 1,500-feet north of the end of the concrete apron provided a cribbed sand core embankment 54-feet long and 20-feet high for firing in of machine guns and airplane aprons. Two concrete aprons with chock holes are placed ahead of the embankment for lining and holding planes. Access to the target butt is provided by a 50-foot bituminous surfaced taxi strip from the east-west runway in line with the concrete taxi strip. A latrine and target houses are also provided adjacent to the ramp of the target butt.

Construction of Rapid City Army Air Base

51. *8 buildings for training facilities.* The training section includes two Base Operations buildings, one Administration building, one Parachute building, one Bombsight storage building, one Arms and Instrument Inspection building, one Link-trainer building, and one Bomb-trainer building.

52. *Administration building.* The Administration building, type BH-1, similar to Headquarters building, has office, and classroom space totaling 9,500 square feet, with steam heat, hot and cold running water, toilet facilities and electricity.

53. *2 school buildings.* The two base Operations buildings, located one on each side of the Administration building, are Type A-10 buildings with 2,160 square feet of floor space to be used as classrooms.

54. *Parachute building.* The Parachute building, with 1,900 square feet of floor space, also has a tower 12-feet square and 32-feet high for hanging parachutes, for airing and drying.

55. *Bombsight storage.* The bomb sight storage building has six vaults of heavy reinforced concrete construction, air conditioned with steel combination lock doors. The walls and ceiling are lined with acoustical title. An alarm system is provided for this building.

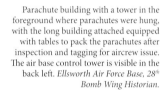

Parachute building with a tower in the foreground where parachutes were hung, with the long building attached equipped with tables to pack the parachutes after inspection and tagging for aircrew issue. The air base control tower is visible in the back left. *Ellsworth Air Force Base, 28ᵗʰ Bomb Wing Historian.*

56. *Armament and Instrument, Inspection and Adjustment building.* The Arms and Instrument Inspection and Adjustment building, type A & I, I & A, is of wood construction similar to other buildings described in this report. It has a total of 7,260 square feet of floor area with repair and adjustment equipment and office space. The building is heated by a steam radiation system complete with ventilation and air conditioning equipment.

57. *Link-trainer building.* The Link-trainer building, type SH-9, complete with hot air heating equipment, has facilities for nine Link-trainers.

58. *Communications center.* The Communications center includes a telephone and telegraph building, a fire station, a post office, a post Quartermaster office. A Wing Headquarters building, type T3A-6, is also provided in this block.

59. *Telegraph-telephone.* The telephone and telegraph building, type TT-4, is a one-story wood frame structure 25-feet by 90-feet, having switchboards, teletype machines, telegraph office, and Signal Corps office with toilet facilities and hot air heating system.

60. *Fire station.* The fire station, type F-2, is a one-story, frame construction building, 38-feet by 107-feet, with three overhead doors and concrete floor apparatus room with facilities for housing three fire trucks and equipment. Housing facilities with bunks and toilets are also provided for the firemen. The relative general alarm system and hospital alarm systems were included.

61. *Post Office.* The Post Office building, type PO-1, is a one-story frame structure 25-feet by 87-feet with 38-feet of dispensing windows.

Construction of Rapid City Army Air Base

62. *Post Quartermaster office.* The Post Quartermaster office is a one-story frame structure, 25-foot by 110-foot with open office space and two private offices.

63. *Service Group.* The Service Group includes four utility shops, one warehouse, two motor repair buildings, one maintenance shop, wash and grease racks and gasoline and oil storage, plumbing and Post Engineering office units. The two Motor Repair buildings, type SP-2, 84-feet by 37-feet, have six stalls each with concrete floors. Each stall has a 12-foot by 36-foot by 13-foot height clearance with double swinging wood doors in one end. Office space and heater rooms are provided in one end of the building. The warehouse, type SH-18, has 8,320 square feet of floor area with office space. Eight foot loading platforms are provided full length on each side of the building. The maintenance shop, type SP-14, has facilities for motor repair, painting, washing and greasing. Two outside wood constructed two-car grease racks are located adjacent to the motor repair shops. A ten-car concrete ramp is provided for outside washing with two hundred fifty pound water pressure available from a booster pump located in building T-864.

64. *24,800 gallons Quartermaster Gasoline.* Underground storage for Quartermaster gasoline with 12,000 gallon capacity underground tanks and two pump islands, one on each side of Third Street near the repair shops is provided.

65. *2 Air Corps warehouses.* The two Air Corps warehouses, each with storage space of 20,000 square feet, are located adjacent to the railroad unloading spur. Platforms are provided full-length along both sides of these buildings. A concrete ramp is also provided along the truck unloading platform. The floors are reinforced concrete. Heating is accomplished with steam boilers and unit heaters.

21

66. 5 SH-18 warehouses. The total of five type SH-18 buildings are provided for Quartermaster warehousing, four of these are designed as Quartermaster storage and the fifth is designated as a Commissary warehouse with meat and vegetable cold storage facilities. An overhead track for transferring the meat from the railroad car to the refrigeration room is provided. The five buildings are provided with 8-foot loading platforms for the full length, on each side with one platform adjacent to the tracks and the other adjacent to a concrete ramp for truck loading. The floors are constructed of wood for 150 pound per square foot floor load.

67. 8 buildings in ordnance. The Ordnance area embodies eight buildings, including for four magazines, one warehouse, one ammunition assembly, one paint and oil storage, and one Auxiliary Light Plant building. The magazine group includes one barricaded igloo, non-reinforced concrete circular arch with 1,075 square feet storage and an earth embankment barricade. It is furnished with overhead track and a bomb handling device. Two magazines, one SH-10 and one SH-11, corrugated metal, wood frame with concrete floors, are provided for miscellaneous storage. A six-bay reinforced tile structure with external metal doors is provided for assembled ammunition storage. Building No. T-878 is provided for ammunition assembly. The building is a tile structure, 33-feet by 97-feet with an overhead metal door in each end of a reinforced concrete floor. Heating is accomplished with a hot water boiler and unit heaters. The heater room is completely isolated from the main building and explosion proof fixtures are installed for fire protection. Building No. T-877 houses the Auxiliary Light Plant for use in case of power failure. Water service is not provided for this group due to remote location. An 8-foot chain link fence encloses the entire area.

Hangar under construction along the flight line. *Ellsworth Air Force Base, 28th Bomb Wing Historian.*

68. Radio transmitter building. The Radio Transmitter building, located in an isolated ten acre tract, southwest of the air base, provided facilities for radio transmission. The building is of wood construction with hot air heat. Office and equipment space totaling 575 square feet is provided.

69. Standard skeet range provided. A standard skeet range, including a high house, a low house, a control house, a shelter shed and a latrine are located north of well number 2 and southwest of the northwest-southeast runway. A 300 yard danger zone is adequately provided.[13]

From the *Daily Journal*:

The tempo of activity at Rapid City's new air bombing base is increasing daily, looking toward the time, within the next few weeks, when it may reach a climax of several thousand workmen throwing up the many buildings and other facilities to be needed there. Workmen under the direction of the US Army Corps of Engineers are rushing an office building for the Corps, which on completion of the project, will be one of many barracks on the grounds. Foundations have been completed and the building itself is beginning to take form. Altogether, at this time, about 50 men are at work at the field.[14]

22

Although the drilling rig by itself, which will seek the required 1,000,000 gallons of water per day, will not arrive until the first of the week [second week in February 1942], a crew of men of the Manning and Martin Company, Denver, Colorado, are erecting a 136-foot steel derrick. The rig, weighing over 200 tons, was to have arrived from the Lance Creek oil Fields in Wyoming on Saturday, but was delayed a few days in starting from Rapid City. Charles E. Basket of the company's Casper, Wyoming office, is in charge of the crew here at present.[15]

Bids for the construction of the railroad spur from Box Elder to the Army air base, and for a fence around the air base, will be opened here Saturday, it was announced Wednesday by Major Elmer H. Oechsle, Area Engineer, U.S. Army Corps of Engineers. The bid opening was originally set for February 26th, but was delayed. The Government has ties on the ground for the approximately three miles of the rail lines from the Chicago & Northwestern Railroad mainline and has also purchased other materials. Bidders on the fence job, however, will be bidding on materials as well as work. That includes posts, gates and about 58,000-feet of barbed wire. The fence will be a five-strand barrier, four feet six-inches high. Bids will be opened at the Hotel Alex Johnson at 11 a.m.

Meanwhile, rotary bits of the Manning and Martin Company of Casper, Wyoming are making progress downward in a search for water and Wednesday afternoon were at approximately 3,000-feet and expecting to reach the Dakota Sands at any time. Glen N. English, Superintendent of drilling, said latest estimates had been that the Dakota Sands would be reached about 3,000-feet and so far only comparatively loose shale has been encountered, which made for fast progress. The best time made so far was about 300-feet in 8-hours, or nearly 40-feet per hour. A thin layer of hard sand was encountered at about 2,810-feet. Some idea of the depth of the hole may be gleaned from the fact that within the next 300-feet the bits will be at sea level.

What will be found in the way of water in the Dakota Sands remains to be seen, English said. Either the Dakota or Lakota Sands, or even the Sundance Sands, might result in a large amount of water, or the drilling may be continued to the Minneslusa Sands, probably about 4,000-feet down. The hole was started at 16-inches in diameter, but at 400-feet was changed to 9 7/8-inches, which will be the size to the bottom. The top 400-feet will be used to hang pumping equipment in. Tests will be made in all sands.

Operations Group hangar. The gravel road along the hangar has an oil applied surface to cut down on dust during dry and hot weather and keep it from becoming muddy during rain and winter weather. *Ellsworth Air Force Base, 28th Bomb Wing Historian.*

23

Operation of the crew on the Manning and Martin Company's rig is about the smoothest ever seen in this section of the country. Changing of tools, for instance, is as neat a piece of teamwork as one could see anywhere. In order to change the drilling tool of all 3,000-feet of pipe, which turns the drilling tool or bit, has to be drawn to the surface, uncoupled in 90- foot lengths and stacked inside the 160- foot derrick. When the bit has been changed, the sections have to be hoisted into the air, recoupled and relowered into the well. It takes about a minute for the crew to hoist a length into the air, couple it onto the piece below with power, lower it into the hole and make ready for a new section. Three five-man crews are pushing the work forward-or rather, downward, 24-hours per day.

Major Oechsle and his office force are now practically settled in their new office headquarters at the air base site. Among officers here now or recently on special phases of the project were Edgard A. Jitson, Real Estate Section, Army Engineers Office at Kansas City, Missouri, who is looking after land transactions; L.E. Johnston, District Office, Fort Peck, Montana and Allen L. Snyder, Utilities Power Board, Divisional Office, Kansas City, Missouri. The latter completed negotiations with Black Hills Power and Light Company for what facilities will be needed at the base.[15]

The United States Engineers' Office, Rapid City, Monday announced the award of contracts for construction of the Rapid City Air Base project.

The Northwestern Engineering Company, Rapid City, was awarded a contract for the construction of cantonment roads, water supply and water storage facilities, sewage disposal, sewage treatment plant, airfield grading, drainage, paving and apron facilities.

Workers were needed to build the Rapid City Army Air Base. With Army Air Force bases going up throughout the United States, construction workers were in short supply. Patriotic advertisements such as this one from the *The Rapid City Daily Journal*, were common. This one appeared on March 30, 1942.

24

United Construction Company, consisting of Henry Carlson Company, Sioux Falls, SD.; S.W. Johnson Company, Aberdeen, SD.; R.A. Mark Company, Brookings, SD.; and H.H. Hacke Construction Company, Rapid City, SD., were awarded contracts for construction of temporary buildings, including barracks, mess halls, administration buildings, shops, warehouses, school building and hospital.

Clifton Engineering Company, Three Rivers, Michigan was awarded a contract for electrical distribution system, including switch house, primary and secondary electrical distribution systems and street lighting.

Work on all phases of construction will begin immediately, the Engineer's office said. No amounts of contracts or details of construction were announced.[16]

Rapid City Airfield is the temporary name for the local base, now under construction, Army engineers announced Friday on receipt of a statement from the District officers at Fort Peck, Montana. That name will be used in designation of the field until the Air Corps Naming Board in Washington gives it an official name. Some suggestion has been heard that it be named for Orvil Anderson, stratosphere high-altitude balloon pilot, or for the late Colonel Karle Lewis, local National Guard officer. Gravel trucks whizzing up and down the road from Old Highway 14/16 to the airfield, spreading gravel on the road…That is the county's contribution. At the airfield itself, the beehive tempo is increasing rapidly…looks like a *swarm* is near. Machinery is spreading oil all over the place and groups of men are at work on various jobs. Army engineers Friday reiterated their warning that no cameras are to be taken onto the site and no pictures are to be taken, either close to the field or at a distance, showing the site. A trailer camp *city* is springing up across the highway from Box Elder and about a quarter mile west. What will we call it, Box Younger?

The new red, white and blue license plates being assigned to all governmental automobiles are really something. A few have appeared to be local officer's cars. They boast a red shield on a white-black background, plus blue numbers and a lettering designating the department. In the meantime, a local metal shop turned out some neat temporary numbers for trucks and other vehicles, pending the arrival of national plates from Washington, so airfield men would be different. There does not seem to be any agitation as to who should have No. 1.

Work has been started on the disposal plants and also on footings for several buildings, starting in the northeast corner, near the Administration building, and including a guard house. Workmen hope the well will soon be in production, as hauling water is expensive and not entirely satisfactory. A popular soft drink, on which there was supposed to be a shortage is *tops* with office workers at the Administration building. The railroad spur is being worked. Grading now extends far onto the airfield site and steel rails are stretching along the grade. Motorists who were worried can now ease off for the grade crossing is flush with the highway. Old Highway 14/16, incidentally, is very busy. Many cars are on the road all the time between Rapid City and the airfield junction. Several small *cafes* have sprung up along the road. Hundreds of cars are parked practically all the time around the air base Administration area. They belong to Army engineers, workers office and airfield visitors, both business and sight-seeing. Inside there are so many workers that one might wonder if they know what they are doing, but they do. So far the policy of having the highway to the Administration area and the area itself open to the public at all times is working out well.[17]

Word was received here Thursday from the office of Congressman Francis Case, Washington, D.C. that the…*War Department unofficially announced approval of an airbase transmitter station at Rapid City Army Airfield.* The station will occupy a 10-acre tract and local officials said it will be located southwest of the airbase. It will be used for all radio communications in connection with the base, between the base and planes and in other ways.[18]

A high-ranking senior officer, who first saw service with the U.S. Army as a Buck Private, is the first commander of the Rapid City Airbase. He is Colonel Charles B. Oldfield, who arrived Monday from Gowan Field at Boise, Idaho, where he had been in command since November 1941.[19]

Rapid City Airbase, that is now the official name, is springing up rapidly.

After six weeks of rain that retarded progress in May, construction is back on schedule, or as near as possible except for a shortage of certain types of workmen.

There is a shortage of carpenters and cement finishers, as well as concrete mixer operators, United States Army engineers said Monday. The shortage is due to the large number of defense projects in this part of the country.

Construction work is in progress in one of the barracks areas. A grader is leveling an area prior to the construction of additional barracks. Concrete block foundation piers allow rapid construction of two-story and one-story wood frame buildings. *Ellsworth Air Force Base, 28th Bomb Wing Historian.*

At that, every building on the reservation is at least started and some quarters are nearing completion. Some streets have been black topped and all buildings finished on the outside are painted. A modern city, completely equipped with everything from hospital to fire department, has sprung up almost overnight in a cornfield.

While carpenters, tinsmiths, plumbers, electrician, painters and bricklayers work on the containment half of the air base, hundreds of trucks and modern machines are making rapid progress with a vast concrete apron before the hangars-to-be and with the blacktop runways. Huge dirt moving machines tore away an entire hillside to give proper clearance on one side of the field and leveled off the field in general.

A construction worker monitors a trench digging machine to install water and sewer lines throughout the base construction. Two story barracks are visible in the background. The first barracks on the left, behind the trencher, still has workers installing the siding. To the left building position stakes are in place, noting the future construction of more buildings. *Ellsworth Air Force Base, 28th Bomb Wing Historian.*

The paving operation is one of the most interesting and fastest-moving *industries* on the project with an equivalent of 21-miles of 20-foot wide pavement being laid to form the apron. The work is about an automatic as it can be made.

Steel rails are laid down for the concrete finishing machines to run on; are laid by instrument to the exact height to a fraction of an inch. A unique machine that runs on each rail tamps the earth firmly under the rail.

Earth moving scrappers knock down high spots on the designated runway and taxiways of the air base. *Ellsworth Air Force Base, 28th Bomb Wing Historian.*

Construction of Rapid City Army Air Base

A bulldozer pulls a power harrow north of the north-south runway to remove excess dirt, which is then loaded into dump trucks for movement to other locations requiring fill material. *Ellsworth Air Force Base, 28ᵗʰ Bomb Wing Historian.*

Trucks rush loads of mixed ingredients from a giant hopper to the paving mixer where water is added and the concrete is mixed. The paving mixer moves on treads alongside the strip being laid, while concrete goes out on a boom in a *go-cart* and is dumped between the forms.

The first machine to go over the rails is a sub-grader, which scrapes the dirt out between the rails, or forms, to the exact depth needed and tosses the excess out to one side. The sub-grader pulls itself along by a steel cable and drum.

Earth scrapper dumps removed dirt from the runway high areas onto low spots as marked by engineers. Once dumped the road grader in the middle of the photograph levels the dirt, followed by a bull dozer pulling spiked sheep foot roller spikes to compact the ground to the proper compression ratio. The process is repeated as necessary until the desired level and grade is reached. *Ellsworth Air Force Base, 28ᵗʰ Bomb Wing Historian.*

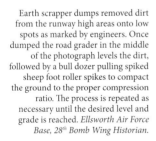

Construction of Rapid City Army Air Base

One of the largest paving jobs ever carried out in South Dakota, boasting the largest paving machine manufactured (as of 1942), which worked night and day at the air base. The concrete paving operations are as automatic as they can be made. This machine is a sub grader, operating on steel rails that are laid ahead by instrument to a specified height. Then a slab of concrete for the huge apron in front of the hangars-to-be is laid. The edge of the finished strip is used as one rail. The sub grader is moved as prongs loosen up the dirt in front and an endless belt scoops out the surplus soil and throws it outside, thereby leveling the grade inside to the exact depth desired. *Ellsworth Air Force Base, 28th Bomb Wing Historian.*

The first machine after the mixer pushes the bulk of the wet concrete ahead of it in a rough distribution, while behind it tamps and works it in firmly. One of the most unusual machines is the joiner, with a long saw blade the width of the strip being laid. At specific intervals, it saws a narrow grove in the wet concrete, in which is placed a two-inch strip of bituminous joint material to take care of expansion and contraction with changes in climate. At the same time, it lays a continuous strip of the same material down the center.

Loading dump trucks with concrete ingredients is a job that never ends, because concrete mixing machines have appetites that are never satisfied. Day and night, the giant hoppers pour ready-mixed ingredients into dump trucks waiting below, and they rush off to dump their contents, with water added, into mixers that pave the large concrete apron. *Ellsworth Air Force Base, 28th Bomb Wing Historian.*

28

The next and last machine to move over the rails is a hand-float, on which workman complete the job of smoothing the surface by hand and also insert recessed iron loops to which airplanes will eventually be anchored. Cement crews worked around the clock, seven days a week, and at night, the work areas are brilliantly lighted by huge flood lights on portable towers that are moved with the work.

Stationary bituminous mixing plant. *Ellsworth Air Force Base, 28th Bomb Wing Historian.*

Construction work is also progressing on the water storage pressure tank, an elevated steel structure that will serve the entire area. The main runways extend north-south and northwest-southeast, while a third runs east-west at the north edge of the field.

Sixth Avenue, the main thorough fare, running north and south, between the cantonment and hangar areas, has been black topped within the military area and the county will prepare the stretch between the military area and Old Highway 14/16. Sixth Avenue will eventually be the main entrance to the air base and is used now as a workmen's entrance. Bus schedules have re-arranged so that workmen now enter on Sixth Avenue, but only workmen with badges are allowed in the bus into the area.

So far, there have been no serious accidents on the job and the accident rate has been remarkably low, engineers announced. There was on fire, a small blaze in a lumber pile, extinguished by workmen. The new fire station is nearing completion and one truck is already on hand. Another is expected later. The hospital section of the area is also nearing completion and all buildings will be connected by covered passageways.[20]

The airfield was oriented along a north/south flight line, where the original hangars were built. To the west of the line lay the ramps, taxiways, and runways. There were three runways (north-south, east-west, and northwest-southeast). To the east of the hangar line lay the administration buildings, barracks, and so on. When Rapid City Army Airfield was built, like most of the other WWII bases it was considered to be only a temporary facility. Consequently, the buildings, even the control tower, were built of wood, with only a few years of planned useful life. The only building on the base made of bricks was the hospital power plant. It provided the only practical way of heating the sprawling, spread out, one-story wood frame hospital wards.

29

Enlisted barracks were typical two-story, 40-man style with open bays, and latrines in the ends. Officers' quarters were the same type of construction, with partial panels separating one or two bed bays. Mess halls were one story wood frame structures; Group Headquarters and Base Headquarters faced each other across an open field with a slough down the middle of it.

Paving the runways at the air base was one of the largest construction projects, and was pushed as rapidly as possible to completion. At the rear of the dump truck is a spreader, into which the dump truck is dumping a load of prepared surfacing material. Following these two pieces of equipment is a steam roller, which compacts the surface material to the desired level and hardness to allow the operation of heavy bomber aircraft. Due to the size of the dump truck, which was standard in the 1940s (small when compared to modern day heavy haul dump trucks), many trucks were needed to dump sufficient surfacing material to pave the three runways and connecting taxiways on the air base. *Ellsworth Air Force Base, 28th Bomb Wing Historian.*

Squadron hangars were also built of wood, each of them large enough to hold one B-17. Other maintenance facilities were equally temporary in nature and in construction, as was the Base Operations building and its associated facilities. Other than the hospital, each building had its own furnace room with a coal burning furnace. Water came from non-flowing artesian wells at about 140-degrees. It had to be cooled for some uses: wooden cooling towers were used until the water was piped from Rapid City when Renel Heights was built in the middle of the 1950s, after the base started to be upgraded from its WWII appearance, as Strategic Air Command improved living conditions on the air base. At first there was no family housing, Married personnel had to find housing off base, and it was scarce. The theory was, that if the Army wanted a man to have a wife, it would issue him one. The service clubs for Officers, NCOs, and enlisted men were the centers of social activity. While adequate for WWII, training single men in B-17s and sending them off to war around the world, in the post-war military, the base was ill-fitted for the larger airplanes that began operating from the air base (first the B-29), and housing for married personnel became critical. As a stop-gap measure, some of the WWII barracks were converted into multi-family barracks and apartments. They were far from satisfactory, and were replaced when better quarters became available. (Although they were still far below civilian housing as available in Rapid City).[21]

Construction of Rapid City Army Air Base

CHAPTER TWO

WORLD WAR II B-17 AIRCREW TRAINING

A Close Look at B-17 Aircrew Positions and Responsibilities to Fight an Air War

Main gate, Rapid City Army Air Base. The large water tower on the base is visible in the back right section of the photograph. The brick chimney for the hospital is visible to the left rear of the photograph. The base was home to the Second Air Force's Combat Crew Training School. In this posed photograph two Army guards stand on each side of the guard entrance building along with a jeep parked in the middle of the entrance with another Army security force member. *Ellsworth Air Force Base, 28th Bomb Wing Historian.*

The *pioneers* of Rapid City Army Air Base were on the job Friday, July 2, 1942. A cadre or nucleus of unrevealed strength arrived in the City last week and after becoming established in quarters began active organizational work. The cadre, consisting of both officers and enlisted men, does not have a flying role. Its work eventually will be to *keep' em flying*, but the group is now here to pave the way for air operations. Major John R. Bollinger, Adjutant of the base by Colonel Charles B. Oldfield, Base Commander…*Our job is administrative. We relieve flying officers of ground work and that means speeded up training.*[1]

Second Air Force's mission was to train the absolute best mission-ready graduates to support combat readiness and fight a world war. It was activated on October 19, 1940, at McChord Field, Washington, as the Northwest Air District of the General Headquarters Air Force. It was redesignated Second Air Force on March 26, 1941, primarily assigned the mission to provide defense of the Pacific Northwest and Upper Great Plains of the United States. During the months after Pearl Harbor, Second Air Force organized air defense for the Northwest Pacific Ocean coastline of the United States (1940-1941), also flying anti-submarine patrols along the coast until October 1942. By this time any threat from Japanese attack had ended, especially after the Imperial Japanese Navy's defeat at Midway Island (four fleet carriers and their aircrew) and heavy air, naval, and ground force losses at Guadalcanal.

Second Air Force started supporting AAF's Training Command prior to October 1942, training AAF units, crews, and personnel for bombardment, fighter, and reconnaissance combat operations. Second Air Force became primary for training Boeing B-17 Flying Fortress and Consolidated B-24 Liberator bombardment groups (Heavy or H) and Rapid City Army Air Base training B-17 aircrews.[2]

Boeing B-17B. This aircraft is pictured after roll out from the Boeing factory in Seattle, Washington. It had a very streamlined appearance, but lacked defensive armament to fend off attacking enemy fighters. This aircraft was not equipped with machine guns (1937). Operational bombers were defended by five .30 caliber machine guns: one on each side of the fuselage aft of the wing, one top fuselage machine gun behind the pilots, one downward firing machine gun aft of the wings, and one mounted in the nose. The aircraft was not equipped with a tail gun, so on early operations enemy fighters attacked from the rear. *Courtesy Milton Kuhl of Fort Bragg, California, to Lt. Col. George A. Larson, USAF (Ret.)*

Probably the most famous American-produced bomber to fly during WWII was the Boeing B-17 Flying Fortress, although late in the war it was eclipsed in performance and size by the Boeing Superfortress during combat operations against Japan. B-29s dropped the two atomic bombs on Japan, ending the fierce fighting in the Pacific. Although it was used in all Combat Theaters of Operations during WWII, B-17s assigned to the Eighth Air Force became known as the bomber which flew attack missions in daylight, at high altitude, and unescorted – until the introduction of the North American P-51 Mustang fighter – deep into Nazi controlled Europe and Germany. Combat experience over Europe led to many improvements of the B-17 throughout the European war, especially to increase defensive armament against attacking German fighters. The final mass-produced variant, the B-17G, accounted for over two-thirds of Flying Fortress production.[3]

Boeing B-17F *Homesick Angel* on static display inside the main gate at Offutt Air Force Base, United States Strategic Command, across from the Officer's Club. *Lt. Col. George A. Larson, USAF (Ret.)*

The B-17F equipped most of the early trained bomb groups in 1942, such as those at Rapid City Army Air Base. Later bomb groups were equipped with the more capable B-17G, which accounted for over 60 percent of the Flying Fortresses built by America's bomber production plants. The B-17F was a production improvement over the B-17E. Defensive armament increased to eleven .50 caliber machine guns, with added cheek guns to provide better production against head-on attacking German fighters. More powerful engines and improved propeller design increased maximum bomb loads to 8,000 pounds, although if this load was carried, it was at a cost of a decrease of 70 mph in maximum cruising speed. Therefore, the B-17F usually carried less than a maximum capable bomb load to maintain speed, reducing vulnerability to attacking German fighters and 88mm flak guns.

BOEING B-17F SPECIFICATIONS AND PERFORMANCE INFORMATION

Wingspan	103 feet 9-inches
Wing area	1,420 square feet
Tail height	19 feet 1-inch
Power plant	(4) 1,200 hp Wright R-1820-97 Cyclone radial engines
Take-off power	Rated 1,000 hp at 25,000 feet Rated 1,000 hp at 25,000 feet
Maximum speed	300 mph at 25,000 feet 325 mph on emergency power
Cruising speed	160 mph at 5,000 feet 200 mph at 25,000 feet
Landing speed	90 mph
Service ceiling	37,500 feet
Combat range	1,300 miles with 6,000 pound bomb load
Ferry range	2,880 mile
Empty weight	34,000 pounds
Loaded weight	40,437 pounds
Combat weight	56,500 pounds
Fuel	2,530 gallons
Tokyo tanks	3,612 gallons
Defensive guns	(3) .30 caliber machine guns in nose, left and right nose cheek (2) .50 caliber machine guns in Sperry power turret, dorsal (2) .50 caliber Browning M2 machine guns in Sperry power turret, ventral (1) .50 caliber Browning M2 machine gun left waist gunner's position (1) .50 caliber Browning M2 machine gun right waist gunner's position (1) .50 caliber Browning M2 machine gun roof of radio operator's compartment (2) .50 caliber Browning M2 machine guns tail gunner's position
Number built	Boeing (2,300) Douglas (605) Lockheed-Vega Burbank (500)

33

Boeing B-17G on display at the Castle Air Museum at Atwater, California. Because of losses from forward attacking German fighters against daylight raids by Eighth Air Force B-17s, Boeing engineers modified the bomber's design. A forward-firing power-operated Bendix chin turret was added, equipped with two Browning M2 .50 caliber machine guns. This increased the amount of machine gun fire which could be brought to bear on forward attacking German fighters. *Lt. Col. George A. Larson, USAF (Ret.)*

34

BOEING B-17G SPECIFICATIONS AND PERFORMANCE INFORMATION

Wingspan _____ 103 feet 9-inches

Wing area _____ 1,420 square feet

Length _____ 74 feet 9-inches

Tail height _____ 19 feet 1-inch

Powerplant _____ (4) 1,200 hp Wright R-1820-97 Cyclone radials

Take-off power with _____ 1,000 hp at 25,000 feet

GE B-22 turbo

superchargers

Emergency Power _____ 1,380 hp

Maximum speed _____ 260 mph at 25,000 feet

300 mph on emergency power

Cruising speed _____ 150 mph at 25,000 feet

Landing speed _____ 90 mph

Service ceiling _____ 35,000 feet

Combat range _____ 1,850 miles with 4,000 pound bomb load

Number built _____ Boeing (4.035)

Douglas Long Beach (2,395)

Lockheed-Vega Burbank (2,250)

Empty weight _____ 32,750 pounds

Loaded weight _____ 55,000 pounds

Combat weight _____ 72,000 pounds

Fuel load _____ 2,520 gallons

Tokyo tanks _____ 3,612 gallons

Defensive guns _____ (2) .50 caliber Browning M2 machine guns. One in each side of nose, cheek positions

(2) .50 caliber Browning M2 machine guns in Bendix chin turret under nose

(2) .50 caliber Browning M2 machine guns in Sperry power turret, dorsal position

(2) .50 caliber Browning M2 machine guns in Sperry power turret, ventral position

(1) .50 caliber Browning M2 machine Gun on right side, waist gunner's position

(1) .50 caliber Browning M2 machine gun on left side, waist gunner's position

(2) .50 caliber Browning M2 machine guns in tail gunner's position [4]

The standard B-17 combat crew consisted of 10 crewmen (four officers: Pilot, Copilot, Navigator, and Bombardier; and six enlisted: Flight Engineer, Radio Operator, Ball Turret Gunner, Left Waist Gunner, Right Waist Gunner, and Tail Gunner).[5]

World War II B-17 Aircrew Training

The Eighth Air Force prepared a detailed statement of the duties and responsibilities of the Pilot/Airplane Commander:

Your assignment to the B-17 airplane means that you are no longer just a Pilot. You are now an Airplane Commander, charged with all the duties and responsibilities of a command post.

You are now flying a 10-man weapon. It is your airplane, and you crew. You are responsible for the safety and efficiency of the crew at all times-not just when you are flying and fighting, but for the full 24-hours every day while you are in command.

Your crew is made up of specialists. Each man-whether he is the Navigator, Bombardier, Engineer, Radio Operator, or one of the Gunners-is an expert in his line. But how well he does his job, and how efficiently he plays his part, as a member of your combat team, will depend to a great extent on how well you play your part as Airplane Commander.

Photo of the pilot's station in a B-17G. *Lt. Col. George A. Larson, USAF (Ret.)*

Get to know each member of your crew as an individual. Know his personal idiosyncrasies, his capabilities, his short comings. Take a personal interest in his problems, his ambitions, his need for specific training.

See that your men are properly quartered, clothed, and fed. There will be many times, when your airplane and crew are away from the home base, when you may even have to carry your interest to the extent of financing them yourself. Remember always that you are the commanding officer of a miniature army-a specialized army; and that morale is one of the biggest problems for the commander of any army, large or small.

Crew Discipline

Your success as the Airplane Commander will depend in a large measure on the respect, confidence, and trust, which the crew feels for you. It will depend also, on how you maintain crew discipline.

Your position commands obedience and respect. This does not mean that you have to be stiff-necked, over bearing, or a loaf, such characteristics most certainly will defeat your purpose. Be friendly, understanding, but firm. Know your job; and, by the way, you perform your duties daily, impress upon the crew that you do know your job. Keep close to your men, and let them realize that their interests are upper most in your mind. Make fair decisions, after due consideration of all the facts involved; but make them in such a way as to impress your crew that your decisions are to stick. Crew discipline is vitally important, but it need not be as difficult a problem as it sounds. Good discipline in an aircrew needs comradeship and high morale, and the combination is unbeatable.

You can be a good CO, and still be a regular guy. You can command respect from your men, and still be one of them.

To associate discipline with informality, comradeship, a leveling of rank, and at times a shift in actual command away from the leader, may seem paradoxical, says a Brigadier General, formerly a group commander, VII Bomber Command. *Certainly, it isn't down the military grave. But it is discipline just the same-and the land of discipline that brings success in the air.*

Crew training

Train your crew as a team, keep abreast of their training. It won't be possible for you to follow each man's courses of instruction, but you can keep a close check on his record and progress.[6]

Second Air Force, which conducted of heavy bombardment training in the United States, divided training into three phases. Until the end of 1943, each of the phases was usually given at a different base, but that arrangement was then abandoned in favor of giving the entire program at one Operational Training Unit (OTU) station. During the first phase, individual crew members received instruction in their specialties, particular attention being given to instrument and night flying exercises for Pilots, cross-country tests for Navigators, target runs for Bombardiers, and air-to-air firing for Gunners. During the second phase, teamwork of combat crew was stressed. Rapid City Army Air Base primarily conducted second phase OTU training. This included bombing, gunnery, and instrument night missions. The third phase aimed at developing effective unit operation, the goal of the entire program. It included extensive exercises in high-altitude formation flying, long-range navigation, target identification, and simulated combat missions.

When the individual Pilot, Gunner, or other flying specialist arrive at the OTU or Replacement Training Unit (RTU) station, his main concern was the character of his crew. The crew was the family of the Army Air Force; each member knew that long hours of work, play, anxiety and danger would be shared. Naturally, each man hoped to be assigned to a crew in whose members he had confidence and with whom he would be congenial. The assignment process was almost entirely a matter of checking names from alphabetical rosters, but the men so assigned generally accepted each other and adjusted gradually to the mixture of backgrounds and temperaments. If trouble flared, reassignment of individuals could always be made. To each member of the crew a vital part of the operational training period was learning about the personalities, as well as duties of his crew mates. Much of the OTU-RTU instruction was given on the ground in classrooms, hangars and on gunnery ranges. Air training was conducted chiefly through informal supervision of flight operations.[7]

The Pilot must get to know each man's duties and problems to know his job, and they try to devise ways and means of helping him to perform it more efficiently.

Each crew member naturally feels great pride in the importance of this particular specialty. You can help him to develop his pride to include the manner in which he performs that duty. To do that you must possess and maintain a thorough knowledge of each man's job and the problems he has to deal with in the performance of his duties.

Duties and Responsibilities of the copilot.

The Copilot is the executive officer-your chief assistant, under study, and strong right arm. He must be familiar enough with everyone of your duties-both as Pilot and as Airplane Commander-to be able to take over and act in your place any time.

He must be able to fly the airplane under all conditions as well as you would fly yourself.

He must be extremely proficient in engine operation, and know instinctively what to do to keep the airplane flying smoothly even though he is not handling the controls.

He must have a thorough knowledge of cruising control data, and know how to apply it at the proper time.

He is also the Engineering Officer aboard the airplane, and maintains a complete log of performance data.

He must be a qualified instrument pilot.

He must be able to fly good formation in any assigned position, day or night.

He must be qualified to navigate by day or at night by pilotage, dead reckoning, and by use of radio aids.

He must be proficient in the operation of all radio equipment located in the Pilot's compartment.

In formation flying, he must be able to make engine adjustments almost automatically.

He must be prepared to take over on instruments when the formation is climbing through an overcast, thus enabling you to watch the rest of the formation.

Always remember that the Copilot is a fully trained, rated Pilot, just like yourself. He is subordinate to you only by virtue of your position as the Airplane Commander. The B-17 is a lot of airplane; more airplane than any one pilot alone can handle over a long period of time. Therefore, you have been provided with a second pilot who will share the duties of flight operation.

Copilot's position on a B-17G. _Lt. Col. George A. Larson, USAF (Ret.)_

38

Treat your Copilot as a brother Pilot. Remember that the more proficient he is as a Pilot, the more efficiently he will be able to perform the duties of the vital post he holds as your second in command.

Be sure that he is allowed to do his share of flying in the Pilot's seat, on takeoffs, landings, and on instruments.

The importance of the copilot is eloquently testified to by Airplane Commanders overseas. There have been many cases in which the Pilot has been disabled or killed in flight and the Copilot has taken full command of both airplane and crew, completed the mission, and returned safely to the home base. Usually, the Copilots who have distinguished themselves under such conditions have been Copilots who have been respected and trained by the airplane commander as Pilots.

Bear in mind that the Pilot in the right-hand seat of your airplane is preparing himself for an Airplane Commander's post too. Allow him every chance to develop his ability and to profit by your experience.

Duties and Responsibilities of the Navigator.

The Navigator's job is to direct your flight from departure to destination and return. He must know the exact position of the airplane at all times.[8] An experienced Navigator, during training at the OTU, would accompany a new team on a practice mission. During the course of the trip, he would observe the recently graduated Navigator, check his techniques, and offer suggestions for improvement. At the conclusion of each mission, the *instructor* would file a report on the progress of the *student*. Informal teaching of this kind was the rule for the other crew positions involving the coordinated use of crews, or large elements, were often demonstrated by experienced crews before the new units attempted them. Teaching methods in the operational programs, both on the ground and in the air, were not strictly standardized.[9]

Navigator's position and table in a B-17G. *Lt. Col. George A. Larson, USAF (Ret.)*

Navigation is the art of determining geographic positions of (a) pilotage, (b) dead reckoning, (c) radio, or (d) celestial navigation, or any combination of these four methods. By any one or combination of methods, the Navigator determines the position of the airplane in relation to the earth.

(a) Pilotage

Pilotage is the method of determining the airplane's position by visual reference to the ground. The importance of accurate pilotage cannot be over-emphasized. In combat navigation, all bombing targets are approached by pilotage, and in many theaters, the route is maintained by pilotage. This requires not merely the vicinity type, but pin-point pilotage. The exact position of the airplane must be known not within 5 miles but within 1/4 of a mile.

The Navigator does this by constant reference to ground sheets and ETA's established for points ahead, the ground, and to his maps and charts. During the mission, so long as he can maintain visual contact with the ground, the Navigator can establish these pinpoint positions so that the exact track of the airplane will be known when the mission is completed.

(b) Dead Reckoning.

Dead reckoning is the basis of all other types of navigation. For instance, if the Navigator is doing pilotage and computes ETA's for points ahead, he is using dead reckoning.

Dead reckoning determines the position of the airplane at any given time by keeping an account of the track and distance flown over the earth's surface from the point of departure or last known position.

Dead reckoning can be sub-divided into two classes.

(1). Dead reckoning as a result of a series of known positions obtained by some other means of navigation.

For example, you, as Pilot, start a mission from London to Berlin at 25,000-feet. For the first hour, your Navigator keeps track by pilotage; at the same time recording the heading and air speed, which you are holding. According to plan, at the end of the first hour the airplane goes above the clouds, thus losing contact with the ground. By means of dead reckoning from his last pilotage point, the Navigator is able to tell the position of the aircraft at anytime. The first hour's travel has given him the wind prevalent at altitude, and the track and ground speed being made. By computing track and distance from the last pilotage point, he can always tell the position of the airplane. When your airplane comes out of the clouds near Berlin, the Navigator will have a very closed approximation of his exact position, and will be able to pick up pilotage points quickly.

(2). Dead reckoning as a result of visual references other than pilotage.

When flying over water, desert, or a barren land, where no reliable pilotage points are available, accurate DR navigation can be performed. By means of drift meter, the Navigator is able to determine drift, the angle between the heading of the airplane and its track over the ground. The true heading of the airplane is obtained by application of compass error to the compass reading. The true heading, plus or minus the drift (as read on the drift meter), gives the track of the airplane. At a constant airspeed, drift on two or more headings will give the Navigator information necessary to obtain the wind by use of his computer. Ground speed is computed easily once the wind, heading, and air speed are known. So by constant recording of true heading, true airspeed, drift, and ground speed, the Navigator is able to determine accurately the position of the airplane at any given time. For greatest accuracy, the Pilot must maintain constant courses and air speeds. If course or airspeed is changed, notify the Navigator so he can record these changes.

(c). Radio.

Radio navigation makes use of various radio aids to determine position. The development of many new radio devices has increased the use of radio in combat zones. However, the ease with which radio aids can be jammed, or bent, limits the use of radio to that of a check on DR and pilotage. The Navigator, in conjunction with the radio man, is responsible for all radio procedures, approaches, etc., that are in effect in the theater.

(d). Celestial.

Celestial navigation is the science of determining position by reference to two or more celestial bodies. The Navigator uses a sextant, accurate time, and many tables to obtain what he calls a line of

40

position. Actually, this line is part of a circle on which the altitude of the particular body is constant for that instant of time. An intersection of two or more of these lines gives the Navigator a fix. These fixes can be relied on as being accurate within approximately 10 miles. One reason for inaccuracy is the instability of the airplane as it moves through space, causing acceleration of the sextant bubble (a level denoting the horizontal). Because of this acceleration, the Navigator takes observation over a period of time so that the acceleration error will cancel out to some extent. If the Navigator tells the Pilot when he wishes to take an observation, extremely careful flying on the part of the pilot during the few minutes it takes to make the observation will result in much greater accuracy. Generally speaking, the only celestial navigation used by a combat crew is during the delivering flight to the theater. But in all cases celestial navigation is used as a check on dead reckoning and pilotage except where celestial is the only method available, such as on long over-water flights, etc.

Instrument Calibration.

Instrument calibration is an important duty of the Navigator. All navigation depends directly on the accuracy of his instruments. Correct calibration requires close cooperation and extremely careful flying by the Pilot. Instruments to be calibrated include the altimeter, all compasses, airspeed indicators, alignment of the Astro compass, astrograph, and drift meter, and check on the Navigator's sextant and watch.[10]

Requirements for Navigators on or before a mission.

1. Attend briefing: understand flight plan, know the targets, draw necessary maps, and plot enemy flak positions. Also, check individual Navigators' equipment to see that it is complete.

2. Between stations and take-off time he will:

a. Brief crew on fighter rendezvous, flak concentrations, best known route if forced to leave formation, and friendly airfields should they make an emergency landing on the continent or England.

b. Check his guns.

c. Check for flare pistol and Aldis lamp.

d. Run the Flux-Gate compass through the caging cycle after engines have run for 10 minutes.

3. During the mission, the Navigator will:

a. Notify Pilot of all turns and climb at least two minutes in advance.

b. Notify the crew when friendly aircraft should be seen.

c. Notify Radio Operator when camera should be turned on and make entry of the same in log.

d. Notify the crew of how time schedule is being followed and deviations noted in log.

e. Notify Pilot when flak concentrations may be expected.

f. Co-ordinate bomb run with Pilot.

g. Continue with navigation until actual fighter attacks occur, keep track of approximate position during attack.

h. Keep a complete log of which includes the following:

(1). Time, altitude and position of assembly with other squadrons and groups.

(2). Continuous record of plane's heading, altitude and airspeed.

(3). Time and altitude of crossing the English and Continent coasts.

(4). Position report on landfalls, IP and at least once every 30 minutes during the entire flight.

(5). Wind vector recordings at least once over England, one over the water, one over target area, and one on return.

4. Recordings for Intelligence:

a. Accurate time entries made of flak, the intensity and accuracy of fire and if possible the position of batteries.

b. Fighter attacks should be logged as to time and place.

c. Enemy shipping, camouflage, smoke screens and other details of military value.

d. Any news concerning friendly aircraft in distress should be accurately given to Hot News before interrogation.

5. Recording for Air Force:

a. Time of bombs away to the half minute.

b. Magnetic heading over target.

c. Indicated airspeed and altitude over target.

d. Wind vector determined near target.

6. Recording for Meteorology:

a. Weather report must be taken by each Navigator on mission noting height, and nature of low, medium, high clouds, also recording of vapor trails, temperature.

7. Recording for Signals:

a. Splasher and fix beacons should be checked on the radio compass and characteristics noted.

b. Beaconing and jamming should be noted on all beacons used. Enemy interference can be noted in one of these ways:

(1). Wide swinging of the compass needle when the identifying letters are sent out.

(2). More than two oral nulls are registered when the loop antenna is rotated through 360 degrees.

(3). A series of dots super-imposed on the call sign by the British implies that they have detected interference on Splasher beacons (known as mutilating).

c. Gee should be checked for jamming, the last fix in enemy territory and the first fix possible on the way out.

d. Care should be taken in tuning radio beacons and bunches because of the great numbers of stations in England.[11]

Pilot-Navigator Preflight Planning.

1. Pilot and Navigator must study flight plan of the route to be flown and select alternate airfields.

2. Study the weather with the Navigator. Know what weather you are like to encounter. Decide what action is to be taken. Know the weather conditions at the alternate airfields.

3. Inform your Navigator at what airspeed and altitude you wish to fly so that he can prepare his flight plan.

4. Learn what type of navigation he intends to use: pilotage, dead reckoning, radio, celestial, or a combination of all methods.

5. Determine check points; plan to make radio fixes.

6. Work out an effective communication method with your Navigator to be used in flight.

7. Synchronize your watch with your Navigator's.

Pilot-Navigator in Flight.

1. Constant course for accurate navigation, the pilot-you-must fly a constant course. The Navigator has many computations and entries to make in his log. Constantly changing course can make his job more difficult. A good Navigator is supposed to be able to follow the pilot, but he cannot be taking compass readings all the time.

2. Constant airspeed must be held as nearly as possible. This is as important to the Navigator as is a constant course in determining position.

3. Precision flying by the Pilot greatly affects the accuracy of the Navigator's instrument readings, particularly celestial readings. A slight error in celestial reading can cause considerable error in determining position. You can help the Navigator by providing as steady a platform as possible from which he can take readings. The Navigator should notify you when he intends to take readings so that the airplane can be leveled off and flown as smoothly as possible, preferably by using the automatic pilot. Do not allow your Navigator to be disturbed while he is taking celestial readings.

4. Notify the Navigator of any change in flight, such as change in altitude, course or airspeed. If change in flight plan is to be made, consult the Navigator. Talk over the proposed changes so that he can plan the flight and advise you about it.

5. If there is doubt about the position of the airplane, Pilot and Navigator should get together, refer to the Navigator's flight log, talk the problem over and decide together the best course of action to take.

6. Check you compass at intervals with those of Navigator, noting any deviation.

7. Require your Navigator to give position reports at intervals.

42

8. You are ultimately responsible for getting the airplane to is destination. Therefore, it is your duty to know your position at all times.

9. Encourage your Navigator to use as many navigation methods as possible as a means of double-checking.

Post-Flight Critique.

After every flight get together with the Navigator and discuss the flight and compare notes. Go over the Navigator's log. If there have been serious navigation errors, discuss them with the Navigator and determine their course. If the Navigator has been at fault, caution him that it is his job to see that the same mistake does not occur again. If the error has been caused by faulty instruments, see that they are corrected before another navigation mission is attempted. If your flying has contributed to inaccuracy in navigation, try to fly a better course next time.

Miscellaneous Duties.

The Navigator's primary duty is navigating your airplane with a high degree of accuracy. But as a member of the team, he must also have a general knowledge of the entire operation of the airplane.

He has a .50 caliber machine gun at his station, and he must be able to use it skillfully and to service it in emergencies.

He must be familiar with the oxygen system, know how to operate the turrets, radio equipment, and fuel transfer systems.[12]

The experiences of the B-17 aircrew at Ellsworth, or at least of one airman in particular, can be recounted in detail by Paul Morcello.

Paul Marcello is the current Ellsworth Air Force Base 28th Bomb Wing Historian, assigned to this position as of February 2011. History is one of his passions, and his career path was set while a young child rummaging around his grandfather's house. "When I was a child, I found my grandfather's uniform and was immediately fascinated by the ribbons and patches. I wanted to learn what they meant. Since then, I've dedicated my life to learning all I can about the military and its unique past." This same grandparent significantly links him to the history of Ellsworth Air Force Base (Rapid City Army Air Base in WWII). "My grandfather Al (Alfred) Marcello, is not only responsible for my interests in history, but it just so happens he was here at Ellsworth some 70 years ago to receive aircrew training before shipping out to England with the 100th Bomb Wing during World War II. It is almost as if my coming here was meant to be. What I do today to record events around the base benefits the people of tomorrow. When future generations are interested in knowing what the 28th BW was doing at this point in history, or they're researching a grandparent's military contributions, they can use the records I'm keeping now."[13]

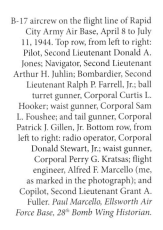

B-17 aircrew on the flight line of Rapid City Army Air Base, April 8 to July 11, 1944. Top row, from left to right: Pilot, Second Lieutenant Donald A. Jones; Navigator, Second Lieutenant Arthur H. Juhlin; Bombardier, Second Lieutenant Ralph P. Farrell, Jr.; ball turret gunner, Corporal Curtis L. Hooker; waist gunner, Corporal Sam L. Foushee; and tail gunner, Corporal Patrick J. Gillen, Jr. Bottom row, from left to right: radio operator, Corporal Donald Stewart, Jr.; waist gunner, Corporal Perry G. Kratsas; flight engineer, Alfred F. Marcello (me, as marked in the photograph); and Copilot, Second Lieutenant Grant A. Fuller. *Paul Marcello, Ellsworth Air Force Base, 28th Bomb Wing Historian.*

43

During WWII Alfred Marcello was assigned to a B-17 aircrew training at Rapid City Army Air Base, commanded by Second Lieutenant Donald (Don) Jones. The members of Don Jones' crew met each other for the first time in the Badlands of South Dakota in the summer of 1944. From April 8 to July 11, 1944, the crew trained at Rapid City Army Air Base. It was here that each man had the opportunity to hone his skills in the positions which the USAAF had schooled them in and assigned to them. The crew members began to familiarize with their jobs as well as each other. Each member would be relying on the other nine for his very life; it was imperative that they develop an *esprit de corps*.

For the pilots there was an emphasis on both ground school and flying proficiency. They spent no time on formation flying. However, the combination of air to ground gunnery flights, navigational flights, practice bombing, and night navigation assured that they gained considerable time in the B-17.[14] It was here that Don Jones proved himself to be a natural leader. He was a student pilot, but understood what was necessary to make his crew successful.

Practice bombing occurred over targets set up in the South Dakota Badlands. On one such flight, Art Juhlin remembered being terribly low as the plane dipped and climbed over the rocky landscape. As he sat at the Navigator's table, he thought that Jones was a bit scary, but quite a pilot. Just then he was brushed back and noticed Don Jones behind him. Now thinking it was Grant Fuller flying the plane, Juhlin again had the same thought; this is some great flying. Upon getting up and peering into the astrodome of the nose he noted, "Al was flying it! The flight engineer was flying the plane. It was a wild ride, but he was pretty good!"[15]

Since most of the crew had some pilot training, Don Jones permitted this privilege from time to time. It suggests his approach to leadership, qualities that nurture a successful crew and unit cohesion. Usually the crew members focused on their specialty, but Jones generously helped crew members put aside hard feelings about being washed out of pilot schools. His egalitarian approach to leadership encouraged the crew to work as a team and to rely on each other to accomplish the job at hand.

(Author's note: the following account summarizes how trained aircrew from Rapid City Army Air Base made their way from South Dakota to their Eighth Air Force assignment in England for air combat missions against Nazi held territory and Germany.) Once crew training was completed, the Jones crew prepared to go overseas and begin their combat tour. Grant Fuller remembered:

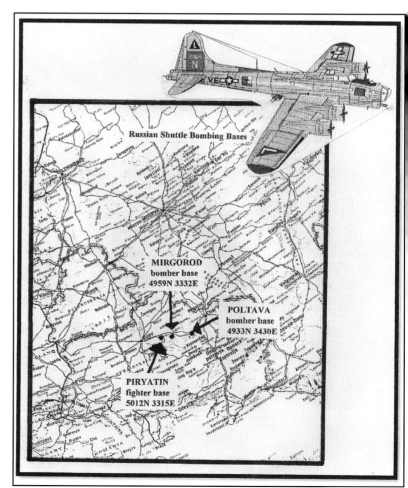

Russian Shuttle Bombing Bases

MIRGOROD
bomber base
4959N 3332E

POLTAVA
bomber base
4933N 3430E

PIRYATIN
fighter base
5012N 3315E

When we left Rapid City we went by train to Kearney, Nebraska where we checked out a new plane. Jones had to sign for the B-17 and the cost listed was $125,000 so we knew right then we had better get it over to England safely. We left Kearney one night and flew to Grenier Field, Manchester, New Hampshire (7 hours and 40 minutes). The next day we flew during the day to Gander Bay, Newfoundland (5 hours and 5 minutes). We were weathered in there for nine days and on July 29, 1944 at midnight we left for Ireland. Twelve hours and thirty minutes later we landed at Shannon. We turned the plane over to the proper authorities and never saw it again. We were sent to a replacement depot just west of London and from there, were assigned to the 100th, arriving there August 16, 1944.[16]

Locations of the three Russian airfields behind the front lines of the Russian Army facing German Army troops. *Drawing: Lt. Col. George A. Larson, USAF (Ret.)*

44

It is impossible to cover all of this single crew's combat missions after leaving Rapid City Army Air Field. I have selected their participation in the historic shuttle bomb missions from England to targets in Nazi held territory, landing at airfields in Russia and then returning, and bombing German targets on their return flight to England, as an example.

Eighth Air Force planners believed that to maximize bomber operations against Germany, shuttle bombing missions flown from widely scattered USAAF bases would place the greatest possible strain on Luftwaffe fighter resources. It was reasoned that if shuttle bases could be established in Russia, USAAF B-17s could range across all strategic targets located in Germany and occupied areas, with many refits.

The Allies gave the Russian shuttle bombing proposal the code name *Operation Frantic*, and in October 1943, USAAF General Henry H. Arnold convinced the U.S. Combined Chiefs of Staff in Washington to send a delegation to Russia to seek permission from Soviet leaders for B-17 and long-range fighter escort recovery and reconstitution airfields to be prepared. These were needed to support refueling, rearming, and repairing of the aircraft for return bombing missions over Germany. In November 1943, the American Military mission to Russia departed from Washington to talk to the Russian military and political leadership in the Kremlin about the viability of such flights. It was led by Major General John R. Deane. At the meeting the Soviet leadership agreed to the concept, but did not give formal approval until after the Teheran Conference (November-December 1943), at which point Stalin authorized his Army officers to prepare six airfields for USAAF B-17s and their fighter escorts.

In February 1944, the Russian Army provided three airfields for USAAF use instead of the six previously agreed to by Stalin. These comprised B-17 airfields at Poltava and Mirgorod, and a fighter base at Piryatin. All three were located near Kiev (spelled Kijev in Russian during WWII), north of the Black Sea. Their proximity to Russian sea ports facilitated supply support, but the airfields were farther behind Russia's front lines than USAAF Generals wanted, increasing flying distances, and all three locations required extensive repairs and reconstruction.

Route of the last shuttle bombing mission flown from England on September 18 and 19, 1944. *Drawing, Lt. Col. George A. Larson, USAF (Ret.)*

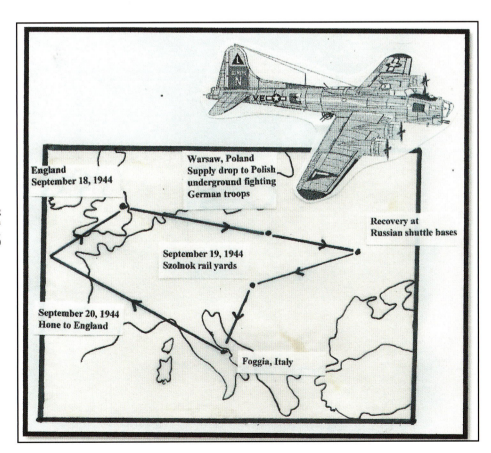

This meant supplies, equipment, and personnel had to be brought in. Supplies sent to Black Sea ports and Persian Gulf ports in Iran enabled Russian construction workers, most of which were women, to lengthen and strengthen the runways at the two proposed B-17 bases, and to improve facilities at all three. This included laying down steel Marston matting on muddy taxiways and parking areas and constructing maintenance hangars and control towers, as well as setting up open-air supply depots for fuel drums, bombs, ammunition, food, and aircraft spares. Guards had to be posted to protect these supplies from theft by Russian Army troops and local citizens who stole anything not nailed down or guarded, then sold it on the black market. It was the cost of doing business in Russia during the war.

Second Lieutenant Donald A. Jones and his crew, assigned to 100th BG, 418th BS, flew the September 18, 1944, B-17 shuttle bombing raid from its base at Thorpe Abbotts, England, Eighth Air Force, to Mirgorod, Russia. On August 1, 1944, the Polish underground in Warsaw radioed Allied Command Headquarters in London, England, for an emergency airdrop of food, ammunition, weapons, and medical supplies. The resistance movement went on the offensive against the Germans in their city as Soviet troops approached, believing they would soon receive their assistance. However, Soviet leaders had a different agenda, and were planning to allow these fighters to lead that country after the Germans had been defeated. The Russian Army halted its advance short of the city, allowing time for the Germans to put down the uprising. The Russians indicated the delay was necessary to replenish and restock their troops to resume the attack. In any event, they did not approve the airdrop until September 11th. This event caused further souring of relations between the Russian political leadership (Stalin) and the Allies (British Prime Minister Winston Churchill, who was already suspicious of Russian occupation of Eastern Europe and President Roosevelt).

Two Russian Army troops assist two USAAF mechanics in turning a B-17's propellers prior to take-off from Poltava Airfield for the return shuttle bombing mission back to England. *Dick Drain, 99th BG Historian to Lt. Col. George A. Larson, USAF (Ret.)*

It took time to organize and prepare the B-17s to drop supplies instead of bombs, and it was not until September 18th that 107 Flying Fortresses were loaded with specially prepared bundles fitted with parachutes. The bombers were escorted from their bases in England by a large escorting force of 150 P-51s. Shortly after take-off three B-17s aborted, returning to England. Two B-17s were lost to Luftwaffe fighters before the attack force began circling over Warsaw. For a period of two hours or so they dropped 1,284 supply containers, but it is estimated that only between 130-to-300 actually reached their intended recipients. The rest were retrieved by German troops. After the airdrop, 64 escorting P-51s flew back to England, while 86 continued on with the B-17s to the Russian airfields for recovery.[17]

The following is from Second Lieutenant Art Juhlin, Navigator on Second Lieutenant Jones' B-17 crew, mission number 5, September 15, 1944, Warsaw, Poland:

Purpose of mission was to drop supplies and arms to Polish partisans in Warsaw. Took off in very adverse weather. Shortly after leaving the English coast we hit the soup and couldn't even see our wingman. Were recalled, just before crossing the Danish coast. Really seated out the trip back, as you couldn't see a thing and it was every man for himself. Had a little excitement on our landing. Brakes didn't hold and so we ground looped [A violent turn of an airplane while taxiing, or during a landing or takeoff run. A ground loop may result from loss of control, or it may result from an intention to avoid an obstacle.] at the end of the runway, narrowly missing the other airplane and scattering ground personnel all over the place. No serious damage. If mission had come off as planned, we would have continued onto Russia after dropping our chutes. Mission Number 6, September 19, 1944, shuttle mission from England [parachute drop to partisans in Warsaw] to Mirgorod, Russia…Purpose of mission was to drop food and arms to the beleaguered Poles in Warsaw. Flew the Northern Sea en route to target. Weather over target was very poor and we spent an hour circling over the target, looking for openings in the overcast. At this time, some of our groups were attacked by ME-109s and several planes were lost. Finally, went over target at 13,500-feet and Jerry threw up everything he had at us. Between dodging flak, parachutes and other planes we had a rough time of it. Picked up a Russian fighter escort at the target and one of them put on quite an aerobatic show for us. Enroute to base crossed the Dnyfer River at Kiev and landed at Mirgorod, Russia. Time of flight was 10 1/2 hours.[18]

The following day they all took off from their Russian bases to attack the Szolnok Marshalling Yard. Afterwards they were received in southern Italy, and after crew rest and aircraft servicing, the B-17s and P-51s departed for England on September 22nd. The era of Russian shuttle bombing missions had come to an end.[19]

Juhlin now describes mission number 16, September 19, 1944, a shuttle raid from Mirgorod, Russia, to Foggia, Italy:

The target was a railway marshalling yard at Szolnok, Hungary which is about 45 miles east of Budapest. Passed over target at 18,000-feet and really plastered it. Flak over target was light but encountered moderate and extremely accurate flak while crossing Drina River at Brod, Yugoslavia on way to Italy. Time of flight was 6 1/2 hours.

September 22, 1944, Foggia, Italy, to Diss, England:

Flew back home without a bomb load and supposedly was to receive credit for a mission, as had previous groups. However, for some unknown reason they decided not to count it this time. This made our average time for each mission on the shuttle raid 12 3/4 hours. Trip back was regular sightseeing tour. Passed by Rome, Corsica, Nice Alps and Paris. Ran into quite a bit of bad weather after getting over the Channel, but didn't have too much difficulty getting back to base. Time of flight was 8 1/2 hours.[20]

Equipment on a B-17 was vital to each crew's survival on high-altitude, long-range missions over Nazi controlled Europe and into Germany:

I. Oxygen, General.

1. Anoxia is the lack of sufficient supply of oxygen to meet the body's needs. Anoxia overtakes one without warning and will result in anything from slight inefficiency to death. It is nearly always caused by carelessness on the part of the individual concerned. Remember, that an individual at 30,000-feet has useful consciousness for only 48 seconds if additional oxygen is not supplied.

2. Oxygen will be used above 10,000-feet in day time flying and from ground up on night flights.

3. An *oxygen check* will be made every five minutes by the Bombardiers as long as the crew is on oxygen.

4. Do not allow oil or grease to come in contract with oxygen equipment or fittings. The results may be a serious fire.

II. Oxygen Supply.

1. Eighteen G-1 cylinders per aircraft at 400 PSI equals 8 1/2 crew hours of oxygen if used properly.

2. The oxygen systems are grouped as follows:

a. Pilot, Navigator and Engineer.

b. Copilot, Bombardier and Engineer.

c. Left Waist Gunner, Radio Operator, Ball Turret Gunner and Bomb Bay.

d. Right Waist Gunner, Tail Gunner and radio room.

III. Oxygen Equipment.

1. Walk around cylinders are provided at each crew position.

a. D2 cylinders (long yellow) and A2 high pressure cylinders (short green) have endurance of approximately 30 minutes.

b. The Ball Turret Gunner will always have an A2 cylinder with A&B mask in the ball ready for immediate use.

2. Mask

a. Test for fit and disconnect tension in equipment room.

b. Modifications required: rubber patch at inlet ports, safety clip for disconnect.

c. Do not allow ice to cake over inlets or outlets.

d. A spare mask will be in the spare equipment bag.

3. Regulator

a. *ON* is the normal position for the auto mix. *OFF* is used only in testing or emergencies as at some altitudes oxygen will be used six times the normal rate and will cut down the endurance.

b. Tighten the knurled collar on the regulator after changing the position of the elbow.[21]

As much detail, training and preparation to be devoted to clothing, parachutes, flak suites and life vests.

I. Clothing

1. Dressing with F-2 electric suit.

a. Feet: One pair on medium weight wool socks, electric shoe, fleece line boots.

b. Waist: Winter underwear, electric trousers.

c. Chest: Winter underwear, OD shirt, electric jacket.

d. Hands: Silk liner, electric gloves.

2. Dressing with F-3 suit.

a. Underclothes: Use medium weight long woolen underwear and socks.

b. Use OD shirt and Trousers.

c. F-3 electric suit complete.

d. Outer clothing: A9 trousers and B-10 jacket.

48

3. Precautions

a. Adjust rheostat for minimum energy output.

b. Do not connect or disconnect shoes or gloves with power on.

c. Prevent sleeve tabs for shoes and gloves from touching metal on aircraft.

d. All electric clothing must be hung in the drying room when not in use.

4. Miscellaneous clothing

a. Tail gunners will wear either hoods or face and neck protectors.

b. Scarfs may be drawn from the equipment sections.

c. Four hand muffs are furnished for each aircraft.

d. A limited number of aircraft are equipped with electrically heated casualty bags.

e. A parachute bag containing blankets and spare clothing is provided each aircraft to use in emergencies only.

f. B-8 goggles (plastic lens) are recommended for use in combat.

g. Copilots will draw one pair of field glasses from the equipment section before each operational flight.

II. Parachutes

1. Handle the parachute pack gently and do not allow it to get wet or greasy.

2. It is advisable to have one side of the parachute pack strapped to the harness when in immediate danger.

3. Jumping suggestions

a. Make delayed jumps.

b. Dampen oscillation.

c. Face down wind.

d. Keep feet together.

e. Unhook snaps during descent or over water.

4. Use static lines to bail out wounded personnel.

5. Three short rings on alarm signal indicates *Prepare to bail out.*

III. Life Vests

1. To be used on all flights.

2. Inspect vest for full CO_2 cylinders, cap screwed down tight over CO_2 cylinder, closed mouth valves, and serviceable sea markers.

3. Do not inflate life vest until in the water and in need of additional buoyancy.

4. Use sea marker when rescue parties are in the area as the sea marker lasts for two hours and is visible for several miles.

IV. Be familiar with the location of first aid kits and fire extinguishers on the aircraft.[22]

Without the Bombardier properly trained and proficient in accurately dropping bombs on target, all efforts of a 10-man B-17 crew would be wasted. While the Air Corps had a splendid heritage of experience to build upon in the wartime training of pilots, it had no such advantage in developing instruction for bombardiers. Before 1940, the training of bombardiers was limited to on-the-job instruction in operational bombardment groups, both in the United States and at overseas bases. A small number of the trainees were pilots, but the majority were non-pilot enlisted men. Plans had been underway as early as 1939 for the establishment of a specialized school for individual bombardier training, and a course for instructors was established at Lowry Field, Colorado, in July 1940. It was not, however, until late in 1941 that training of students in specialized bombardier schools began. At first, the effort was made to conduct this training at fields where advanced pilot training was being given, but by the end of 1942, it was generally agreed that it was more satisfactory to conduct only one kind of training at one station. Specialized schools were established as rapidly as facilities would permit, and they expanded to meet the rising flow of students during 1942 and 1943. The students who had received pre-flight training before entering these bombardier schools were given a twelve week course until the middle of 1943; as soon as combat requirements for bombardiers declined the course was lengthened to eighteen weeks. By 1945, it was possible to provide twenty-four weeks of advanced individual training.

49

In addition to this specialized instruction, bombardier students from 1942 on were required to take a regular six-week flexible gunnery course. This training was received either before or after the bombardier course, depending upon when it could best be scheduled by the gunnery schools. Upon completion of training in gunnery and bombardiering, the students, who had become rated officers upon graduation, were assigned to one of the continental air forces for crew and unit training. When there was a shortage of bombardier graduates coming into the operational units, the continental air forces completed their crews with enlisted men who had received individual bombardment training in their own units. (This was part of aircrew training at Rapid City Army Air Base.) This program distracted from the primary mission of the air forces, which was unit training, and it was generally considered to be less efficient than the individual instruction conducted in regular specialized bombardier schools.

Flexible training gun. The gun did not fire live ammunition, but scored hits when the gun camera (mounted on top of the gun) film was developed and scored. This allowed instructors to review the gunner's proficiency and make suggestions for corrective procedures before going on to live firing at sleeve targets towed by aircraft. This gun is on display at the Commemorative Air Force Museum, Midland, Texas. *Lt. Col. George A. Larson, USAF (Ret.)*

Although most of the bombardiers who were trained up to the middle of 1943 were qualified only in that specialty, plus flexible gunnery, plans persisted for dual Bombardier-Navigator training. It was believed more efficient to carry as part of the bombardment team a man who was both Navigator and Bombardier, and the crews specified for the very heavy bomber (B-29 and B-32) called for two fully qualified Bombardier-Navigators. The chief obstacle to this training goal during the early part of the war was the chronic shortage of men trained in one or the other of the specialties, it proved difficult to keep for supplementary instruction a man already qualified and needed elsewhere. Another hindrance was the difficulty of securing men who were capable of being trained in both bombardment and navigation. As a result of these major deterrents, the AAF repeatedly postponed inauguration of its dual training plans, which had been in existence since September 1941. Insistent demands continued to be made, however, for some type of Bombardier-Navigator course. During 1942, small numbers of qualified Navigators were given an abbreviated course on the D-8 (low-altitude) bombsight, and were then assigned to medium bombardment units as Navigator/D-8 Bombardier. In the early part of 1943, a short supplementary course in dead-reckoning navigation was given to a small number of qualified Bombardier graduates. Complete dual training to produce precision Bombardier/celestial Navigator crewmen was initiated in January 1943 and was continued on a limited scale. In July, the regular Bombardier curriculum was enlarged to include dead-reckoning navigation. This change appeared to be the most feasible answer to the demand for some form of dual training.

One of the principal criticisms of early training was the lack of attention given to analysis of the causes of bombing errors. Bombardier graduates appeared, in general, to be deficient in knowledge of their equipment and inaccurate in making fundamental computations. Shortages of qualified instructors and proper equipment and lack of training time were considered to be the chief causes for the deficiency.

After numerous conferences, a standard eighteen-week course was adopted in June 1943 and continued in effect, with minor variations, until the end of the war. The ground instruction, which was the foundation for air exercises, consisted of some 425 hours. The largest portion of this time was given to critiques of air missions and preflight inspection of bombing equipment. Other parts of the ground training dealt with bombing and bomb equipment. The most important of these were on basic theory and the bombsight; students became acquainted with the forces acting upon a falling projectile and with the principles of a bombsight. Both Norden and Sperry sights were studied until 1944, but beginning in that year instruction was limited to Norden equipment.

Another subject was bombing accessories, which included the study of bomb racks, fuzes, aerial cameras, and radio. Separate courses were given on flight instruments and the elements of bombsight maintenance and calibration. In the course on bombing procedure, most of the student's time was spent on a synthetic bomb trainer, which served as a device for transition from ground to air training. A special course was also

50

devoted to the C-1 automatic pilot, which the Bombardier normally operated during actual bombing runs. Analyzing results of practice bombing and explaining scoring methods made up the bombing analysis course, and the basic AAF doctrines of employment of air power were expounded in a course called bombardment aviation. In addition to the subjects related only to bombing, some dealt with elementary navigation. Training was given in pilotage, dead-reckoning, and computers; the graduate Bombardier was therefore qualified in elementary navigation, as well as in his primary specialty. Several courses common to the preparation of other aircrew members – weather, flight planning, oxygen indoctrination, aircraft and naval recognition, and continuing practice in radio code – rounded out the ground school curriculum. Military instruction and physical condition were practically the same as in pilot training.

The Norden bombsight was named after its inventor, C.L. Norden. It is a gyroscopically-stabilized synchronizing bombsight used mainly for synchronous bombing, but was also useful for fixed-angle bombing. Utilizing preset data and manual operation by the Bombardier, the Norden bombsight computes the correct dropping angle and, in connection with an automatic pilot or pilot direction indicator, determines the proper course of the aircraft required to maintain the necessary line of sight to the target. This Norden bombsight is on display at the Commemorative Air Force Museum, Midland, Texas. *Lt. Col. George A. Larson, USAF (Ret.)*

Air training, eagerly anticipated by the Bombardier students, did not begin until after three weeks of preparatory ground instruction. It consisted of a total of some 120 hours and was divided into two parts: the instructional and qualification stage, and the combat stage. On his first several flights the student learned the feel of the bombsight in the air. He made several dry runs over simulated targets without having dropped any bombs from the plane. For about a week he flew over practice ranges, developing his ability to sight by aiming at various ground objects, such as bridges, farmhouses, roads, factories, and the like.

A one hundred pound inert practice bomb, AAF designation M38A2, on display at the Commemorative Air Force Museum, Midland, Texas. *Photograph: Lt. Col. George A. Larson, USAF (Ret.)*

51

World War II B-17 Aircrew Training

Practice bombs were then dropped under close supervision by the instructor. If the student's progress was considered satisfactory, he was permitted to bomb for record and qualification; otherwise he was subject to elimination or further preparatory training. Ordinarily about 100 bombs were dropped during the qualification phase of the Bombardier's training. At least seven recorded missions, four by day and three by night, were required; in order to qualify for the combat stage, the student's average circular error for these missions could not exceed 230-feet. Bombing runs during the qualification phase were generally from two to three minutes in duration.

The combat stage of training tried to improve bombing proficiency, and to simulate the diverse conditions which could be expected in combat. Some sixty bombs were generally dropped during this phase, thirty of them without the instructor present in the airplane, and all releases were scored for the record. Circular error was the basis for scoring until early 1943, when the method was changed to counting only hits and misses. It was believed that the latter method provided a stronger incentive for the student to strive for the greatest possible accuracy. The required standard of proficiency, established in 1943, was a minimum of 22 percent hits. Combat training missions were exacting; continuous evasive action was exacting; continuous evasive action by the pilot of the aircraft was required within a ten-mile radius of the target; and final approaches had to be straight and level and could not exceed sixty seconds.

Most combat missions were flown with the aid of the C-1 automatic pilot, a device for keeping the aircraft on a set course. The Bombardier, as he prepared to make his run, engaged a clutch on his bombsight that gave him control of the plane's direction. As he made adjustments on the sight in order to hold the hairline on target, the aircraft was automatically guided to the required course. A certain number of manual missions was also flown. These were accomplished with the aid of the pilot's directional indicator, which was first zeroed by the Bombardier when the plane was lined up on a collision course with the target. The pilot then made the necessary adjustments of stick and rudder to hold the indicator on zero. Not until 1944 did it become possible to provide enough cameras to permit all scoring to be done by photographic record. During 1943 and 1944 interest was stimulated by bombing contests among the several schools.

The Bombardier schools had difficulty, as did practically every military training program during the war, in procuring adequate training staff. Instructors were needed for ground classes as well as for the air phase training. Some military personnel with flying experience were made available for these assignments, but most of the instructors were commissioned directly from civilian life, or were drawn from the ranks of graduating classes at the schools themselves. Equally as important as the instructors were the bomb-approach pilots. The effectiveness of the Bombardier was dependent in large measure upon the proficiency exercised in piloting the airplane; this held true not only when the pilot controlled the plane manually, but even when the automatic pilot was used. In both instances, the pilot needed to understand the bombing problem and how to operate the equipment successfully. A general policy was developed of rotating bomb-approach pilots to operational units and replacing them with new graduates of advanced two-engine pilot schools.

Training manuals, books, films, and mock-ups were produced with considerable success in the Bombardier program. Most of the materials, as well as ground school teaching outlines, were at first prepared locally; during the latter part of the war numerous aids were produced at higher headquarters for distribution to the schools. One of the most generally used mock-ups was a device to demonstrate bombing theory; it showed the influence of airspeed, altitude, drift, and other factors upon the course of a falling projectile. While devices such as this one had the advantage of arousing student interest, many instructors believed that the same principles could be taught as easily by using the blackboard or other simple techniques. Instruction on particular items of equipment, such as the C-2 automatic pilot, was generally conducted on working models of the equipment involved. Giant-size mock-ups were sometimes built so that large groups of students could be taught simultaneously.

The most important synthetic device use in the Bombardier program was the A-2 bomb trainer. This consisted of a steel scaffold about twelve feet high. It was mounted on wheels and could be electrically propelled across the floor of a hangar, where the trainers were usually kept. The top of the structure represented the Bombardier's compartment of an airplane, including the actual bombsight, and was large enough to accommodate the instructor, student Bombardier, and another student who acted as bomb-approach pilot. A third student sat in the lower section of the scaffold and operated a movable bug or electric motor on wheels at which the Bombardier aimed his sight. The bomb release was a small plunger that struck a paper target on the bug, thereby registering the student's accuracy.

52

Bombardier A-2 bomb trainer on display at the Commemorative Air Force Museum, Midland, Texas. *Lt. Col. George A. Larson, USAF (Ret.)*

As in the pilot program, the need was recognized for training of instructors in desired techniques. Eventually, a central instructor's school (Bombardier) was established at Carlsbad Army Air Field, New Mexico, and later moved to Midland Army Air Field, Texas. It conducted advanced courses for regular Bombardier instructors and refresher courses for Bombardier returnees, some of whom were assigned to teaching duties. Although some attention was given to teaching methods, the principal part of these courses treated the subject of the standard Bombardier curriculum with particular emphasis. The work of the central school was supplemented throughout the war by local training courses at the Bombardier schools.

At Rapid City Army Air Base bombardiers flew in the Lockheed Ventura, modified and given the designation B-34 Lexington by the AFF. Lockheed identified this aircraft as the Model 137. It carried no defensive armament, but was fitted with a central bomb bay that carried 100-pound practice bombs for training Bombardiers. The AAF also took the Ventura for additional training roles, referred to as the B-34B when B-43As produced for the Royal Air Force (RAF) under Lend Lease were returned to the AAF for navigation training.

Lockheed Ventura on display at the Commemorative Air Force Museum and aircraft operations facility, Midland, Texas. *Lt. Col. George A. Larson, USAF (Ret.)*

The B-34 twin-engine light bomber was more capable than the previous AAF Bombardier trainer, the Beech AT-11. It could fly higher, had an increased range, and had a degree of handling characteristics closer to that of the B-17 and B-24 into which the Bombardiers would be assigned and fly into combat. The AAF bombsight, rather than the training plane, was the cardinal item in the Bombardier schools, and it was required both for air and ground instruction. Several types of sights were used: the Norden M-1 and Sperry S-1 precision sights, and the D-8 and T-1 non-precision types, but the Norden was the sight in most general use. Although several kinds of practice bombs were used (as noted previously), the standard was the 100-pound, sand filled M38A2. Bomb targets were of various outline shapes until 1944, when all except circular targets were discontinued.

Looking north along Rapid City Army Air Base's concrete aircraft parking apron. The aircraft to the left is a Lexington B-34. Bombardier students are gathered beside the aircraft in their pre-takeoff briefing for that day's training. Two B-17s are father down the apron, in from one of the hangars. The top of the control tower is visible above the bombers and controlled daily flight activities at and around the airfield. Army trucks are parked to the right of the apron. *Ellsworth Air Force Base, 28th Bomb Wing Historian.*

The number of students assigned to specialized Bombardier instruction in the Flying Training Command increased rapidly from 1941 until the middle of 1943; following a slump in enrollment for several months, the number moved up again to a wartime monthly peak of over 2,000 students in training in September 1944. After that time the number in training declined sharply until the end of the war. During the entire period, a total of over 45,000 Bombardiers graduated. This figure does not include several thousand who were given instruction outside Training Command in OTUs (as happened at Rapid City Army Air Base).

B-17s were a common sight over Rapid City Army Air Base, over the Badlands Bombing Range, and throughout the area as aircrew trained for combat in Europe, preparing to destroy Germany's war industries and defeat the German Air Force to clear the way for the planned aerial and amphibious invasion of France (June 6, 1944, commonly known as D-Day, the invasion of Normandy). Here a B-17 makes a low-level pass over the Rapid City Regional Airport in July 1977. *Lt. Col. George A. Larson, USAF (Ret.)*

Procurement of qualified trainees was one of the most difficult problems in the Bombardier program. Hardly any of the applicants for aircrew training, most of whom desired to be pilots, wanted to become Bombardiers. In order to rectify this unbalanced situation, the AAF conduced a publicity campaign to glamorize the position of Bombardier and other members of the combat crew; the role of the Pilot was de-emphasized. By means of press releases, books, films, and radio broadcast, the campaign produced favorable results by 1943. In the meantime, however, those eliminated from Pilot training were the principal source

54

of students. These men, most of whom had been washed out in primary flying, were offered the option of entering bombardier instruction. Although this policy was necessary in order to fill training quotas, it was strongly criticized by many Air Corps authorities. The eliminated Pilot trainee often had a serious morale problem because of failure to attain their first goal, and because they were inclined to regard their new status with resentment. By the time these students reached operational training, they had usually become adjusted to the role of Bombardier, but their efficiency in training was lessened because of the initial attitude.

No rigid policy for elimination of unsatisfactory students was established in the Bombardier schools. The principal cause for elimination was unsatisfactory performance in the qualification stage of air training; students who were deficient in ground instruction, but who could bomb well, were usually held over at the school for additional ground classes. The method of measuring bombing proficiency was more objective than that for flying proficiency in the pilot program. Various factors, including the immediate demand for graduates, influenced the elimination rates in Bombardier training. They fluctuated from a low of 3.3 percent for one class to a high of 26.1 percent for another. The average for all graduating classes during the war was about 12 percent.[23]

Following are several excerpts from various training manuals and specific Bomb Wing instructional manuals:

Duties and Responsibilities of the Bombardier.

Accurate and effective bombing is the ultimate purpose of your (Pilot) entire airplane and crew. Every other function is preparatory to hitting and destroying the target.

That's your Bombardier's job. The success or failure of the mission depends upon what he accomplishes in that short interval of the bombing run.

When the Bombardier takes over the airplane for the run on the target, he is in absolute command. He will tell you what he wants done, and until he tells you *Bombs Away*, his word is law.

A great deal, therefore, depends on the understanding between the Bombardier and the Pilot. You expect your Bombardier to know his job when he takes over. He expects you to understand the problems involved in his job, and to give him full cooperation. Team work between Pilot and Bombardier is essential.

Under any given set of conditions-ground speed, altitude, direction, etc.-there is only one point in space where a bomb may be released from the airplane to hit a predetermined object on the ground.

There are many things, which a bombardier must be familiar with in order to release his bombs at the right point to hit his predetermined target.

He must know and understand his bombsight, what it does, and how it does it.

He must thoroughly understand the operation and upkeep of his bombing instruments and equipment.

He must know that his racks, switches, controls, releases, doors, linkage, are in first class operation conditions.

He must understand the automatic pilot as it pertains to bombing.

He must know how to set it up, making any adjustments and minor repairs while in flight.

He must know how to operate all gun positions in the airplane.

He must know how to load and clear simple stoppages and jams of machine guns while in flight.

He must be able to load and fuse his bombs.

He must understand the destructive power of bombs.

He must understand the destructive power of bombs and must know how vulnerable the spots of various types of targets.

He must understand the bombing problem, bombing probabilities, bombing errors, etc.

He must be thoroughly versed in target identification and aircraft identification.

The Bombardier should be familiar with the duties of all members of the crew and should be able to assist the Navigator in case the Navigator becomes incapacitated.

For the Bombardier to be able to do his job, the pilot of the aircraft must place the aircraft in the proper positions to arrive at point on a circle about the target from which the bombs can be released to hit the target.

Consider the following conditions, which affect the bomb dropped from an airplane.

1. Altitude is controlled by the Pilot is sustained in flight and affected by atmosphere conditions, thus affecting the range (forward travel of the bomb) and deflection (distance the bomb drifts in a crosswind with respect to the airplane's ground track).

2. True airspeed controlled by the Pilot. The measure of speed of the airplane through the air. It is the speed which is imparted to the bomb and which gives the bomb its initial forward velocity and, therefore, affects the trail of the bomb, or the distance the bomb lags behind the airplane at the instant of impact.

3. Bomb ballistics is the size, shape and density of the bomb, which determines its air resistance. Bombardier uses bomb ballistic tables to account for type of bomb.

4. Trail is the horizontal distance the bomb is behind the airplane at the instant of impact. This value, obtained from bombing tables, is set in the sight of the Bombardier. Trail is affected by altitude, airspeed, bomb ballistics and air density, the first three factors being controlled by the pilot.

5. Actual time of fall is the length of time the bomb is sustained in air from instant of release to instant of impact. Affected by altitude, type of bomb and air density. Pilot controls altitude to obtain a definite actual time of fall.

6. Ground speed is the speed of the airplane in relation to the earth's surface. Ground speed affects the range of the bomb and and varies with airspeed, controlled by the Pilot. Bombardier enters ground speed in the bombsight through synchronization on the target. During this process the Pilot must maintain the correct altitude and constant airspeed.

7. Drift is determined by the direction and velocity of the wind, which determines the distance the bomb will travel downwind from the airplane from the instant the bomb is released to its instant of impact. Drift is set on the bombsight by the Bombardier during the process of synchronization and setting up course.

Prior to takeoff, the Pilot must ascertain that the airplane's flight instruments have been checked and found accurate. These are altimeter, airspeed indicator, free air temperature gauge and all gyro instruments. These instruments must be used to determine accurately the airplane's altitude.

The Pilot's Preliminaries.

The autopilot and Pilot's Direction Indicator (PDI) should be checked for proper operation. It is very important that PDI and autopilot function perfectly in the air; otherwise it will be impossible for the Bombardier to set up an accurate course on the bombing run. The Pilot should thoroughly familiarize himself with the function of both the C-1 autopilot and PDI.

If the run is to be made on the autopilot, the Pilot must adjust the autopilot before reaching the target area. The autopilot must be adjusted under the same conditions that will exist on the bombing run over the target. For this reason, the following factors should be taken into consideration and duplicated for initial adjustment.

1. Speed, altitude and power settings at which the bomb run is to be made.

2. Airplane trimmed at this speed to fly hands off with the bomb-bay doors opened.

The same condition will exist during the actual, except that changes in load will occur before reaching the target area because of gas consumption. The Pilot will continue making adjustments to correct for this by disengaging the autopilot elevator control and re-trimming the airplane, then re-engaging and adjusting the autopilot trim of the elevator.

One of the most important items in setting up the autopilot for bomb approach is to adjust the turn compensation knobs so that action made by the bombardier will be coordinated and at constant altitude. Failure to make this adjustment will involve difficulty and delay for the Bombardier in establishing an accurate course during the run with the possibility that the Bombardier may not be able to establish a proper course in time, the result being considerably large deflection errors in point of impact.

Uncoordinated turns by the autopilot on the run cause erratic lateral motion of the crosshair of the bombsight when sighting on target. The bombardier in setting up course must eliminate any lateral motion of the fore-and-aft hair in relation to the target before he has proper course set up. Therefore, any erratic motion of the cross hair requires an additional correction by the Bombardier, which would not be necessary if autopilot was adjusted to make coordinated turns.

56

During WWII, it was not unusual for residents living around Rapid City Army Air Base to see B-24s overhead, flying into the state from their bases in Nebraska, as their B-17 counterparts were at the base. Here a B-24 makes a low altitude pass over the Rapid City Regional Airport in July 1997. *Lt. Col. George A. Larson, USAF (Ret.)*

Use of the PDI. The same is true if the PDI is used on the bomb run. Again, coordinated, smooth turns by the pilot become an essential point of the bomb run. In addition to added course corrections necessitated by uncoordinated turns, skidding and slipping introduce small changes in airspeed affecting synchronization of the bombsight on the target. To help the pilot flying the run on PDI, the airplane should be trimmed to fly practically hands off.

Assume that you are approaching the target area with autopilot properly adjusted. Before reaching the IP there is evasive action to be considered. Many different types of evasive tactics are employed, but from experience it has been recommended that the method of evasive action be left up to the Bombardier, since the entire anti-aircraft pattern is fully visible to the Bombardier in the nose.

Evasive Action. Changes in altitude necessary for evasive action can be coordinated with the Bombardier's changes in direction at specific intervals. This procedure is helpful to the Bombardier since he must select the IP at which he will direct the airplane onto the briefed heading for the beginning of the bomb run. Should the pilot be flying the evasive action on PDI, at the direction of the Bombardier, he must know the exact position of the IP for beginning the run, so that he can fly the airplane to that point and be on the briefed heading. Otherwise, there is a possibility of beginning the run too soon, which increases the airplane's vulnerability, or beginning the run too late, which will affect the accuracy of the bombing. For best results the approach should be planned so the airplane arrives at the IP on the briefed heading, and at the assigned bombing altitude and airspeed.

At this point, the Bombardier and Pilot as a team should exert an extra effort to solve the problem at hand. It is now the Bombardier's responsibility to take over the direction of flight, and give directions to the pilot for the operations to follow. The Pilot must be able to follow the Bombardier's directions with accuracy and minimum loss of time, since the longest possible bomb run should seldom exceed 3 minutes. Wavering and indecision at this moment are disastrous to the success of any mission, and during the crucial portion of the bomb run, flak

57

and fighter opposition must be ignored if bombs are to hit the target. The Pilot and Bombardier should keep each other informed of anything which may affect the successful completion of the run.

Holding a Level. Either before or during the run, the Bombardier will ask the Pilot for a level. This means that the Pilot must accurately level his airplane with his instruments, ignoring the PDI. There should be no acceleration of the airplane in any direction, such as an increase or decrease in airspeed, skidding or slipping, gaining or losing altitude. For the level, the Pilot should keep a close check on his instruments, not by feel or watching the horizon. Any acceleration of the airplane during this moment will affect the bubbles, primarily through centrifugal force, on the bombsight gyro, and the Bombardier will not be able to establish an accurate level.

Holding Altitude and Airspeed. As the Bombardier proceeds to set up his course, called synchronize, it is absolutely essential that the Pilot maintain the selected altitude and airspeed within the closest possible limits. For every additional 100-feet above the assumed 20,000-foot bombing altitude, the bombing error will increase approximately 30-feet, the direction of error being over. For erroneous airspeed, which creates difficulty in synchronization on the target, the bombing error will be approximately 170-feet for a 10 mph change in airspeed. Assuming the airspeed was 10 mph in excess, from 20,000-feet, the bomb impact would be short 170-feet.

The Pilot's responsibility to provide a level and to maintain a selected altitude and airspeed within the closest limits cannot be over-emphasized.

If the Pilot is using PDI, at the direction of the Bombardier, instead of the autopilot, he must be thoroughly familiar with the corrections demanded by the Bombardier. Too large a correction or too small a correction, too soon or too late, is as bad as no correction at all. Only through prodigious practice flying with the PDI can the Pilot become proficient to a point where he can actually perform a coordinated turn, the amount and speed necessary to balance the Bombardier's signal from the bombsight.

Erratic airspeeds, varying altitudes, and poorly coordinated turns make the job of establishing course and synchronizing doubly difficult for both Pilot and Bombardier, because of the necessary added corrections required. The resulting bomb impact will be far from satisfactory.

After releasing the bombs, the Pilot or Bombardier may continue evasive action-usually the Pilot, so that the Bombardier may man his guns.

The Pilot using the turn control may continue to fly the airplane on autopilot, or fly it, manually with the autopilot in a position to be engaged by merely flipping the lock switches. This would provide potential control of the airplane in case of emergency.

Reducing Circular Error. One of the greatest assets towards reducing the circular error of a bombing squadron lies in the Pilot's ability to adjust the autopilot properly, fly the PDI, and maintain the designed altitude and airspeeds during the bombing run. Reducing the circular error of a bombing squadron reduces the total number of aircraft required to destroy a particular target. For this reason, both Pilot and Bombardier should work together until they have developed a complete understanding and confidence in each other.[24]

The Bombardier's training is highly significant to the B-17's crew operations, but is not often reflected in the history of stateside training and combat operations during the war. It was more than just climbing into the bomber and allowing the Pilots to get the aircraft to the designated position so the bombs could be released onto the designated target. It was always more glamorous to talk about the fighter pilots, and the bomber crews often made headlines for the losses suffered during daylight bombing missions against attacking German fighters.

58

1. Lead and Deputy Lead Bombardiers will be called (2) hours before breakfast and will report to the S-2 Intelligence Library.

a. They will fix their mission folders in complete order.

b. All aiming points of chosen target will be properly marked.

c. Orientation spinners for obscured bombing procedure will be put on the necessary target pictures and used.

d. Headings to the target will be drawn on the target pictures and also on the night target maps.

e. A line will be drawn from the IP to the target and definite, outstanding pilotage points will be encircled in the area flown from the IP to the target.

f. All Lead and Deputy Lead Bombardiers will check out maps from the map room and will draw the course from the enemy coast to the target on them. They will do pilotage during the entire mission in addition to studying the target, etc.

g. Secondaries and last resort target, after leaving the primary, is to be number one.

1. All Bombardiers will attend the Bombardiers briefing.

2. All Lead and Deputy Lead Bombardiers will report to Group Operations after the main briefing.

3. All Bombardiers will thoroughly check their equipment before take-off.

4. Lead or Deputy Bombardiers will perform all sighting operations, i.e., sight for both range and deflection. Flight leaders will drop their bombs when the leader drops. All Wing Bombardiers will drop their bombs on their respective Section Leader's release.

5. The leader of the Low Squadron will be the Deputy Leader if the Leader aborts or cannot bomb before the IP. If the Deputy Leader cannot takeover, the Leader of the High Flight will takeover. This refers to a 12-ship group. In an 18-or-21 ship group, the Deputy Lead will be the Leader of the High Section and the second Deputy will be the Leader of the Low Section. If the Leader cannot bomb during the bomb run, the number four man in the Lead Section will do the sighting.

6. The Deputy Leaders will have their sights and equipment set up to bomb in case the Leader aborts or cannot bomb.

7. A series of red *dits* from the Aldis Lamp in the tail gunner's position of the Lead ship will be the signal to the High Squadron that the Lead ship cannot bomb and the High Squadron to take the Lead. A steady green light from the Deputy Lead ship will be the signal that this Bombardier is prepared to assume the Lead.

8. Leaders will attack only assigned targets and assigned MPI's of those respective targets. The target of opportunity will be attacked when the assigned targets cannot be bombed due to inclement weather, etc. Lead and Deputy Lead Bombardiers will be sure that they know where the bomb line is located before departing on a mission.

9. After take-off.

a. Pins will be pulled after an altitude of 5,000-feet is reached. In case bombs are returned, the pins will be replaced before landing.

b. Lead Bombardier will have his stabilizer warmed up and his PDI held center so that the Pilot can set up AFCE immediately after take-off.

c. Bombardiers will know their flare colors and will keep on alert for them and give the Pilot all the cooperation they can in forming with the Group.

d. Bombardiers will act as Oxygen Officer and Gunnery Officer during the entire mission and will order the crew to check in every five minutes when at high altitude. When under attack by enemy aircraft or in flak, crews will be checked every three or four minutes, or as often as possible.

e. The Lead and Deputy Lead Bombardiers will take several drift runs on each leg of the flight and notify their respective Navigator of the proper drift. The Navigator will use this drift in conjunction with Metro Data or Gee Box data. It is absolutely impossible to get a more accurate drift than that obtained from the bomb sight.

Drift readings will be taken with a dead sight.

f. The bomb run will be flown on AFCE, after bombing altitude is reached. The Lead Bombardier will check the ratio of the automatic pilot by making a turn with the secondary clutch and then engaging it while still in the turn. If the auto pilot brings the PDI back past center, or stops short of center, the ratio is not correct and the Pilot is not correct and the Pilot can make the necessary adjustments.

g. If the auto pilot fails to function, the Leader will not relinquish his position to the Deputy Lead, but will make the run on PDI.

h. After bombing altitude is reached and just prior to the IP, the Bombardier will obtain from the Navigator the exact drift for the bomb run and pre-set his sight. He will also get a good stabilizer level and check it while on the bomb run.

10. After the mission.

a. In case of a *Tactical Mission,* there will be a Bombardier from each squadron in the respective squadron equipment rooms to take the bomb count. On a *Strategic Mission*, the Bombardiers will report to the Group Bombardier's table immediately after the mission and turn in their reports. They will tell him whether they released their bombs on their respective leader, on time, and as briefed. In case of any malfunction of any sort, they will report this to the Group Bombardier and the Armament man of their ship.

b. After being interrogated, the Lead Bombardier will report to the Group Bombardier's Office and write their narrative and give a complete account of their mission.

c. All Bombardiers, except Lead Bombardiers will clean their guns, on the same day of the mission. Lead Bombardiers will not be required to install or clean their guns.

Procedure on the Bomb Run.

1. The Lead Navigator will have the Pilot swing wide on the IP and come directly over it.

2. When the formation is headed for the target, the Lead Bombardier will take over the ship and the rest of the crew will stay off the interphone during the entire bomb run. Only the Pilot, Navigator and/or Bombardier will do any talking over the interphone during the bomb run.

3. The Lead Bombardier will check his pre-set drift, obtained from the Navigator, at the beginning of the bomb run. The Lead Navigator will stand directly behind him and do pilotage. After the Bombardier has pre-set his drift and killed his rate, he will show on his map where he is in relation to the ground directly in front of him.

4. The Lead Bombardier will notify the Pilot when he has located the target and when he is coming up on it. About 20 or 30 seconds before *Bombs Away* the Bombardier will notify the Pilot and leave his airspeed and altitude exactly as it is. This will enable him to kill his rate perfectly.

5. After *Bombs Away* the Pilot will take over the ship and the Bombardier will take down his necessary date and close his doors and shut off the power switches.

6. It is the Bombardier's duty at *Bombs Away* to make sure all cameras are turned on.

7. The Bombardier will be sure that the intervalometer counter knob is on zero and the select switch on train after leaving the target and while enroute to it.

8. Procedure if interphone is out.

a. Making run on C-1 auto pilot.

1.) Two minutes before the IP, the Bombardier will attract the Pilot's attention by flashing bomber-call Light, two long flashes.

2.) Upon receiving this signal, the Pilot will stand by to turn at IP.

3.) When IP is reached, Bombardier will indicate this by one long flash of bomber call.

4.) Pilot starts turn toward, and levels out when Bombardier gives too long flashes of bomber call.

5.) At this time Bombardier takes control of aircraft and makes normal run on target.

6.) As the bomb release light flashes, Pilot again assumes control of aircraft.

b. Making run manually.

1.) If C-1 auto pilot ceases to function properly, a manual run will be made. Bombardier will notify Pilot of this turning on PDI.

2.) Two minutes before IP is reached, Bombardier will attract the Pilot's attention by two long flashes of bomber call light.

3.) Upon receiving this signal, the Pilot will stand by to turn at IP.

4.) When IP is reached, Bombardier will indicate this by one long flash of bomber call.

5.) Pilot turns toward target as Bombardier indicated direction of turn with PDI. Pilot levels out when PDI is centered.

6.) From this point, the bomb run is made with the Pilot following PDI until the bomb release light flashes.[25]

Duties and Responsibilities of the Engineer.

Size up the man who is to be your Engineer. This man is supposed to know more about the airplane you are to fly than any other member of the crew.

He has been trained in the Air Force's highly specialized technical schools. Probably he has served some time as a Crew Chief. Nevertheless, there may be some inevitable blank spots in his training which you, as a Pilot and Airplane Commander, maybe able to fill in.

Think back on your own training. In many courses of instruction, you had a lot of things thrown at you from right and left. You had to concentrate on how to fly; and where your equipment was concerned you learned to rely more and more on the enlisted personnel, particularly the Crew Chief and the Engineer, to advise you about things that were not taught to you because of lack of time and the arrangement of the training program.

Both Pilot and Engineer have a responsibility to work closely together to supplement and fill in the blank spots in each other's education. To be a qualified Combat Engineer a man must know his airplane, his engines, and his armament equipment thoroughly. This is a big responsibility: the lives of the entire crew, the safety if the equipment, the success of the mission depends upon it squarely.

He must work closely with the Copilot, checking engine operation, fuel consumption, and the operation of all equipment. He must be able to work with the Bombardier, and know how to cock, lock, and load the bomb racks. It is up to you, Airplane Commander, to see that he is familiar with these duties, and, if he is hazy concerning them, to have the Bombardier give him special help and instruction.

He must be thoroughly familiar with the armament equipment, and know how to strip, and re-assemble the guns.

He should have a general knowledge of radio equipment, and be able to assist in tuning transmitters and receivers.

Your Engineer should be your chief source of information concerning the airplane. He should know more about the equipment than any other crew member-yourself included.

You, in turn, are his source of information concerning flying. Bear this in mind in all your discussions with the Engineer. The more complete you can make his knowledge of the reasons behind every function of the equipment, the more valuable he will be as a member of the crew.

In emergencies, the Engineer will be the man to whom you turn first. Build up his pride, his confidence, his knowledge. Know him personally; check on the extent of his knowledge. Make him a man upon whom you can rely.[26]

Duties and Responsibilities of the Radio Operator.

There is a kit of radio equipment in a B-17. There is one man in particular who is supposed to know all there is to know about this equipment. Sometimes he does, but often he doesn't. And when the Radio Operator's deficiencies do not become apparent until the crew is in the combat zone, it is then too late. Too often, the lives of Pilots and crew are lost because the Radio Operator has accepted his responsibilities indifferently.

The Arizona Wing of the Commemorative Air Force's B-17G taxiing to the old passenger terminal at Rapid City Regional Airport in July 1997. *Lt. Col. George A. Larson, USAF (Ret.)*

61

Radio is a subject that cannot be learned in a day. It cannot be mastered in 6 weeks, but sufficient knowledge can be imparted to the Radio Operator during his period of training in the United States if he is willing to study. It is imperative that you check your Radio Operator's ability to handle his job before taking him overseas as part of your crew. To do this you may have to check the various departments to find any weakness in the Radio Operator's training and proficiency and to aid the instructors in overcoming such weaknesses.

Training in the various phases of the heavy bomber program is designed to fit each member of the crew for the handling of his jobs. The Radio Operator will be required to:

1. Render position reports every 30-minutes.
2. Assist the Navigator in taking fixes.
3. Keep the liaison and command sets properly tuned and in good operating order.
4. Understand from an operational point of view.
a. Instrument landing
b. IFF
c. VHF and other navigational aids and equipment in the airplane.
5. Maintain a log

In addition to being a Radio Operator, he is also a gunner. During periods of combat he will be require to leave his watch at the radio and take up his guns. He is often required to learn photography. Some of the best photographs taken in the Southwest Pacific Theater of Operations were taken by B-17 and B-24 Radio Operators. The Radio Operator who cannot perform his job properly may be the weakest member of your crew-and the crew is no stronger than its weakest member.[27]

Duties and Responsibilities of the Gunners.
The B-17 is a most effective gun platform, but its effectiveness can be either applied or defeated by the way the gunners in your crew perform their duties in action.

Your gunners belong to one of two distinct categories: turret gunners and flexible gunners.

The power turret gunners require many mental and physical qualities similar to what we know as inherent flying ability, since the operation of the power turret and gun sight are much like that of airplane flight operation.

While the flexible gunners do not require the same delicate touch as the turret gunner, they must have a fine sense of timing and be familiar with the rudiments of exterior ballistics.

All gunners should be familiar with the coverage of all gun positions, and be prepared to bring the proper gun to bear as the conditions may warrant.

They should be experts in aircraft identification. Where the Sperry turret is used, failure to set the target dimension dial properly on the K-type sight will result in miscalculation of range.

They must be thoroughly familiar with the Browning aircraft machine gun. They should know how to maintain the guns, how to clear jams and stoppages, and how to harmonize the sights with the guns. While participating in training flights, the gunners should be operating their turrets constantly, tracking with the flexible guns even when actual firing is not practical. Other airplanes flying in the vicinity offer excellent tracking targets, as do automobiles, houses, and other ground objects during low-altitude flights.

The importance of teamwork cannot be overemphasized. One poorly trained gunner, or one man on the alert, can be the weak link as a result of which the entire crew may be lost.

Keep the interest of your gunners alive at all times. Any form of competition among the gunners themselves should stimulate interest to a high degree.

Finally, each gunner should fire the guns at each station to familiarize himself with the other man's position and to insure knowledge of operation in the event of an emergency.[28]

I. Combat Gunners.
1. All Gunners, Navigators and Bombardiers will be responsible for cleaning their guns.
a. The guns will be cleaned not later than the evening of the day on which they are fired. Never, under any circumstances, permit a gun to be set without cleaning it after it has been fired.

b. When guns are being fired, the pores of the metal tend to open and expand. With this expansion, a certain amount of moisture is absorbed. When the guns have been cleaned and stored away, the metal pores of the guns contract and force a slight amount of moisture to the surface and the rust starts to form. One cleaning is not sufficient to prevent this rust, and the guns must be cleaned every day for at least seven days and left with a thin coating of oil after each cleaning.

c. All working parts of the gun will be disassembled and cleaned thoroughly with gasoline. Dry the parts immediately after removing the gasoline. Then the parts will be wiped with an oiled rag, oiled with AXS 777.

d. The bore will be cleaned with bore cleaner; saturate a clean patch with the bore cleaner and run through the bore. Repeat with clean patch saturated with bore cleaner two or three times until a clean patch picks up no foreign matter. Then use a clean dry patch to remove all the cleaner before applying lubrication oil Spec. No. AXS 777.

e. Proper oiling is second in importance only to intelligence cleaning. All working parts of the gun must be oiled, but sparingly.

(1). In oiling guns, be careful not to use too much oil. In low temperatures the oil will congeal and cause malfunctions. It is best to oil by wiping with an oil rag.

(2). Lubricating oil, Specification No. AXS 777 will be used on all machine guns. The oil is not considered a suitable preservative. It must be renewed every 48 hours.

f. The oil buffer will be filled while cleaning the gun after it has been fired. It will be checked again before assembly to see that it is full and has not leaked.

(1). Recoil Oil Specification No. 2-36D will be used in the oil buffer.

(2). The oil buffer will be set at two notches from full open.

g. On damp and rainy days, the receivers, barrel jackets, and barrel jacket bearing will be wiped free from moisture. This must be done to prevent freezing of the gun at altitude.

h. After guns have been assembled, the gunners will preflight their respective positions.

(1). Turret Gunners will make sure their ammunition cover is fastened securely.

(2). Ball Turret Gunners will make sure their ammunition cover is fastened securely.

(3). For correct procedure to adjust G-11 Solenoid, the Turret Gunners will check their turret sections.

2. All ordnance and armament functions will be left intact for investigation by the Armament officer if it is unrepairable in the air.

3. All guns will be test fired at least fifty rounds every third combat mission or whenever briefed to do so. Airplane Commanders will insure that the guns are fired only in such places where they will not endanger friendly troops and installations. This is very important because negligence in handling the caliber .50 machine gun has resulted in needless deaths as well as damage to aircraft.

4. Bendix chin turret guns will be stowed in the extreme right position at a minus three degrees elevation to prevent moisture draining down the barrels into the receiver group.

5. Modification of combat loading.

a. The radio hatch gun with its accessories and ammunition will be removed.

b. Only one Waist Gunner will be carried and will ride in the radio compartment from time of take-off until altitude is reached, or within thirty minutes of the enemy coast, whichever is first.

c. Ammunition load will be limited to 800 rounds in the tail gun position and 350 rounds at each waist gun position, with spare ammunition not to exceed two cases, divided between the nose and radio compartments.

6. Modification for preparation of oxygen fires in the upper turret.

a. Installation of emergency oxygen shut-off valve on the rear of the Pilot's seat to permit Pilot or Copilot to immediately shut-off the upper turret oxygen supply in case of emergency.

b. Installation of Emergency Power Switch on the same panel behind the Pilot's seat to permit immediate shut-off of power supply to upper turret in case of emergency.

7. Sights

a. The fact this Group does not use tracer ammunition, makes the use of the gun sights imperative. In order to use the sights the Gunner must understand the theory of sighting, and know how to use them.

63

(1). Turret Computing Sights will be used or operated in compliance with existing sight manuals or Gunner's Information File.

(2). The Ring and Post Sights and Optic Sight will be used in accordance with letters of instructions issued by this Headquarters. The sighting theory reduced and explained in the most simple manner Possible in these Letters of Instructions, dealing with the HEWITT Sighting System. The Gunner's Information File also deals with sighting theory for the Iron and Optic Sights. Each Squadron Gunnery Officer has a supply of these instructions to be issued and explained to each combat Gunner. Any difficulties in sighting techniques experienced should immediately be taken with the Squadron Gunnery Officer.

b. Flexible guns, including the chin turret are harmonized for 1,000 yards,

c. All sights are checked and harmonized every third mission.

8. Auxiliary equipment.

a. Combat Gunners cannot operate at maximum efficiency unless they understand and know how to operate their equipment. This includes radio interphone equipment, and emergency medical supplies. Full information on these subject can be obtained from current publications, technical orders, and Squadron Equipment Officers.

II. Zones of Search.

1. All turrets will search 360 degrees with added emphasis in the area of most likely attack according to present enemy tactics.

2. Zones of search for all other gun positions will be maintained in the area covered by the traverse of the particular gun or guns concerned.

3. The zone of search on any given turret depends on only one thing and that is the position of the aircraft in Squadron or Group formation.

4. It is imperative that all combat crew personnel know and understand formation, the position in any formation, and search accordingly.

5. The efficiency of this system depends completely upon the absolute cooperation of each and every gunner. Gunners will maintain an alert search in their assigned areas at all times. Gunners will be in their assigned positions at all times over the enemy territory.

6. Bombers and their crews have been needlessly lost because Gunners were not watching and let fighters sneak in on them

7. The clock system is the approved method of fighter location in Group and will be used.

8. Further indication of exact position on enemy aircraft may be made by calling *High-Low or Level*. Example: *Fighter-11-o'clock-low*. This system tends to eliminate unnecessary interphone conversation. In addition, Gunners should always keep the pilot informed when any enemy aircraft commits himself to a direct attack on the Gunner's aircraft. This enables the pilot to take individual evasive action.[29]

During Group formation on combat missions an aerial observer was assigned to maintain tight formation flying, improving effectiveness of combined B-17 Gunners:

Aerial Observers.

I. In each Squadron Lead Aircraft, an officer will ride in the Tail Gunner's position to act as Aerial Observer, usually a Pilot or Copilot.

II. His primary function is to give the Squadron Leader a running account of the number and condition of the aircraft in his formation throughout the mission.

1. Keep the Squadron Leader posted on the Squadron assembly and notify him when all aircraft are in proper Assembly. The Aerial Observer in the Group Leader's aircraft will, in addition, keep the Group Leader Informed as to position of the High and Low Squadrons during assembly and particularly as the time for departing the assembly point approaches.

2. Keep the Squadron Leader informed as to any changes in formation, aircraft leaving the formation, straggling, etc.

64

3. Notify the Squadron Leader of the position and distance of all fighter aircraft, friendly or enemy. If the fighter Aircraft is enemy or it is suspected to be enemy the Observer will keep the Pilot constantly advised as to the fighter's actions to enable the Pilot to take evasive action. Fighter aircraft generally begin their attacks at a range of 1,500 yards and evasive action must be taken before the fighter commits himself to the attack.

4. The greatest emphasis is to be placed on the compactness of the formation during the bomb run and it is of prime importance that the Leader be given a running account of the formation at this time. He also will make note of all aircraft dropping bombs early or late.

III. His secondary function is to make a report after the mission on the formation flown that day.

IV. The purpose of the that report is:

1. To establish a numerical rating for each Pilot in the formation.

2. To make constructive suggestions as to how the individual Pilots can improve their formation flying by pointing out the errors made.

Photograph of the flight line at Rapid City Army Air Field. Alfred Marcello is at the right side of the front row, sitting on the parking apron. He was undergoing continuing training while at the base before flying to England, assigned to a replacement crew, and eventually onto the 100th BG. A B-17F is in the background, with one of the base's hangars visible under the wing of the B-17F. Directly in the middle of the photograph, underneath the B-17F, is another B-17F parked on the ramp. *Paul Marcello, Ellsworth Air Force Base, 28th Bomb Wing Historian to Lt. Col. George A. Larson, USAF (Ret.)*

World War II B-17 Aircrew Training

V. The report will be made as follows:

1. The Lead Squadron Aerial Observer will rate the Lead aircraft of the High and Low Squadrons on the quality of their Squadron leadership in relation to Group formation on the basis of excellent, good, or fair or poor.

2. The Aerial Observer in each Squadron will rate all the Pilots in that Squadron on a numerical basis with the Pilot flying the best formations as #1 Lead so on through #12.

3. The Pilots are to be rated on their formation flying throughout the mission with the greater emphasis placed on the correctness of formation flying during the bomb run. If any aircraft is forced to lag because of battle damage which hinders the normal efficiency of the aircraft, it will be rated on its performance up to the time it suffered the damage.[30]

During B-17 training operations conducted at Rapid City Army Air Field, nine B-17 BGs trained as a unit, in addition to those crews trained and assigned as replacement crews after reaching England that were assigned to the Eighth Air Force. The following is a series of short group histories for these units:

96th Bombardment Group (H).

Constituted as the 96th BG on January 28, 1942. It was activated on July 15, 1942. It was the first B-17 unit to train at Rapid City Army Air Base, September 30 to October 30, 1942. While at the base it conducted third phase OTU training. The training emphasized navigational flights, formation flights and general polishing for combat efficiency. Flights over large bodies of water and long distances were also taught. Much practice bombing was on the daily training schedule. The bomb group was told that it would remain in the states for more than a year in order to train combat crews for overseas combat. The bomb group arrived Snetterton, England, assigned to Eighth Air Force on June 12, 1943.[31]

88th Bombardment Group (H).

Constituted as 88th Bombardment Group (H) on January 28, 1942. Activated on July 15, 1942, but not manned until September. Trained at Rapid City Army Air Base, October 28 to November 28, 1942. Served for a short time as an OTU then RTU. First assigned to Second Air Force, alter to Third Air Force. Inactivated on May 1, 1944.[32]

A B-17F crew at Rapid City Army Air Base, from left to right: Pilot, 2nd Lt. Don Jones; Right Waist Gunner Cpl. Sam Foushee; Navigator, 2nd Lt. Arthur Juhlin; and Bombardier, 2nd Lt. Ralph P. Farrell, Jr. *Paul Marcello, Ellsworth Air Force Base, 28th Bomb Wing Historian to Lt. Col. George A. Larson, USAF (Ret.)*

World War II B-17 Aircrew Training

383rd Bombardment Group (H).
Constituted as 383rd Bombardment Group (H) on October 28, 1942. Activated on November 3, 1942. Trained at Rapid City Army Air Base, November 12, 1942 to June 20, 1943. Equipped and served as an OTU and RTU Squadron. Inactivated on April 1, 1944.[33]

95th Bombardment Group (H).
Constituted as 95th Bombardment Group (H) on January 28, 1942. Activated on June 14, 1942. Trained at Rapid City Army Air Base, from December 14 1942 to March 11, 1943. The bomb group moved to England, March-May 1943 (ground echelon by train to east coast port of embarkation and transport to England, with air echelon flying, assigned to Eighth Air Force).[34]

452nd Bombardment Group (H).
Constituted as 452nd Bombardment Group (H) on May 14, 1943. Activated on June 1, 1943. Trained at Rapid City Army Air Base from June 15 to October 11, 1943.[35]

398th Bombardment Group (H).
Constituted as 398th Bombardment Group (H) on February 15, 1943. Activated on March 1, 1943. Started training of overseas combat, but interrupted from July to December 1943 to train replacement crews for other organizations.[36]

During training at Rapid City Army Air Base there were frequently B-17 crashes. The following is one documented crash, showing the inherent dangers of Stateside training flight operations prior to entering the active combat theater of operations in Europe with the Eighth Air Force:

A B-17F, 398th BG, 603rd BS on January 17, 1944, took off on a night bombing training mission. Once airborne the Pilots' encountered engine problems, forcing the Pilots to attempt an emergency landing in an empty farm field. The Pilots over shot a grassy area, plowing into the soft ground of a wheat field. The crash killed the four officers (Pilot, Copilot, Navigator and Bombardier) and three enlisted crew members, with the resulting fire completely destroying the aircraft. This was just one of many Stateside training crashes of military aircraft during the war, before the crew and aircraft could reach their respective combat theaters. Preparing for war was a dangerous and serious mission.[37]

447th Bombardment Group (H).
Constituted as 447th Bombardment Group (H) on April 6, 1943. Activated on May 1, 1943. Trained at Rapid City Army Air Base from July 1 to August 1, 1943.[38]

457th Bombardment Group (H).
Constituted as 457th Bombardment Group (H) on May 19. 1943. Activated on July 1, 1943. Trained at Rapid City Army Base from July 9 to October 29, 1943.[39] In august 2003, former members of the bomb group returned to their WWII training site. These veterans talked about the changes to the military facility. In 1943, the primary runway was one-half the length of the present 13,500-foot Ellsworth Air Force Base runway. There were two additional runways, one of which is now used as a street for base support activities. B-17 trainees came to Rapid City Army Air Field from other training bases after completing schooling in their main aircrew functions and responsibilities. As Pilots practiced their formation flying skills; the Navigators, Bombardiers, Gunners and specialized crew members trained on their assigned functions during day and night training missions.

Norman Frantz…*We had little buses that took us from our barracks to the flight line. During one of the South Dakota winter January 1944 blizzards, the bus couldn't make it all the way, and two guys tried walking back. They were found a day later, frozen to death about halfway back to the barracks. I had a great deal of respect for the B-17 pilots who were learning how to fly the planes in formation. They practiced at some low-altitudes and had to deal with a lot of winds that could move the planes up and down a lot.*

67

Albert Porta…*I was a new Pilot. I had always wanted to be a pilot and started flying when I was 16. B-17 controls were all mechanical then, we didn't have hydraulics. It was a lot of work to keep the big bombers just yards apart. We could lose as much as five pounds during an eight-to-ten hour bombing mission over enemy targets. The planes flew in tight formations so each bomber could provide covering fire for the rest of the aircraft and make it more difficult for enemy fighter planes to single out one target. The downside to the State side formation training was that at the local airfield, the bombing missions were 30-seconds long. Bombing missions over enemy targets could take three to four minutes, and you had to keep your plane level despite flying through heavy flak.*[40]

463rd Bombardment Group (H).

Constituted as 463rd Bombardment Group (H) on May 19, 1943. Activated August 1, 1943. Trained at Rapid City Army Air Base for August 1 to November 5, 1943.[41] By August 17th, fifteen officers and eighteen enlisted men had reported for duty at Camp Rapid, Rapid City. A small tent city was servicing as a gathering point for newly activated groups for the Second Air Force. There the Groups received their key personnel who were later sent to AAFSAT for a 30-day period of combat training. During the last week in August, 135 officers and enlisted men who made up the key personnel of the group, had reported for duty, and on September 1st left Rapid City to move to the AAF School of Applied Tactics at Orlando, Florida. While at Orlando, each squadron planned general simulated raids over occupied Europe and the Southwest Pacific. Briefings were held to acquaint the men with standard procedure for the planning of heavy bombardment missions. Upon completion of the ground school program, the Group moved to Montbrooke, Florida, a small satellite base approximately 127 miles north of Orlando. Four B-17s were picked, one for the model crew of each squadron. The model crews continued their training throughout the month of September, flying simulated bombing missions over cities throughout the southwest. They returned to Camp Rapid on September 29th. More personnel had continued to arrive and by October 23rd, there were 83 officers and 707 enlisted men in the Group. During this time at Camp Rapid, an intensive Ground School program was conducted. Model crews continued to fly their planes which were based at Rapid City Army Air Base.[42]

CHAPTER THREE

THE B-29 SUPERFORTRESS YEARS

Toward the end of WWII, Rapid City Army Air Base remained an operational training center for weather reconnaissance squadrons, as well as combat fighter squadrons flying the Northrop P-61 Black Widow, North American P-51 Mustang, and North American B-25 Mitchell.

When one thinks of the long bomber flight line now in operation at Ellsworth AFB, the operation of smaller military aircraft is overlooked. During the closing weeks of WWII, the 55th Reconnaissance Squadron was formed on July 21, 1945, at Will Rogers Field, Oklahoma, using personnel from the 369th Fighter Group. By the time they moved to Rapid City Army Air Base on July 28, 1945, Germany had surrendered. However, fighting in the Pacific appeared to have no end in sight, with Japanese troops, airmen, and pilots killing as many U.S. military personnel as possible on Iwo Jima and Okinawa, and preparing for all-out suicide defense of the Japanese home islands. An important aspect of any future campaign would surely be aerial bombardment.

The Boeing B-29A Superfortress. *Ellsworth Air Force Base, 28th Bomb Wing Historian.*

Although used sparingly in the European Combat Theater of Operations, bombing by means of radar and radio navigation had not been perfected. Using the aircraft's optical (Norden) bombsights was preferred. In the Pacific, over the Japanese home islands, weather conditions over a planned target were crucial to any bombing formation, particularly the high-altitude bomb releases by B-29s, before switching to the low-level incendiary night bombing missions of Japanese cities. Bombardment wing commanders required exact weather information upon which to base their decision to strike the primary target or switch to an alternate target.

The Army Air Forces looked for a special aircraft to provide weather information for the Superfortresses attacking the Japanese home islands. The aircraft that the Army selected was a stripped down variant of the Northrop P-61 Black Widow night fighter. Its armament was removed to accommodate the installation of electronic weather collection equipment and camera. Operating out of airfields on Iwo Jima and Okinawa, it could cover proposed targets in the Japanese home islands. The plan was for the aircraft to precede the B-29s and radio back last minute weather information over the selected target areas in Japan.

Northrop P-61 Black Widow on display at the National Museum of the United States Air Force. *Lt. Col. George A. Larson, USAF (Ret.)*

Although considered a viable mission, WB-29s were used to perform this mission in the interim while the weather version of the P-61 was modified. The 20th Air Force had to use what equipment was available for the assigned weather mission. The 58th Reconnaissance Squadron did not see any combat with their P-61s after the Japanese government and military accepted the Allied terms of unconditional surrender after the 509th Composite Group dropped two atomic bombs (Hiroshima on August 6th and Nagasaki on August 9, 1945). At the time, the 58th Reconnaissance Squadron had completed preliminary training at Rapid City Army Air Base for its small cadre of 250 and one P-61. After the end of the war, many aspects of the unit were returned to Will Rogers Field, and on May 31, 1946, the unit was inactivated.[1]

Rapid City Army Air Base became Rapid City Army Air Field on September 1, 1946, with military missions ending at the base and the base closed until March 1947, at which time it came under control of the newly created United States Air Force.[2] The National Security Act of July 26, 1947, created a National Military Establishment of the Army, Navy, and Air Force.[3] The base was renamed Rapid City Air Field on November 28, 1947.

Air Force and local dignitaries dedicated Rapid City Air Force Base on June 24, 1948. *Ellsworth Air Force Base, 28th Bomb Wing Historian.*

The B-29 Superfortress Years

To understand the 28th BW's history, one must first understand its relationship to the 28th BG. The 28th BG was the Wing's primary operational element when the War Department first organized the Wing in August 1947. The Bomb Group is seven years older than the Wing and was already a WWII veteran before the Air Staff created the larger organization. Nevertheless, in tribute to the Group's gallant exploits, a review of its early history provides a background in explaining how it later fit into the Wing's structure.

On December 22, 1939, the U.S. Army Air Corps originally constituted the 28th Composite Group. Two months later, the Group saw its first active duty as the 28th BG (Composite), assigned to March Field, California. After a short assignment at Moffett Field, California, the Group moved to Elmendorf Field, Alaska, where its assigned squadrons operated primarily from bases in the Aleutian Islands. They flew various aircraft (B-17s, B-18 Bolos, B-24 Liberators) and compiled an impressive combat record against the enemy in the Northern Pacific. After the war, the Group inactivated temporarily before coming back as the 28th BG Very Heavy (VH) under the new Strategic Air Command (SAC). This time the Group's assigned squadrons flew the B-29 while they conducted six months of post-WWII Arctic operations in Alaska.[4]

Strategic Air Command shield on display at the SAC Museum. *Lt. Col. George A. Larson, USAF (Ret.)*

Three B-29s on a training mission in formation. The photograph is taken from the left B-29 in formation, showing the other two B-29s close by. *Ellsworth Air Force Base, 28th Bomb Wing Historian.*

The B-29 Superfortress Years

SAC was established on March 21, 1946, as one of three major commands of the United States Air Force. The command emblem is on a sky-blue shield over two clouds: one in the upper left and one in the lower right extending to the edges of the shield. A cubit arm in armor issues from the lower left and extends toward the upper part of the shield, and the hand grasping a green olive branch, along with three red lightning flashes. On the shield, in heraldic terms, it is azure, two clouds proper, one issuing from sinister chief and one issuing from the dexter base, a cubit arm in armor in bend issuing from the sinister, the hand grasping a branch of olive proper, and three lightning flashes or glues (the end of lightning bolts).

The blue sky is representative of Air Force operations. The arm and armor is a symbol of strength, power, and loyalty, and represents the science and art of employing far reaching advantages in securing the objectives of war. The olive branch, a symbol of peace, and the lightning flashes, symbolic of speed and power, are qualities underlying the mission of SAC.[5]

The B-29 was born in 1938, created when Boeing Aircraft Company submitted an aircraft design proposal to the United States Army Air Corps (USAAC) for an upgraded long-range bomber equipped with a pressurized cabin and crew compartments for high-altitude combat operations. However, nothing was done until February 1940, when the USAAF authorized the Boeing Aircraft Company to submit a formal design proposal for a high-altitude bomber. This design was referred to by Boeing engineers as Model 345, under military designation XB-29. The XB-29 prototype first flew on September 21, 1942, a little more than nine months into WWII and the U.S.' active belligerent status on the Allied side (England, Russia, and China). The USAAF funded a production run of 1,664 B-29s, a commitment to build a complex aircraft that had not been fully tested prior to full/mass production at multiple production facilities in the United States, which had never been done before. The demand to fight a long-range air war overrode concerns over the viability of the bomber before testing could be completed. This resulted in many engineering changes to the bomber after aircraft left the production facilities and the deaths of many aircrew members. But it was war, and the U.S. was facing long distances in the Pacific and needed the bomber to attack the Japanese home islands. Another important factor was that the B-29 was the only aircraft capable of carrying the top secret weapon under development by the Manhattan Project, the atomic bomb.[6]

In May 1947, the 28th BG and its subordinate units relocated to Rapid City Army Air Field and awaited activation of the new headquarters to which they would be assigned under the 28th BW. The Wing traces its history to July 28, 1947, when the USAAF first established it as the 28th BW VH, a unit that, until that time, existed only on paper. On August 15, 1947, SAC organized the Wing at its first duty station: Rapid City Army Air Field, South Dakota, 15th Air Force. This was when the 28th BG and its operational squadrons first joined the 28th BW.[7]

At a Dakota Thunder Air Show (Ellsworth AFB's open house), the lineage of strategic bombers assigned to the base once again was parked on the aircraft parking apron, except for the massive Convair B-36 Peacemaker. From bottom left to top right: Boeing B-17 Flying Fortress, Boeing B-29 Superfortress (at that time Confederate Air Force-today Commemorative Air Force), Boeing B-52H Stratofortress, and the Rockwell International-Boeing North American B-1B Lancer. This was a historic gathering of past and present bombers at the base. *Ellsworth Air Force Base, 28th Bomb Wing.*

The B-29 Superfortress Years

98th Bomb Group B-29 airborne during the Korean war. This aircraft operated late in the war, as indicated by the lower fuselage and wing sections, painted black in an effort to hide the standard silver-colored B-29s in the night sky. This was protection against ground based, radar-guided searchlights and airborne MiG-15s guided to the B-29s first by ground-based radar controllers, and then visually by the searchlights on the aircraft. *98th Bomb Group Historian.*

B-29A SPECIFICATIONS AND PERFORMANCE INFORMATION

Wing span	141 feet 3-inches
Wing area	1,736 square feet
Length	99 feet
Tail height	27 feet 9-inches
Empty weight	70,140 pounds
Gross weight	120,000 pounds
Maximum	weight 135,000 pounds
Powerplant	(4) 2,200 hp Wright, Double-Eagle Cyclone 18-cylinder air-cooled Wright R-3350-23 engines with a Pair of GE-13-11 superchargers
Max speed	375 mph
Cruising speed	220 mph
Ferry range	5,830 miles
Combat range	3,250 miles
Service ceiling	31,850 feet
Bomb load	20,000 pounds
Armament	(10) .50 caliber machine guns double turrets Top forward Top rear Bottom forward Bottom rear Tail (some equipped with one 20mm cannon)
Crew	Pilot (Aircraft Commander) Copilot (Tail gunner) Navigator (Left gunner) Bombardier (Right gunner) Flight engineer (Central Fire) Radio operator (control gunner) Radar operator[8]

The B-29 Superfortress Years

SAC was created on March 21, 1946, and was one of three major commands of the USAAF on September 18, 1947. Its mission was to be prepared to conduct long-range offensive operations in any part of the world, either independently or in cooperation with land and naval forces; to conduct maximum range reconnaissance over land and sea, either independently or in cooperation with land and naval forces; to provide combat units capable of intense and sustained combat operations employing the latest and most advanced weapons; to train units and personnel for further maintenance of the strategic forces in all parts of the world; to perform such special missions as the Commanding General, Army Air Forces, may direct. In developing an atomic bombing force, SAC relied heavily upon the men and equipment of the 509th CG, redesignated a VH BB on July 30, 1946.[9]

This only looked good on paper, as operational plans. General Curtis LeMay was assigned Commander SAC on October 19, 1948. When he took command, General LeMay was not satisfied with the operational combat status. I was able to obtain a paper copy (photocopy) of General LeMay's strategic development war-fighting plan for SAC in 1948. I kept it on my desk as a Second Lieutenant and until I was Captain, when I was assigned to SAC as a photographic intelligence officer; first as general photo interpreter, and then to the Office in Charge of Middle East Section, which was very busy during the 1973 Arab-Israeli conflict. It gave me insight into the historical SAC mentality and the vigorous and intense planning, preparations, and culture of submission that led to the personal fear of being evaluated by SAC Operational readiness Inspection (ORI). One could not make a mistake or one's career effectively ended, but it was hard not to when there were so many obstacles in the way:

1. The 509th BG (VH) was not combat ready to deliver atomic bombs, SAC's primary mission.
2. Operational training had been neglected, so crews and support personnel could not conduct SAC's primary mission, which was nuclear deterrence.
3. No target lists or routes planned to get the bombers to targets inside the Soviet Union to provide a credible nuclear shield and deterrence for the United States.
4. In 1948, SAC had to get its atomic weapons from storage facilities controlled by the civilian Atomic Energy Commission (AEC).

As a nuclear Weaponeer and Nuclear Staff Officer on the Joint Strategic Target Planning Staff (JSTPS), I was responsible for the nuclear weapons on the SAC base I was assigned to. All our training, inspections, and readiness were established many years previously by General LeMay. As a non-rated SAC intelligence officer assigned to a SAC BW in a more modern era, my responsibilities went far beyond SAC's early years. I was a *Cold War Warrior,* carrying on the legacy of nuclear deterrence to protect my country. I often stated, *I helped win the Cold War.*[10]

B-29 assigned to the 28th BW (Rapid City Air Field and then Weaver Air Force Base on January 13, 1948).
Ellsworth Air Force Base, 28th Bomb Wing Historian

The B-29 Superfortress Years

On July 12, 1948, after the first of SAC's many name changes, SAC activated the organization as the 28th BW, Medium. Just one week later, the wing deployed the 28th BG and its 77th, 717th, and 718th BS to England for a 90-day SAC B-29 show-of-force during the Soviet Union's military blockade of Berlin.[11]

The Berlin Airlift, which took place between June 24, 1948, and September 30, 1949, kicked off the first round of the Cold War between the United States and the Soviet Union. Some Cold War historians, with whom I share my own conclusions, state that when the Soviet Union closed railroad, highway, and canal routes from Ally occupied West Germany into Ally occupied (controlled) West Berlin, the two former WWII allies escalated their postwar confrontation from political to military provocation. The Cold War was primarily characterized by an atomic arms race, followed by a more deadly thermonuclear arms race that lasted into the 1990s. In 2010, the arms race appeared to be rising again with the delay by the U.S. Congress in ratifying the updated START with Russia (eventually voted on and passed by the U.S. Senate, voted on and passed by the Russian legislative branch, and then exchanged by both governments, thus activating the treaty), the rise of the Peoples Republic of China's Army's (PLA) intercontinental ballistic missile capabilities, and the nuclear threat from North Korea. This does not ignore the nuclear weapons capabilities of India, Pakistan, France, England, or those available but not disclosed by Israel, or the imminent threat that is Iran.

On June 24, 1948, the Soviet Union closed overland transportation routes from West Germany through Soviet-occupied East Germany into West Berlin, isolating the city except for the use of air transport. Fortunately, a November 1945 Allied Control Agreement, which the Soviet Union was a member of, allowed three twenty-mile wide air corridors from Ally-controlled West Germany into West Berlin. The Soviet Union had a historical record and tradition of abiding by its signed international legal agreements, such as the non-aggression Pact with Germany and the one with the Japanese Empire during WWII, and most importantly, Stalin's agreement at Yalta to enter the war in the Pacific against Japan three months after Germany's unconditional surrender, because this allowed the three air corridors to remain open to Allied air operations. However, the Soviet military still did not want the Allies to supply West Berlin by air, and determined they could wait for the residents of the city to surrender because of a lack of fuel, food, and basic living necessities.

U.S. Army Lt. General Lucius D. Clay, Deputy Commander, American Occupation Zone Berlin, only had the air transport option to supply Western Berlin and keep the city from falling under Soviet military control. He ordered USAF Europe (USAFE) transport aircraft to begin preparations to fly supplies into Berlin, code name *Operation Vittales*. This was a difficult assignment because of the limited number of transports available: USAFE's Douglas C-47s, two four-engine Douglas C-54s, and smaller twin-engine Fairchild C-82 Packets (forerunner of the C-119 Boxcar) were used to move supplies by air into West Berlin. The RAF contributed to the effort with a small number of Short Sunderland Flying Boats to land on a lake in West Berlin to deliver high-priority cargo for the city.

The first Berlin airlift flight was conducted on June 26, 1948. On May 12, 1949, the Soviet military acknowledged defeat in trying to starve the city into submission and ended its blockade. The Allies continued to fly in supplies to the city until September 30th to build up a level of emergency supplies. During 276,926 transport flights the USAF lost seventeen aircraft and the RAF seven. After the landing of the final transport flight in West Berlin, the Allies had delivered 17,835,727 tons of cargo, keeping the city alive and free of Soviet military control.

B-29s on the flight line at Weaver Air Force Base. *Ellsworth Air Force Base, 28th Bomb Wing Historian*

The B-29 Superfortress Years

One little known action taken by the U.S. in response to the Berlin Crisis, as it was officially called, was the deployment of B-29s to air bases in England and Europe. President Harry Truman hoped this implied threat to unleash SAC's B-29s, armed with atomic weapons, against targets inside the Soviet Union would deter that country from further aggressive action.

In 1948 the Soviet Union was developing, but had not detonated, an atomic weapon. President Truman wanted to use America's atomic bomb monopoly to keep the Soviet Union's expansion in check. This was an early implementation of America's *Containment Policy*, which became the primary strategy used against the Soviet Union. The Berlin Crisis, which spawned the extensive airlift into West Berlin, brought the U.S. and Soviet Union close to WWIII. After WWII the U.S. demobilized most of its military and supporting military industrial base, maintaining a small force in Europe opposing a huge Soviet military force in occupied Eastern Europe, which demobilized none of its war time forces. The U.S. could only threaten the Soviet Union with atomic bombs delivered by SAC's B-29s. A Joint Anglo/American Intelligence Group developed an atomic war plan, code named *Charioteer*, to attack Soviet cities and industrial centers. The plan targeted seventy Russian cities and industrial centers, with eight atomic bombs to be dropped on Moscow and seven on Leningrad. The goal of the plan was to destroy Soviet political and administrative control, the majority of Russia's oil industry, and thirty to forty percent of the country's industrial base. In doing this, it was estimated seven million Russian citizens would be killed in the atomic attacks.

The Berlin Crisis forced U.S. political leadership to contemplate its first atomic military option. President Truman did not want to use ground troops to force open ground routes into Berlin, primarily because he did not want to pit two U.S. Army divisions stationed in Europe (60,000 troops, of which only 10,000 classified as combat capable) against a Soviet Army force estimated at 300,000 to 400,000 troops stationed in East Germany and Eastern Europe. The Berlin Crisis forced President Truman and his military advisors to look at a change in U.S. strategic policy. During WWII thousands of American aircraft were stationed at airfields in England, flying combat bomber and fighter sorties over Nazi-occupied Europe and deep into Germany (Berlin). By the end of 1946 American bombers were not based in England.

During discussions with U.S. Ambassador to England Lewis W. Douglas on June 26, 1948, Ernest Bevin, British Foreign Secretary, indicated the Berlin Crisis called for the stationing of B-29s in England. Bevin reasoned this would signal to Soviet political leadership and military officers that the western alliance would not back down from keeping West Berlin a free city. At the same time Sir Brian Robertson, British Ambassador to West Germany, informed General Clay, in his role as Commander of the Office of Military Government and Military Governor of Germany, that the British government would approve the stationing of SAC B-29s at RAF bases. On June 27, 1948, General Clay submitted a request for B-29s to be deployed to England. General Hoyt Vandenburg supported his request, forwarding it on to President Truman. On June 28th, President Truman signed off on the request. However, it took until July 13th for the British Cabinet to sign the formal agreement allowing SAC B-29s to be flown to England and operate at select RAF bases.

President Truman had three objectives to achieve during the Berlin Crisis:

1. The U.S. and its Allies (England and France) would remain in West Berlin.
2. Supply West Berlin by air transport, not resorting to a ground confrontation with opposing superior Soviet military forces in Eastern Europe.
3. Increase USAF strength in Europe, including the deployment of B-29s.

SAC responded to the Soviet military blockade of Berlin. One B-29 BS from the 301st BG was already on a scheduled rotational training deployment at Furstenfledbruck, West Germany. General George C. Kenny, Commander SAC, ordered the 301st's two remaining squadrons to deploy to Goose Bay, Labrador, in preparation to fly to designated European bases. Although the B-29 was considered a long-range bomber, it still required forward airfields close to Russian targets to be able to recover at airfields in Germany after dropping their atomic bombs. SAC was in the process of integrating a new very heavy, long-range bomber, the Convair B-36 Peacemaker, into its nuclear arsenal. The B-36 was not nuclear capable at this time.

SAC placed the 28th and 307th BGs on alert for deployment to Europe. The 28th BG could deploy within twelve hours and the 307th BG within three hours. On July 15, 1948, the U.S. National Security Council approved sending SAC B-29s to England. The Soviet Union knew SAC's B-29 could carry an atomic weapon because the U.S. dropped two atomic bombs on Japan in WWII. By deploying B-29s to England, SAC aircrew were within range of targets in Moscow. This was an implied threat to the Soviet Union. It held the same military significance as when Secretary of the Navy Theodore Roosevelt sent the U.S. battle fleet around the world as a demonstration of U.S. naval power. SAC ordered other B-29 BGs on twenty-four hour deployment notification. By early July 1948, the 301st BG's two remaining squadrons were on the ground at Furstenfledbruck, Germany. Later in the month the 28th BG deployed from Weaver AFB, SD, to RAF Scampton, England, for a ninety-day show-of-force mission.

SAC-deployed B-29s were bedded down at RAF bases at Lakenheath, Marham, Scampton, and Waddington. In July 1948, B-29s landed at RAF Lakenheath, assigned to the 2nd BG. Nine B-29s assigned to the 304th BS, 97th BG landed at RAF Marham. Between July 1948 and February 1949, RAF Scampton supported 30 B-29s assigned to the 28th BG, later relieved by B-29s from the 301st BG. Other SAC B-29s operated out of RAF Waddington.

28th BG at Weaver AFB, SD. The bomber completed a landing and is heading toward the parking apron. The aircraft at the right edge of the photograph is a SAC C-54 transport, which transported spare parts and ground personnel when SAC bomb groups deployed, as to England during the Berlin Crisis. *Ellsworth Air Force Base, 28th Bomb Wing Historian*

The 3rd Air Division (3AD) Provisional was created to serve as a temporary 30- to 60-day deployment command. From June 1, 1974, to June 15, 1976, I was assigned to the 3AD, Andersen AFB, Guam, as a Wing Intelligence Officer, part of SAC's nuclear force in the western Pacific. Due to the continuing severity of the Berlin Crisis, the provisional designation was dropped on August 23, 1948, with the 3AD organization headquartered at Bushey Park Air Station on September 8, 1948.

The British government gave SAC aircraft and personnel to RAF airfields at no additional expense, as long as operational expenses did not exceed normal RAF operating costs. The agreement was formalized on January 4, 1949, after Major General Leon W. Johnson, Commander 3AD, received the formal financial agreement to support SAC's B-29s in England from the British Air Ministry. This would not have been possible if airfield facilities in England had not been upgraded by the RAF to handle B-29s. This upgrade started in 1946, after British Air Marshall Sir Arthur Redder, Chief of the Air Staff, and General Carl Spaatz, Commanding General USAAF, decided WWII B-17 and B-24 runways at RAF bases required extensive upgrades. Over the next two years, runways were lengthened and widened to handle combat loaded B-29s, which if loaded with a plutonium atomic bomb had a bomb-bay load of 10,300 pounds. B-29s operating at high combat loads would have broken through the existing runways.

USAF and RAF committed transports to fly supplies into West Berlin in a show of united resolve. At the same time, SAC wanted to demonstrate the possible atomic lethality of its deployed B-29s. In reality, SAC aircrew bombing capabilities and accuracy were short of acceptability, with larger CEP (Circular Error Probable), even when considering the destructive capacity of early U.S. atomic bombs. In May 1948, prior to the Soviet Army closing ground routes into West Berlin, Major General Clements McMullen, Deputy Commander SAC, began addressing this deficiency. He started an annual SAC bombing competition to develop accurate delivery and release of atomic bombs. In June 1948, three aircrew from each of SAC's bomb groups flew to Castle AFB, California, to compete in the command's first bombing competition. It was not very sophisticated. Each aircrew dropped three inert practice bombs by visual identification (bombsight) and three by radar from 25,000-feet. The cumulative competition's bombing scores resulted in a CEP of 1,065-feet for visual and 2,985-feet for radar releases. This would destroy a large area, but was indicative of a lack of accuracy to destroy hardened targets inside the Soviet Union.

The deployment by the 28th BG was typical of SAC B-29 groups in support of the Berlin airlift. On July 15, 1948, a detachment of the 717th BS consisting of 79 officers and 139 enlisted took off from Weaver AFB, along with a provisional headquarters squadron and a detachment from the 77th BS for thirty days TDY to Europe. Thirty B-29s from the 28th BG carrying maintenance and medical personnel, in addition to the aircrew, flew via Goose Bay, Labrador, to Prestwick, Scotland, to RAF Scampton, Lincolnshire, England. Supplies and equipment, along with additional support equipment, were flown to England on commercial passenger aircraft. This was before the creation of the Civilian Air Reserve Fleet. Once at RAF Scampton, the 28th BG came under the jurisdiction of the 3AD, Provisional. Upon arriving in England, members of the 28th BG learned their TDY had been extended to sixty days. News of the change from 30 to 60 days, along with overcrowded barracks, badly prepared food, lack of recreational facilities, and other deficiencies brought a sharp decline in bomb group morale.

On July 26, 1948, twenty-six (four were not operational due to maintenance) 28th BG B-29s were dispatched to fly a formation flight witnessed by Mr. Henderson, British Secretary of State for Air. Three days later a division mission flown by the 307th BG displayed B-29 formations over Lincoln, Cambridge, Brighton, Southampton, Plymouth, Bristol, Liverpool, and Cottingham. Well publicized, this gave British citizens their first glimpse of SAC's Superfortresses in massed formation. Equally important and planned, London Soviet Embassy personnel witnessed the flyover of U.S. air power, reporting back to Moscow that SAC was flying massed training missions in England and practicing the bombing of British cities, probably in preparation for conducting atomic bomb strikes against Moscow to destroy the city.

The 28th BG's morale improved in August 1948, when squadron commanders began authorizing personnel with forty-eight hour passes to go into London, though they were briefed and warned by intelligence officers not to talk about any training, preparations, or activities on their base under severe penalties. This was to get away from the constant activities involved in keeping B-29s and their aircrew ready for forward deployment to Germany for pre-strike positioning. August also brought VIPs to visit the 28th BG at RAF Scampton: Stuart Symington, Secretary of the Air Force; General Hoyt S. Vandenburg, USAF Chief of Staff; and Major General Leon W. Johnson, Commanding General, Fifteenth Air Force. In September more VIPs visited RAF Scampton: Major General Johnson returned on September 1st and 2nd; U.S. Senator Chan Gurney of South Dakota made an official inspection trip on September 3rd; and Lord Tedder, RAF, visited on September 4th. Lord Tedder watched as 28th BG B-29s took off on a joint American-British defense exercise. Soviet intelligence officers were in their cars and outside with other British citizens watching and taking photographs of the B-29s taking off from RAF Scampton. This was a sign to Soviet intelligence officers assigned to the London Embassy that the Berlin Crisis threatened Soviet security, with the stationing of SAC B-29s in England within range of the Soviet Union. Lt. General Lauris Norstad, Commanding General USAFE, also visited the 28th BG one more time. Toward the end of 1948 the 28th BG was alerted for another TDY to England, but did not deploy. Starting in 1949, the 28th BG began the conversion from the B-29 to the B-36, as SAC improved its long-range nuclear weapons delivery capabilities.

The 307th BG deployed from MacDill AFB, Florida, to RAF Marham and Waddington. General George Kenny viewed the Berlin Crisis as an opportunity to demonstrate America's willingness and military capability to use air power against the Soviet Union. Soviet political and military leaders were surprised by

78

the rapid American military response to the closing of the ground routes into West Berlin. Soviet intelligence was unable to determine if or how many SAC B-29s deployed to England or West Germany were armed with atomic bombs. B-29s were guarded as if they were armed, carrying on the active deception. They deployed , uin fact, from the United States with atomic bombs.

On October 19, 1948, General LeMay assumed command of SAC, starting a robust wartime training and upgrade of SAC's delivery capabilities, and to be ready to fight a nuclear war against the Soviet Union. Even though the Soviet Union had been America's WWII partner, the U.S. planned to use its atomic monopoly against her if required. On July 19, 1945, President Truman approved a memorandum prepared by Secretary of War Henry Lewis. This memorandum stated that the United States should use its monopoly on atomic weapons to force the Soviet Union to adopt a political system that would suit America, and that the bomb could be used to gain political advantage over the Soviet Union in the post-WWII period. After the end of the war, the U.S. Joint Chiefs of Staff (JCS) recommended the U.S. must be ready to strike the first blow against the Soviet Union (first strike) with atomic weapons if needed. The resulting war plan targeted twenty Russian cities for atomic bombing. Included on this list were Moscow and Leningrad. This first atomic bombing strike against Russia was called *Totality*.

Photograph of original B-29 *Legal Eagle*, assigned the the 28th BG. *Ellsworth Air Force Base, 28th Bomb Wing Historian*

As forward-looking as U.S. atomic war plans appeared on paper, the problem of implementation depended on the number of produced and available atomic weapons. The first three atomic bombs (one uranium and two plutonium) produced by the U.S. during WWII cost over two billion dollars. The *Fat Man* Nagasaki bomb was technically the blueprint for the postwar Mark (MK) III atomic bombs. By the end of 1948, U.S. inventory of MK IIIs was approximately one hundred. The MK III had a variable yield of 18 to 49 kilotons (kt). One B-29 could carry one MK III in its forward bomb-bay. SAC had thirty-eight B-29s modified to upload and deliver the MK III in 1948. By the end of 1949 this number increased to ninety-five. This improvement was because, at the start of the Berlin Crisis in 1948, SAC only had the 509th BG certified to deliver the MK III, but this group was suspect due to a lack of training and access to AEC controlled atomic bombs away from the bombers' home bases. At this time training was a great problem for SAC. This was the time period before the intense, war level training program initiated by General LeMay, which started an evolution in strategic planning and thinking and the transition to a credible nuclear force able to maintain nuclear deterrence, as well as an immediate response.

For the B-29s deployed to West Germany, the on-site Squadron Commander did not know how to prepare for possible combat operations. The Squadron Commander questioned SAC Headquarters as to whether they should be prepared to deliver conventional or atomic bombs. He did not have an operational or contingency plan of operations for striking targets in Russia from forward bases in West Germany. Initially, with many USAF transports being tied up moving supplies into West Berlin, it was assumed SAC's B-29s would also move supplies. This was not done. It was correctly assumed the Soviet military and political leadership would view the presence of B-29s in West Berlin as a pre-set for a first atomic bomb strike against targets in the Soviet Union, thus requiring a Russian Air Force preemptive strike to remove the immediate threat.

The greatest obstacle to arming SAC's B-29s was the assembly of atomic bombs at the designated forward operating bases. In 1948, with one hundred MK IIIs at the ready, SAC munitions personnel could only assemble two MK IIIs per day. This meant not enough bombs could be assembled and delivered simultaneously to destroy selected targets as specified on paper under *Charioteer*. Even though SAC deployed B-29s to England and Germany in 1948, President Truman never authorized the movement of atomic weapons from the United States to England, let alone into Germany. This was the same pattern President Truman would follow during the Korean War, moving SAC atomic bomb certified aircrew to Okinawa without the subsequent movement of atomic bombs from the United States to Andersen AFB, Guam. It was an implied threat, but one which helped bring the Communist Chinese to the armistice table in Korea, ending the fighting. For SAC's B-29s deployed to Europe, they could only be armed with conventional bombs.

Even with reports of massed B-29 formations over England, forward deployment of B-29s into West German bases, and guarded flight lines, Soviet Intelligence (GRU and KGB) probably and correctly assumed the following:

1. Deployed B-29s were not armed with atomic bombs because of the lack of associated special munitions facilities at these bases to store and assemble the bombs, with secured areas heavily guarded.

2. B-29s were not guarded at the same level as those which would be carrying atomic bombs.

3. The 509th BG had not been deployed, which was SAC's primary atomic delivery group in 1948.

After Soviet intelligence confirmed the movement of B-29s from the United States, there was no indication the Soviet political leadership fully discounted the U.S. would deploy SAC bombers without atomic bombs. Regardless, Soviet intelligence officers were constantly probing to collect information on SAC's B-29 operations in England and Europe. This continued during the Korean War, even though by that time SAC was deploying newer and more capable strategic bombers. The configuration of the B-29A on

B-29 on display at the South Dakota Air and Space Museum, outside Ellsworth Air Force Base. After restoration it was marked to represent a B-29 stationed at Weaver Air Force Base called *Legal Eagle*. Lt. Col. George A. Larson, USAF (Ret.)

display at the South Dakota Air and Space museum corresponded to that of the 28th BG's B-29 deployment by SAC to Germany.[11]

Because three battle-damaged 20th AF B-29s diverted to and landed in Siberia during WWII, the aircrew was detained (eventually released) and the aircraft retained for technical evaluation and exploitation by Soviet Aviation engineers. Consequently, Soviet Air Force officers knew of the operational capabilities of the U.S. B-29s. Based on these three B-29s, Soviet aviation engineers developed and produced their own copy of the Superfortress, the Tupolov Tu-4 Bull. These were eventually armed with a near production copy of the *Fat Man* plutonium atomic bomb from plans stolen by a Manhattan Project scientist, Klaus Fuchs, who passed this classified information to Soviet espionage agents working in the United States. Even with investigations by the FBI, Fuchs' spy activities were not uncovered until after the Soviet Union detonated their first atomic bomb.

The deployment of SAC B-29s to Europe was the aviation version of former U.S. gunboat diplomacy, such as the deployment of Nimitz class aircraft carriers into areas of increasing tension throughout the world. This saber rattling gave the Soviet political leadership pause to think about the long-term military ramifications of their blockade of Berlin. General Kenny complained about the inactivity of SAC B-29s deployed to Europe:

> The Russians may of course be worried about our 90 B-29s now in Europe. But they don't seem to be seeing them as a club. Perhaps in time, the Russians will figure that as long as we don't mention them around the green table [negotiations to end the blockade], that they are no good anyhow.

But it does appear Soviet Premier Joseph Stalin viewed U.S. B-29s as a deterrence, since he never tried to force the occupation of West Berlin by available, overwhelming military force. U.S. intelligence reports indicated reoccurring combat training and the manning of levels of Soviet military forces in Eastern Germany were not that common for a major ground attack. There were smaller scale Soviet military unit trainings, but no large scale combined arms training or forward deployment. However, the movement of two Soviet Army divisions toward the West Germany border in September 1948 lacked the necessary logistical support for the initiation of offensive ground operations. On June 30, 1948, the Soviet Politburo discussed preparations and adequacy of Soviet Army anti-aircraft defenses around Moscow and other large industrial/city areas. This reflected the Soviet political leadership's concern over the current U.S. monopoly of atomic weapons and the presence of SAC B-29 bomb groups in England and Europe, including Western Germany.

Despite the lack of atomic weapons for deployed B-29s, there appeared to be an elaborate effort to convince the Soviet leadership that President Truman was prepared to use atomic bombs in the defense of West Berlin. The deployment of SAC B-29s to England was publicized as the movement of atomic bombers from U.S. SAC bases to England. The British government allowed SAC to station B-29s in England. Comments made by officials of the British government indicated the issue that who would control the release of atomic weapons from England was uncertain and under discussion. In reality, SAC would only respond to orders from the President. For Soviet Premier Stalin, the Berlin Blockade and subsequent crisis was a political setback. In the aftermath of the Berlin Crisis responses, many of which had long lasting effect, continued to mature and develop:

1. U.S. President Truman did not and would not recognize Soviet political or military predominance in Germany.

2. The U.S. rapidly turned the Berlin Blockade into a propaganda attack against the Soviet Union.

3. The rapid creation and maintenance of an air transport capability to move supplies into West Berlin was something the Soviet leadership did not believe could be sustained.

4. Premier Stalin did not plan to start a world war over the control of Berlin, which U.S. intelligence accurately determined and reported to President Truman. The deployment of SAC B-29s played a significant role in the peaceful outcome of the Berlin Crisis.

5. By 1949, Premier Stalin had to recognize a permanent Western Allied political right to remain in West Berlin.

6. Countries in western Europe turned to the United States for protection against the Soviet military threat, leading to the creation of the North Atlantic Treaty Organization (NATO) in 1949.

The B-29 Superfortress Years

In April 1949 the United States joined NATO, a European military alliance that was an outgrowth of the Brussels Treaty. A key provision of the treaty was that an armed attack against one or more of the members would be construed as an attack against all members. In this was an implied threat that the U.S. would use its atomic weapons against the Soviet Union.

In 1948, President Truman was told by his intelligence advisors, as well as by General Groves, military director of the Manhattan Project, that the Soviet Union would not have an atomic bomb until 1953 (at the earliest date). On September 3, 1949, a USAF WB-29 collected airborne radiation samples indicating the Soviet Union had detonated a plutonium bomb, a copy of the U.S. plutonium bomb.

To give SAC credibility as a deterrent force, General LeMay held SAC's bombing competition October 3-7, 1949, establishing it as an annual test of the command's readiness, as well as each bomb group's skill. This was a maximum stress environment. The same format was used as during the first competition, later altered to test actual bombing missions. While stationed with the 93rd BW, Castle AFB, 1976-1978, I radar scored two SAC bombing competitions, and was selected to participate due to my performance as a SAC intelligence officer. Interestingly, the 93rd BG, which had to switch back to B-29s after its new B-50s developed fuel leaks and were grounded, won the unit award for SAC bombing. A 28th BG B-36 aircrew carried home the honor for the best individual crew. (The B-36 will be covered in the next chapter.)

Another development of deploying B-29s to England was after the end of the Berlin Airlift. On March 22, 1950, the first group of B-29s was flown from the U.S. to England, transferred to the RAF under the Mutual Defense Assistance Program. The RAF initially accepted seventy B-29s, with another eighteen on a second transfer. A larger transfer to the RAF was originally scheduled, but the USAF had to pull B-29s out of storage at Davis Monthan AFB and from operational units for transfer to the Far Eastern Air Force (FEAF) for strategic and tactical bombing operations during the Korean War. The RAF designated their B-29s as the Washington B-1.

In April 1950 Lewis Douglas, U.S. Ambassador to England, and Aidan Crawley, British Secretary for Air, reached the determination that RAF bases in East Anglia were too vulnerable to possible Soviet Air Force attack. They decided to relocate SAC B-29s to RAF Upper Heyford, Greenham Common, Brize Norton, and Fairford.[12]

On January 13, 1948, the base's name was changed to Weaver AFB; Air Force Chief of Staff General Carl A Spaatz wanted the base name change to honor Brigadier General Walter A. Weaver, one of the pioneers in the development of the Air Force. On June 24, 1948, the name changed again to Rapid City Air Force Base. At the same time, the base was declared a permanent installation.[13]

Even though today's Ellsworth AFB is a bomber base, in its history it was home to the 54th Fighter-Interceptor Squadron (FIS). It was activated on January 15, 1941, at Hamilton Field, California. It deployed to Alaska and fought against Japanese forces then operating in the Aleutians. On December 1, 1952, the 54th FIS replaced the North Dakota Air National Guard's 175th FIS, taking control of the previous unit's North American P-51D Mustangs. The 54th FIS was assigned to the Air Defense Command (ADC) with a mission to provide air defense for the upper Midwest from attacking Soviet Air Force bombers (Soviet Long Range Aviation, or LRA) flying one-way missions over the North Pole to drop atomic bombs on the United States.

P-51Ds assigned to the 54th FIS.
Ellsworth Air Force Base, 28th Bomb Wing Historian

82

The P-51D was the major production variant of the high-performance WWII fighter later used during the Korean War, 1950-1953. It remained in active service after WWII as part of SAC's fighter assets until 1949, acting as part of the transition from piston powered aircraft to the first generation of turbojet aircraft, bombers, and fighters. At this time, SAC still had its own fighters to protect its bombers after takeoff from their bases prior to crossing in Russian air space if activated under the current war plans. The 54th FIS flew the P-51D until 1953.

Diorama of a 54th FIS P-51D on display at the South Dakota Air and Space Museum. *Lt. Col. George A. Larson, USAF (Ret.)*

The P-51D was retained at United States Air Defense Command bases during the Korean War. Even though the FEAF was outnumbered by the Russian MiG-15s supplied to the Communist Chinese Air Force and the re-equipping of the North Korean Air Force, the USAF did not pull F-86s from Europe to South Korea out of a concern that the Russians might decided to initiate air offensive actions against Germany and NATO.

Republic Aviation F-84F Thunderstreak on display at the National Museum of the United States Air Force. *Lt. Col. George A. Larson, USAF (Ret.)*

The B-29 Superfortress Years

The 54th FIS transitioned to Republic Aviation F-84As in July 1953, when they returned from FEAF and combat operations in Korea. The Squadron upgraded to F-84Gs in December 1953. This variant had a maximum speed of 620 mph. The variant was equipped with the most powerful engine of the thunder jet series, fitted for aerial refueling by SAC's KB-29s. A lesson learned during the Korean War was that fighters needed to be air-refueled.

North American F-86D Sabre on display at the National Museum of the United States Air Force. *Lt. Col. George A. Larson, USAF (Ret.)*

The 54th FIS upgraded to the North American F-86D in April 1954. The F-86D, referred to as the *Sabre Dog*, was an all-weather interceptor equipped with a nose mounted radar for interception of enemy bombers trying to penetrate North American air space. It had a maximum speed of 700 mph and a top altitude of 54,600-feet, and was armed with (24) 70mm Mighty Mouse unguided rockets carried in a retractable rocket pod on the centerline of the fuselage.

Northrop F-89J Scorpion on display at the National Museum of the United States Air Force. *Lt. Col. George A. Larson, USAF (Ret.)*

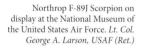

The last fighter used by the 54th FIS was the Northrop F-89J Scorpion in 1957, which was flown until December 25, 1960. The F-89J was a two-seat, all-weather interceptor armed with two Douglas MD-1 Genie atomic warhead rockets (one inert Genie, white in color, is hung under the left wing of the aircraft to the left) designed to knock down large formations of Soviet LRA bombers attacking the United States. These Soviet bombers were the Myasischyev M-4 Bison four-engine turbojet bomber (only produced in limited numbers) and Tupolev Tu-95 Bear four-engine turbo-prop bomber (the Russian Air Force's to replicate the longevity of the USAF's B-52 Stratofortress, currently the B-52H).

The history of the 54th FIS and ADC at Rapid City Air Base is maintained at the South Dakota Air and Space Museum. The concept for the museum (it will be covered in more detail in a later chapter) began when the 15th AF Commander pushed for the establishment and development of an Air Force program at Ellsworth AFB in 1982. The museum began operations in 1942 in an H-shaped wooden building along the east side of the flight line. Initial displays were primarily pictorial, with a few artifacts and tables to display them. In 1987, the current museum director, Ron Alley, began looking at possible options for a new facility to accommodate larger displays, as well as provide adequate space for an outdoor aircraft display area.

Four-bay ADC alert facility moved from the flight line at Rapid City AFB to the location of the new South Dakota Air and Space Museum. *Lt. Col. George A. Larson, USAF (Ret.)*

The Ellsworth Heritage Foundation assisted in the building search project, which focused on an intact, former flight line structure scheduled by the Air Force to be torn down. The former ADC four-bay alert fighter-interceptor hangar facility finished its active duty career by housing four F-89Js. Four pilots and four radar operators, aircraft, and crews were on 24-hour alert, prepared to scramble and takeoff within three minutes to intercept Soviet LRA bombers at high-altitude attempting to penetrate North American air space, thus protecting SAC's bombers on the base. The ADC alert facility was considered to be an ideal structure to house the museum's growing displays. In November 1988, the Ellsworth Heritage Foundation began discussions with Air Force and SAC officials on moving the alert bays and central connecting building from the flight line to a location near the main gate. In August 2010 this gate was closed, after demolition of the North Gate and reconstruction was completed. The Commercial Gate was also rebuilt farther to the south, providing an enclosed vehicle inspection bay able to inspect large commercial vehicles before clearance and entry onto the base. The Main Gate was reopened in the summer of 2011. These upgrades were necessary to

85

meet possible terrorist attack attempts on U.S. military installations. Each road onto the base is protected by an active vehicle barrier that pops up from the road, blocking any unauthorized attempt to break through the three security gates/entry checkpoints. It is a statement to the world which the U.S. military currently operates, even in Stateside installations.

The $60,000 needed to move the ADC alert bays was raised through public non-government donations. The facility was moved one alert bay at a time to the new location and set on a concrete floor (foundation). By June 1989 the museum's new facility was ready to open to the public, although not all the interior had been refurbished.[14]

In July 1949, after runway improvements, the 28th BW began the conversion process from the B-29 to the B-36.[15]

During the conversion from the B-29 to the B-36 at Rapid City AFB, B-29s were flown to Davis-Monthan AFB, Arizona, for storage in the desert. Most were eventually sold to commercial metal scrap companies, chopped up and melted down. Some were retained for displays at museums throughout the world as a tribute to those who flew the Superfortress during WWII, the Korean War, and opening years of the Cold War between the United States and the Soviet Union. The South Dakota Air and Space Museum is fortunate to have one of these magnificent and historically significant bombers on display. *AMARC historical files to Lt. Col. George A. Larson, USAF (Ret.)*

B-29 *Legal Eagle* on display at the South Dakota Air and Space Museum. *Lt. Col. George A. Larson, USAF (Ret.)*

CHAPTER FOUR
SIX TURNING AND FOUR BURNING

SAC's First Long-Range (Very Heavy) Atomic Bomber

B-36D on final approach to runway number 27 at San Diego, California, returning to the Convair San Diego plant after being converted to the "D" variant with the addition of four turbojet engines on February 22, 1952. Milton R. Kuhl, Jr., took the photograph as the large bomber flew overhead. Hence, the phrase as the upgrade became known on the aircraft, *six turning and four burning. Milton R. Kuhl, Jr., to Lt. Col. George A. Larson, USAF (Ret.)*

The existing north-south runway at Weaver AFB is no longer operational because of construction of a new parking apron, maintenance hangars, and support buildings. The east-west runway also became non-operational for the same reason. The current runway (13/31), which is concrete, 13,497-feet long, and elevation 3,276-feet, supports B-1B bomber operations for Air Combat Command (ACC).[1]

The Consolidated (Convair) B-36 Peacemaker was born after the USAAC became concerned that the German Luftwaffe might possibly defeat the RAF during the fierce air war in the summer of 1940 referred to as the Battle of Britain. On paper, the Luftwaffe possessed numerical superiority and should have defeated the RAF to gain air superiority over the English Channel, clearing the way for a German amphibious landing and airborne operations to invade England. The German invasion and losses in men and equipment would have been extensive, but probably less than subsequent losses to Russian forces on the Eastern Front. Operation *Seeloewe* (Sealion) might have succeeded if Adolf Hitler had thrown caution to the wind and bet everything on the invasion.

Weaver AFB flight line with 28th BW B-29s parked, facing north, prior to the transition to the B-36.
Ellsworth Air Force Base, 28th Bomb Wing Historian

A successful German invasion, supported by captured forward RAF airfields, would have denied the USAAC the use of these airbases needed if or when the U.S. entered the European War. American military planners believed it was only a matter of time before entering the war on the side of the British, given the existing military cooperation with England to counter German and Italian aggression. Most importantly, the USAAC needed RAF airfields to attack German controlled areas and gain air superiority to clear operations for an invasion of Nazi-occupied Europe. The loss of England as a forward operating location would force the U.S. military to conduct long-range bombing raids on Nazi-controlled England from airbases on the U.S. east coast. Even if this was only a remote scenario, U.S. military planners had to consider and develop contingency plans.

XB-36 on its first test flight from Convair-Fort Worth, Texas, production facility on March 26, 1947. *Convair historical archives to Lt. Col. George A. Larson, USAF (Ret.)*

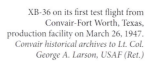

This dramatically altered existing military planning, which had the potential to develop into a serious logistical and strategic problem for the USAAC (which became the USAAF). At this time, neither the four-engine B-17 nor B-24 had the combat range to attack a German occupied England from bases along the U.S. east coast. With the War Department's concern over German military successes over British forces in Norway, France, North Africa, and Greece, preliminary design work began on a new class of very heavy, long-range intercontinental strategic bomber. Its size, bomb load capacity, and range would exceed that of the B-29. Fortunately the RAF narrowly defeated the Luftwaffe, removing the need to initiate another high-priority and expensive war production project under an already strained U.S. war economy. Consolidated engineers continued design work on the large bomber. The B-29 and Consolidated B-32 Dominator long-range bombers were scheduled for Pacific combat operations against the Empire of Japan. The B-29 was planned as the delivery bomber for the top secret Manhattan Project's atomic bomb. Even as the atomic bomb was being developed, Edward Teller, a Manhattan Project nuclear physicist, was working on a thermonuclear (hydrogen) bomb, which would be bigger and heavier, requiring a very heavy bomber to deliver it over intercontinental ranges.

On November 15, 1941, the USAAF signed an aircraft contract with Consolidated Company (Consolidated merged with Vultee Aircraft in 1943, becoming Convair) to design and build two intercontinental bomber prototypes under designation VB-36 and YB-36. Because of overly optimistic range (8,000 miles) and bomb load (10,000 pounds) requirements work progressed slowly. The bomber needed a wingspan of 230-feet with a slight, swept-back wing configuration, and was powered by six 3,000 hp Pratt & Whitney R-4360-25 Wasp Major air-cooled piston radial engines, each driving a 19-foot diameter three-blade Curtiss variable-pitch pusher propeller. It had a maximum speed of 346 mph, cruising speed of 216 mph, service ceiling at 36,000-feet, and a maximum range of 9,500 miles carrying a 10,000 pound bomb load.

B-36H crew (1954). This was not an aircraft assigned to Ellsworth AFB, but shown to depict the crew which had to be trained to fly the large bomber. The 15-man crew was the largest required to operate a SAC bomber. *Photograph sent to author by the Navigator, Chuck McManus, standing second from the right.*

Six Turning and Four Burning

The XB-36 had an empty weight of 131,700 pounds and a gross weight of 276,506 pounds. It far outsized the B-29, which at that time was designated a heavy bomber. Its fifteen man crew consisted of a Pilot, Copilot, Radar Operator/Bombardier, Navigator, Flight Engineer, two Radio Operators, and eight Gunners. A proposed reconnaissance variant would require seven additional crew members to perform duties as observers. To provide crew rest on long-range missions, four crew bunks were installed in the aft compartment.

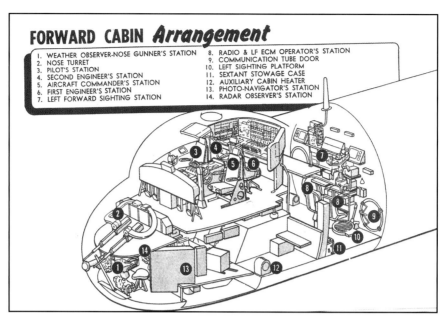

Forward B-36 cabin arrangement. *From B-37 technical manual, provided by (former) SAC Historian Office, Offutt Air Force Base, Nebraska, 1990.*

Defensive armament was to be provided by five 37mm twin cannon turrets, with a .50 caliber machine gun added to the tail turret. Four of these turrets retracted into the fuselage to reduce aerodynamic drag.

Consolidated's design proposal was approved on July 20, 1942 (while WWII raged, and just after the U.S. Navy victory off Midway Island), with the production of the XB-36 moved from Consolidated San Diego to the company's B-24/B-32 production plant at Fort Worth, Texas. If production demand for the B-24 never decreased during the war, complicating production of the B-32, it would have left little resources and production floor space for the XB-36. In 1943, it was determined the B-36 might be needed to attack the Japanese home islands from bases then under Allied control: the Mariana Islands of the Japanese (Guam, Tinian, and Saipan). Plans for its production were increased in priority, with a War Department contract for one hundred issued on July 23, 1943.

B-36 aft cabin arrangement. *From B-36 technical manual, provided by (former) SAC Historian Office, Offutt Air Force Base, Nebraska, 1990.*

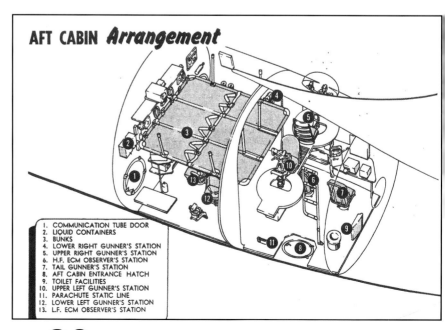

Six Turning and Four Burning

The XB-36 forward crew compartment was redesigned, and the B-24 twin tail was switched to the single B-32 tail arrangement. To reduce weight, the number of defensive guns was reduced. One design on the large bomber almost stopped its production: the landing gear configuration. The design produced a 110-inch diameter, 36-inch wide, 1,300 pound single main (one on each side of the fuselage, under the wing) landing gear and a smaller dual-wheel nose landing gear. This produced a huge load-to-wheel weight, limiting the aircraft's operational locations to: Consolidated Fort Worth, Texas (which became Carswell AFB, across the runway from the production facility); Eglin AFB, Florida; and Fairfield-Suisan AFB (today Travis AFB), California. On the production version B-36A, the landing gear configuration changed to a four-landing-gear-truck arrangement that spread the load-to-wheel weight, allowing the bomber to operate from standard SAC bomber runways.

B-36A at Consolidated Fort Worth. It is equipped with four tires at 56 inches in diameter.
Convair historical archives to Lt. Col. George A. Larson, USAF (Ret.).

It was not until October 7, 1943, that Convair received an official government letter titled *Construction Breakdown for Experimental XB-36*, releasing funds from AAF appropriations to build the XB-36 and YB-36. It took until June 15, 1944, for the AAF to authorize B-36 series production. The AAF letter stated:

As evidence of such a program is the intent of the War Department, the government will place a formal contract with Consolidated for the purchase of 100 B-36s, spare parts, and engineering data, at an estimated cost of $153,750,000, exclusive of the company's fixed fee of 4%.

On March 25, 1946, XB-36 aircraft serial number 42-13570 was towed out of the Fort Worth bomber plant, almost ten months after the end of WWII. On August 25, 1946, Convair's Chief Test Pilot, Beryl Erickson, took off in the XB-36 for the first time. As with all new aircraft, especially one as large as the B-36, there were many problems identified, including: engine overhearing (which the B-29 also suffered from), propeller malfunctions, fuel and oil leaks, airborne fires, and pressurization failures. Production and subsequent modifications resulted in a life cycle cost for 383 B-36s of $3,800,000 per aircraft.

B-36B, aircraft number 44-92033. Some B-36Bs were easily identified by their red tails and wing tips, painted for operations in extreme cold weather during Arctic test flights. The aircraft were assigned to Carswell AFB, Texas. By the end of 1948, Carswell AFB had a complement of 35 B-36Bs. *Ellsworth Air Force Base, 28th Bomb Wing Historian*

Six Turning and Four Burning

CONVAIR B-36 PRODUCTION

Aircraft Designation	Number Built	Modifications
XB-36	1 (F)*	
YB-36	1 (F)	
YB-36A	1 (F)	
RB-36E		
B-36A	21 (F)	
RB-36E		
B-36B	73 (F)	(62)
B-36D	(7)	
RB-36D		
B-36C	(variant cancelled and no airframes built)	
B-36D	22 (F)	62 (M)**From B-36B airframe
RB-36E	22 (M)	(1)
YB-36A	(21) B-36A	
B-36F	34 (F)	
RB-36F	24 (F)	
B-36G (Yb-60)	1 (F)	Swept wing with 8 turbojets
B-36H	83 (F)	
RB-36H	73 (F)	
B-36J	33 (F)	
XC-99	1 (F)	Cargo variant

Notes: (F)* Factory built (M)** Factory modification
(5)   (Corrected space alignment in front of Travis AFB, CA)

General Curtis E. LeMay, SAC's second commander. Painting on display SAC Museum. *Lt. Col. George A. Larson, USAF (Ret.)*

Six Turning and Four Burning

Flight testing at Fort Worth was completed on June 19, 1948, and Convair released the XB-36 to SAC for use as an initial aircrew trainer at Carswell AFB. This aircraft, although considered historically significant, never made it into the safety of an aviation museum. After SAC no longer wanted to retain the aircraft as an aircrew trainer, it was towed across the runway to the Convair plant in 1957. Its engine was stripped and operational equipment removed, and it was used as a fire department trainer until consumed to the point where the hulk was hauled away as scrap.

Although the B-36's range and speed did not meet Air Force specifications, it was SAC's only VH long-range strategic bomber, capable of flying intercontinental atomic missions in the time before the full implementation of aerial refueling tankers (the KC-97 and KC-135) to support B-47s and B-52s, and later B-58s and FB-111s. It was the only bomber able to carry the large, first generation of thermonuclear gravity bombs. It was used by General LeMay to portray SAC as a nuclear-capable deterrent and bridge from all piston bombers to the new turbojets coming into SAC's inventory.

During the Berlin Blockade, SAC withheld the 509th BG because SAC was still building the B-36s into a credible force and needed to train aircrew and support personnel, perfect the new equipment, and hone nuclear bomb delivery procedures. The B-36 was considered to be a transitional bomber, pending the arrival of the B-47 medium and B-52 heavy bombers. Since the B-36 was not aerial refuelable – although one was tested with a drogue refueling modification – it had to be deployed overseas so that it had sufficient range to deliver the heavy and bulky thermonuclear bombs onto targets inside the Soviet Union. No modifications were ever made to allow aerial refueling by the flying boom concept from tankers such as the KC-97.

The first production B-36A (serial number 44-92004) made its maiden flight on August 28, 1947, prior to the second prototype YB-36 (serial number 42-13571). The flight took off from Fort Worth, landing at Wright Field, Ohio. Once on the ground, the B-36A was towed into an aircraft hangar and stress tested until structural failure, allowing engineers to create stress limitations and tolerance tables. A total of 21 B-36As were built, with 19 assigned to the 7th BG at Carswell AFB. These aircraft were not armed, as they were used by SAC until 1949 as airborne aircrew training aircraft. General LeMay clamped a top secret, need-to-know only classification on the defensive armament and delivery capabilities of the B-36A in order to indicate to the Soviet Union that SAC's nuclear delivery capabilities were greater than they really were.

Prior to the outbreak of the Korean War on June 25, 1950, U.S. intelligence agencies began receiving reports that the Soviet Air Force was deploying an advanced turbojet-powered, swept-wing fighter designed by the Mikoyan-Gurevich Aircraft Design Bureau, which became universally known as the MiG-15 Fagot to its Frontal Aviation fighter units. The introduction of MiG-15s into the Korean War prevented FEAF B-29s (pulled from SAC bomber units and storage at Davis Monthan AFB outdoor storage) from operating independently and with impunity during daylight hours. The MiG-15 also was a threat to the B-36 as a deep penetrating nuclear bomber built to destroy strategic and military targets in the Soviet Union. General LeMay believed that at high-altitude, at night, at maximum speed, the B-36 could complete its assigned mission.

The second prototype (YB-36) made its first flight on December 4, 1947. One significant modification was the replacement of the single main landing gear wheels with a four-wheel bogie-type, 56-inch diameter undercarriage. This allowed the bomber to operate from U.S. and overseas bomber-capable runways.[2]

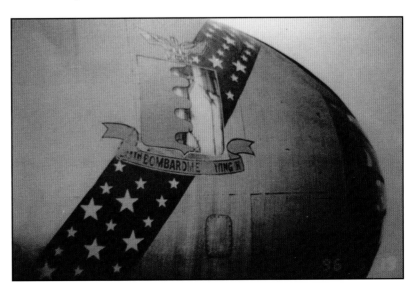

28th BW crest over SAC star band marking on the nose of a B-36. *Ellsworth Air Force Base, 28th Bomb Wing Historian*

SAC B-36 Bomb Wings

Bomb Wing	Bomb Squadrons	Operational Base Assigned
5th	23rd/315th/72nd	Travis AFB, CA.
6th	24th/39th/40th	Walker AFB, NM
7th	9th/436th/492nd	Carswell AFB, TX
11th	26th/42nd/98th	Carswell AFB, TX
28th	72nd/717th/718th	Ellsworth AFB, SD
42nd	69th/70th/75th	Loring AFB, Maine
72nd	60th/73rd/301st	Ramey AFB, Puerto Rico (USA)
92nd	325th/326th/327th	Fairchild AFB, WA
95th	334th/335th/336th	Biggs AFB, TX
99th	346th/347th/348th	Fairchild AFB, WA

28th Bomb Wing (Heavy) crest. *Ellsworth Air Force Base, 28th Bomb Wing Historian*

To improve the bomber's performance the B-36D was born, fitted with four General Electric (GE) J47-GE-19 turbojet engines augmenting the six pusher piston engines. Each turbojet engine was capable of producing 5,200 pounds of thrust. Two turbojets were housed in a pod positioned outboard of the last pusher engine. This improved the aircraft's maximum speed to 435 mph, maximum altitude to 45,000-feet, and a combat radius of 3,530 miles carrying a 10,000 pound bomb load. The addition of two auxiliary fuel tanks in the forward bomb bay, positioned to maintain the aircraft's center of gravity, increased the aircraft's airborne endurance to 48-hours.

To convince the Russians of SAC's delivery capabilities, from December 7-8, 1948, one B-36A flew an 8,000-mile round trip of 35 1/2 hours, flying from Carswell AFB to a designated target area north of the Hawaiian Islands and dropping a boiler plate 10,000 pound bomb (a replica of a thermonuclear bomb) into the water. General LeMay publicly announced the long-distance flight was a thermonuclear bomb training mission.

SAC's first hydrogen bomb, the MK 17 weighed 41,400 pounds with a one megaton yield and was 24.8 feet long and 61.4 inches diameter. Boiler plate replica photograph taken at the original location of the SAC Museum, located outside the east fence of Offutt AFB before moving to the new and indoor museum facility at South Bend, north of Interstate 80, between Omaha and Lincoln, Nebraska. *Lt. Col. George A. Larson, USAF (Ret.)*

94

SAC further demonstrated its capabilities to deliver a thermonuclear bomb over intercontinental range on March 12, 1949, when a 7th BG B-36 flew a 43 1/2 hour training mission over a distance of 9,600 miles.

General LeMay told reporters that SAC's B-36Ds allowed SAC to reach a Soviet target at high altitude on a high-speed bomb run to the target, jamming Soviet radars, and with its presence in Soviet air space not detected until after the bombs were released and the targets destroyed.[3]

B-36Ds on the flight line at Weaver AFB, parked north to south. In the center left background are two Douglas C-124 Globemaster four-engine transports used to support overseas deployments of the bombers. *Ellsworth Air Force Base, 28th Bomb Wing Historian*

At Ellsworth, the first permanent, on-base family housing unit's construction started in 1949, with incremental financing supposedly paying for the project as it was built. There was not enough money in the original funding to furnish them, so they ended up as shells, with no exterior siding and no interior doors, cupboards, etc. The housing need was so desperate, however, that families were moved into the shells anyway. They were later furnished, but became known as Shell Housing.

Wherry Housing on Weaver AFB, November 27, 1951. *Ellsworth Air Force Base, 28th Bomb Wing Historian*

Early in 1951, a new wooden housing complex called Skyway Housing was built just east of the main part of the base, consisting of one- and two-story flat-roofed multiple units with detached multiple garages. It was minimum standard housing, but met the basic requirements of a substantial number of the growing population of military families who gave up their housing allowances to live there. Cold weather brought frozen plumbing and other problems, including ceilings that fell in from inadequate roof loading due to wet snow. There was still not enough housing for families, and many chose to rent or buy in nearby Rapid City or Box Elder, along Old Highway 14/16 down the hill below the base. Mobile home parks became common.

95

Many of the base-assigned personnel became homeowners, particularly in Robinsdale, southeast of Rapid City. [Author's note: Housing was always a problem on Air Force installations. From July 1976 to June 1978, I was assigned to the 93rd BW at Castle AFB, California. The price of off base housing was ridiculous, even then. We lived in a stucco home with a water chiller/evaporator for air conditioning, which did nothing to cool the home. 1960s appliances, vinyl tile floors, car port, hot in the summer and during cooler weather always cold, along with termites that the base entomology department could never eradicate. But as the airmen suffered at Weaver AFB, I did not have any other option.]

A couple of years later Renee Housing, consisting of single-story, ranch-style, multiple family units, was built about one half mile south of the Skyway Housing area. Water and natural gas were piped in from Rapid City. A rental corporation – the successor to the construction firm – took over operation of the project, so families living there retained their rental allowances and paid rent to the corporation, in addition to paying utilities. The first fuel bills that came in with the advent of cold weather were outrageously high. It was discovered that the tenants were being asked to pay off the cost of the natural gas pipeline from Rapid City in a very short time period. A renegotiation of the pipeline contract quickly followed.

Not too long afterward the nice looking units began falling apart. Large cracks appeared in the floors and walls, and garages began sliding away from the houses, all revealing poor site preparation and poor construction. The rental firm was not financially able to keep up with the repairs, and the construction firm had dissolved and disappeared, so there was no one that could be held accountable. The Air Force finally had to buy the whole project and spend large sums on repairs and maintenance. Thereafter, tenants simply gave up their housing allowance to live there. Private enterprise, including investors from among the military population, developed and built a nice and convenient shopping center called Villa Ranchero between the two housing areas. It provided goods and services not available on the base, and an opportunity to purchase things needed on short notice without driving ten miles into Rapid City.

The Vandenburg Building (school) outside and east of the north gate to Ellsworth AFB. The road to the gate is visible in the lower right corner of the photograph. *Lt. Col. George A. Larson, USAF (Ret.)*

Six Turning and Four Burning

The large population of school age children living in base housing led to construction of the Vandenburg School, with high school students being bused to Rapid City. Box Elder eventually built Douglas High School, serving students through the twelfth grade. Because of Box Elder's small tax base, at that time Federal Impact Funds helped finance the school system.

On the operational side of the base, the northwest-southwest runway was lengthened and strengthened to accommodate first B-29s and then B/RB-36s. The aircraft parking ramp was expanded, the north-south and east-west runways were closed, and a new control tower, fire station, and base operations facility were constructed.

Photo of construction workers who built the large aircraft hangar at the north end of the aircraft parking apron, which at that time was publicized as the world's largest concrete hangar. It was built by Steenberg Construction Company of Rapid City. The photograph was taken in July 1948. *Lt. Col. George A. Larson, USAF (Ret.)*

Earlier a very ambitious project was undertaken, and completed in 1949: the largest spanning aircraft hangar in the world, known ever since as the Big Hangar, and now referred to as the Pride Hangar.

Side view of the Pride Hangar during the 2007 Dakota Thunder Open House and Air Show. *Lt. Col. George A. Larson, USAF (Ret.)*

RB-36H and its crew standing at attention for this photograph with the cameras that are carried in the aircraft along the front right side of the bomber, with other equipment underneath the wing. This was similar to what was displayed during periodic base open houses. *Ellsworth Air Force Base, 28th Bomb Wing Historian*

97

Construction was not entirely trouble free. On one occasion a South Dakota breeze (very high wind gusts) struck, blowing down all the work scaffolding. B-36 maintenance personnel were present at a meeting describing the building that was to be built. The architect had a model of the hangar under construction from which the top could be removed, along with B-36 scale models inside. Maintenance personnel looked at the scale construction model, took out the B-36 scale models, set the hangar's roof back on, and commented to the architect "Now, let's see you put three B-36s in there."

During an annual Armed Forces Day open house maintenance scaffoldings were placed along each side of the aircraft so people could look inside the cockpit and even walk into the forward crew compartment. *Ellsworth Air Force Base, 28th Bomb Wing Historian*

The architect had been positioning the B-36 scale models inside the hangar scale model vertically, not thinking that you can't do that with a real airplane. They just could not be maneuvered in that way due to their large wingspan and large propellers. Maintenance personnel could never get more than two B-36s inside.[4]

Two nose docks, one on the left with a B-36 parked inside and the second to the right empty, looking toward the west across the aircraft parking apron. *Ellsworth Air Force Base, 28th Bomb Wing Historian*

98

In 1950, the first annual Armed Forces Day was held at Rapid City Air Force Base. During the Armed Forces Day celebration, the world's largest hangar was dedicated. It cost over two million dollars to construct, and it was 314-feet long, 369-feet wide, and 95-feet high. As previously pointed out, two B-36s (each 165-feet long with a wingspan of 230-feet) could be towed inside the hangar for maintenance work. The hangar remains an impressive and imposing structure on the base, and is now used as a large indoor recreational facility.[5]

The maintenance dock hangars could not hold the entire B-36, with the tail and part of the rear fuselage sticking out into the cold weather. *Ellsworth Air Force Base, 28th Bomb Wing Historian*

At the Fort Worth Convair B-36 production facility weather did not reach the extremes encountered in South Dakota, so they produced and sold to the Air Force portable maintenance structures referred to as Convair Airplane Maintenance Docks. The Docks had heavy canvas curtains which could be closed around the work airplanes. In the harsh weather of South Dakota, especially during below-zero temperatures increased in severity by winds, the resulting high wind chills made the metal dangerous to touch, with exposed skin freezing to metal surfaces.

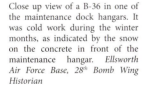

Close up view of a B-36 in one of the maintenance dock hangars. It was cold work during the winter months, as indicated by the snow on the concrete in front of the maintenance hangar. *Ellsworth Air Force Base, 28th Bomb Wing Historian*

99

The portable nose docks were part of the purchase package that came with the RB-36s (the reconnaissance version of the Peacemaker) and provided partial shelter from the exposure to the effects of weather. On a day with the temperature below zero and a northwest wind of 30 mph, however, they were useless. The canvas shields could not be kept closed in a strong wind, and there was just no way of adequately heating the portable work docks. Even the use of Herman Nelson heaters (a portable hot-air heater powered by a gasoline engine primarily used to warm aircraft engines before engine start, used prior to the current hot air blower heaters) did not work.

View of a B-36 inside a Luria Dock, looking back toward the tail. This allowed the majority of the work on the big bomber to be conducted in dry and relative warm conditions. *Ellsworth Air Force Base, 28th Bomb Wing Historian*

There were also canvas shelters which could be set up for maintenance work on individual engines. With a Herman Nelson heater running at full blast, and if the tent could be anchored, some work could be done on an engine when the weather was cold and windy.

Rail access into the south perimeter of the base. The road in front center of the photograph is the old U.S. Highway 14-16. *Lt. Col. George A. Larson, USAF (Ret.)*

Six Turning and Four Burning

Later, large docks were built to cover the nose and wings of the airplane. The tails stuck out unsheltered, but flexible canvas seals could be drawn tight around the fuselages with the doors closed to the fuselage. Cold weather working conditions for maintenance men thus greatly improved. These were called Luria Docks.

Matching the size of the B-36 was that of the crew (15 for bomber operations and 22-23 for reconnaissance variant), and the number of support personnel and equipment required for one mission. This is an excellent photo from back left to right: fuel trucks and a long armament weapon's dollies with conventional bombs that the B-36 could hold, emergency medical vehicle, and oxygen and more fuel vehicles. This required constant ground maintenance efforts. *Ellsworth Air Force Base, 28th Bomb Wing Historian*

One of the key facilities established to support the RB-36 was the engine shop. It was placed in a large hangar with big sliding doors at each end, facing north and south, located near the south end of the old flight line. In it was established what was touted to be the first assembly line operation in South Dakota. Tracks were anchored to the floor to carry dollies on which the R-4360 engines were mounted vertically for movement through the cleaning, disassembly, and servicing line. The engines were initially suspended from an overhead rail. They were then turned vertically and mated with oncoming engine mounts. Work continued at two levels to complete assembly of the Quick Engine Change kits.

Other specialized support functions which had served B-17 operations, such as a parachute shop, sheet metal shop, tire shop, battery shop, and supply, continued in the same functions to support B-29 and RB-36 operations.

During national celebrations, such as Labor Day, Memorial Day, and the Fourth of July, the base's color guard, band, and airmen participated. *Ellsworth Air Force Base, 28th Bomb Wing Historian*

Six Turning and Four Burning

Road access to the base was provided by Old Highway 14/16, a two-lane road which was a half mile south along Box Elder Creek. It was converted to four lanes in the mid-1950s. A two lane county black top road led to the main gate, which was upgraded to the base's commercial gate. In 2010, the gate was upgraded to a security entry point with automatic barriers if someone attempted to force unauthorized entry in a vehicle. The old main gate was also closed in 2010, after the north gate was ungraded and opened the same year as well near the Vandenburg elementary school. The main gate was opened in 2010. As the base's activities and assigned personnel increased, the paved road to town became congested. One partial solution was to stagger duty hours, so some went to work at 6:30 a.m., some at 7:00 a.m., and others at 7:30 a.m.

Rail transportation provided passenger service to Rapid City until the early 1960s, but there was no stop at the airfield. The Chicago & North Western Railroad line paralleled Old Highway 14/16 along Box Elder Creek, and a spur line provided freight service to the base, entering it just east of the main gate. There was also a scheduled bus service between downtown Rapid City and the base.

Radio navigation to the base was by an Adcock Low Frequency Radio Range Station sited on the other side of the ridge south of Box Elder Creek, so as to provide a straight-in final approach path to the north on the north/south runway. It operated on 254kc, call sign RAP. By 1949, the base was also equipped with Ground Controlled Approach, using radar to provide a precision approach landing to the northwest. Like terrain east from the base to the Missouri River, the base itself was practically treeless. Colonel Neil D. Van Sickle, Wing Commander in 1955, thought it should not be so barren, so he directed that each person living on the base had to plant a tree and care for it. This became known as Van Sickle's National Forest. In spite of jokes on base, as well as wind, drought, and frigid winters, a surprising number of trees managed to survive.

One of the traditional fixtures of an air base missing from Ellsworth was an officer's club swimming pool. In fact, there was no swimming pool on the base at all. Some publicly spirited officers, pushed ahead by their wives, proposed that the Officer's Club build a pool for the use of Officer Club members, their families, and guests. With majority member approval, each club member was assessed twenty-five dollars, with the money raised to be used on construction of a swimming pool positioned next to the Officer's Club. In addition, each officer was expected to contribute sweat labor to the project (something which would never happen in today's Air Force environment). The twenty-five dollar fee was not considered a permanent contribution, as each contributor was to receive twenty-five dollars back when transferred from the base. That promise was honored, even for those officers who did not transfer for some time. A contractor was hired and construction started, with officers showing up each day to assist. Contractors were heard to say that the work could have been completed sooner if they did not have the officer help. Regardless, the pool was finished.

RB-36s on the flight line.
*Ellsworth Air Force Base,
28th Bomb Wing Historian*

Six Turning and Four Burning

The WWII wood building construction did not provide a very attractive office environment, and in 1950 there was no money available for improvements. One day a railroad boxcar containing mahogany plywood appeared on the base. There was no address to indicate who sent the wood. It did not take long before offices on the base got a tremendous boost in appearance, paneled in mahogany plywood. The Officer's Club took on an entirely new and attractive look. In the mid-1950s most of the WWII barracks were torn down, providing more comfortable living quarters. Two new mess halls were also built.[6]

On July 13, 1949, the first B-36 arrived at the base. It was assigned to the 77th BS, joining the 717th and 718th BS. The 28th BW was busy transitioning from the B-29 to the B-36. By the end of August only one B-36 had arrived on base. Four more arrived in September, two in October, and one in November. By December all B-36s were grounded because of fuel leaks, requiring extensive maintenance in the B-36 hangar.

By the spring of 1950 SAC was extensively reorganized. The purpose of the reorganization was to make each of the Command's three Air Forces a functionally independent, self-sufficient entity. Up to March 31, 1950, SAC's reconnaissance capability was represented by the three reconnaissance wings of the Second Air Force. The Second Air Force had no fighter or bomber units assigned. The new plan was to assign fighter and bombardment capabilities to the Second Air Force and reconnaissance to the Eighth and Fifteenth Air Forces. At the same time units were to be reassigned, so that all those in the western part of the United States would belong to the Fifteenth Air Force, those in the central part to the Eighth, and those in the east to the Second.

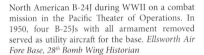
North American B-24J during WWII on a combat mission in the Pacific Theater of Operations. In 1950, four B-25Js with all armament removed served as utility aircraft for the base. *Ellsworth Air Fore Base, 28th Bomb Wing Historian*

The 28th BW was relieved from assignment to the Fifteenth Air Force and transferred to the Eighth. The Fifteenth Air Force was given its reconnaissance capability by being assigned to the 5th Strategic Reconnaissance Wing already stationed in Fifteenth Air Force territory at Fairfield-Suian AFB, California. The 91st Strategic Reconnaissance Wing at Barskdale AFB, Louisiana, remained with the Second Air Force, instead of transferring the other strategic reconnaissance wing (the 9th, stationed like the 5th at Fairfield-Suisan) to the Eighth Air Force and moving it to an Eighth Air Force base. The organization was converted to a bombardment wing. The Eighth Air Force was given its reconnaissance capability by having one of its bombardment wings converted to reconnaissance. On April 1, 1950, the 28th BW (H) became the 28th SRG (H), and the 77th, 717th, and 718th BS became strategic reconnaissance squadrons. With a change in their primary mission from long-range bombardment to long-range reconnaissance, members of the 28th SRG hoped that bombardment would be retained as a secondary mission.

To show the importance of the B-36 mission to Rapid City business and government leaders, the base held a special function inside the Pride Hangar. Maintenance personnel moved a C-124 inside for close inspection. Businessmen and Rapid City officials walk up the lowered cargo ramp of the C-124. *Ellsworth Air Force Base, 28th Bomb wing Historian*

103

Effective July 16, 1950, the 28th BW was redesignated the 28th SRW (H), the 28th Group became the 28th SRG (H), and its three squadrons became the 77th, 717th, and 718th SRS (H). During the second half of 1950 and the first four months of 1951, the 28th Wing accomplished numerous photographic missions. Aircraft strength of the 28th Wing as of January 1, 1951, consisted of (20) RB-36s, (7) C-47s, and (3) B-25Js.

After the tour of the C-124 Globemaster and RB-36, Rapid City businessmen and officials eat in the chow hall. *Ellsworth Air Force Base, 28th Bomb Wing Historian*

A rather extensive reorganization of the 28th Wing took place in February 1951. The chief effect on the 717th BS was that the 28th Group was, for all intents and purposes, abolished; one officer and one airman remained assigned to the Group in order to preserve the unit's legal existence. These men were attached to Wing Headquarters, while other Group personnel were assigned to Wing Headquarters. Thus, the three tactical squadrons were brought under Wing control.

Crew of the first B-36D to fly over 600 hours at Ellsworth AFB. *Russell K. Smith to Lt. Col. George A. Larson, USAF (Ret.)*

104

In April, the 28th Wing flew a mission to penetrate into the Eastern and Western Air Defense Areas and the Albuquerque, New Mexico, Air Defense Identification Zone for reconnaissance purposes. Some of the targets could not be photographed visually because of bad weather; all the rest of the visual photography and all of the radar photography were acceptable. The performance was rated as excellent. Also in April 1951, SAC Headquarters notified the Wing of a requirement to photograph the Mississippi river between Guttenberg, Iowa, and Minneapolis, Minnesota, for flood control purposes. This mission was accomplished on the day the commitment was received. Several other missions were also flown in April.

A C-119 Flying Boxcar, part of 81 aircraft which staged into Rapid City AFB on October 27, 1952, transporting 3,200 paratroopers assigned to the 603rd Regimental Combat Team from Fort Campbell, Kentucky, and selected passengers. This was an intermediated stop for the paratroopers en route to Alaska to participate in a joint Army-Air Force exercise. All exited the aircraft for a meal and rest. This was their only stop on the long flight from Kentucky to Alaska. *Ellsworth Air Force Base, 28th Bomb Wing Historian*

During 1951, the 28th Wing was receiving aircraft to build it up from 18 to a total of 39, and by the middle of the year it reached its authorized strength of 30 RB-36s. An increase in personnel was called for, too, of course; as of June 30, 1951, the Wing was slightly less than 80 percent effectively manned in officers and airmen.

RB-36H, aircraft serial number 51-5751. *Ellsworth Air Force Base, 28th Bomb Wing Historian*

Numerous missions were flown to photograph ZI (installation-Zone of Interior) airfields that had been used in WWII, with a view to enable USAF planners to make appropriate selections of fields to be reactivated or expanded. On September 1, 1951, a violent hail storm struck Rapid City AFB. Out of the Wing's 30 RB-36s, nine were at various Air Material Command (AMC) depots undergoing modifications, one was inside the Pride Hangar, and 20 were parked on the aircraft parking apron. These aircraft were badly damaged by large hail. Hail storms normally hit in the spring and midsummer months, while fall hail storms were rare. Even though AMC depots at Oklahoma City, Oklahoma, and San Antonio, Texas, sent sheet metal workers to Rapid City AFB, the repair work was not completed until September 15th, with fabric repairs taking until September 23rd. It took so long to repair the damage that the Wing's operations were curtailed for lack of flyable aircraft, and it was necessary to ask Headquarters, Eighth Air Force, first on October 1st and then October 31st to suspend dates on all photographic projects. Both requests were granted. As of September 30th, the 28th Wing's strength consisted of 699 officers, 12 warrant officers, 12 nurses, and 4,128 airmen for a 84.56 percent effective manned-officer strength and 80 percent manned-airmen strength.

To improve flight line security, a barbed wire fence, mounted on X-wood 2-by-4s, was set up along the edge of the access road and flight line. The portable maintenance docks are visible on the aircraft parking apron, with one B-36, nose first, into a dock on either side of the aircraft's fuselage. Other B-36s are visible in the top of the photograph. *Ellsworth Air Force Base, 28th Bomb Wing Historian*

Among the missions flown in the spring of 1952 were emergency flood control photography missions on the Mississippi River from the town of Browns Valley, Minnesota, to Minneapolis; on to Hastings, Minnesota, and on to Guttenberg, Iowa; on to Louisiana; and on the Mississippi River from Fort Peck, Montana, to Williston, North Dakota. This was the first time the author saw a B-36 in the clear sky over Iowa. Even at high altitude it was easy to spot because of its large wingspan.

In October 1952, three aircraft with crews from the 717th BS took part in an exercise designed to test the support capability of Thule AFB and capability in night photography and high-altitude gunnery of the 28th Wing. On October 27th, an advance party of administrative and support personnel left rapid City AFB in three C-124s of the 1st Strategic Squadron, Biggs AFB, Texas, and flew to Thule, Greenland, via Goose Bay, Labrador, and Bluie West 8, Greenland.

Close-in security points were set up to control entry to each B-36 on the flight line. The security guard had to check each crewman's security badge to that of his ID. *Ellsworth Air Force Base, 28th Bomb Wing Historian*

The departure of the tactical aircraft assigned to the Thule mission was delayed 24 hours because of adverse weather in Greenland. However, three RB-36s from the 717th BS and two from the 77th BS took off from Rapid City AFB early in the morning on November 1st. One of the 77th BS' aircraft developed propeller problems two and one half hours after takeoff and was forced to return to base. The other four aircraft arrived at Thule after 12 hours. The crew of the aborted reconnaissance aircraft took off in a spare RB-36 on the morning of November 2nd, landing in Greenland later the same day.

Personnel of the 28th Wing involved in the Thule mission were severely critical of the support furnished by the 1st Strategic Support Squadron, but had nothing but high praise for the work of Weather Detachment 824, 8th Weather Squadron, Thule, Greenland.

At the end of December 1952, the 77th and 717th BS had sixteen RB-36Hs each, with the 718th BS equipped with one. The 28th Wing had fifteen additional aircraft, including a number of RB-36Ds and Es assigned to it. The 718th BS was flying Ds and Es. During January 1953 there was a reallocation of RB-36Hs, which resulted in the assignment of eleven to each of the three tactical squadrons.

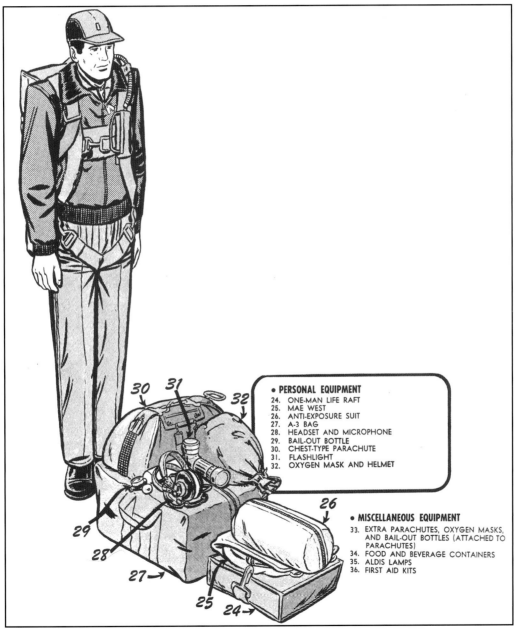

B-36 aircrew personnel and survival equipment. *From B-36 technical manual provided by (former) SAC Historian Office, Offutt Air Force Base, Nebraska, 1990.*

Six Turning and Four Burning

In March 1953, the Wing was directed to fly a simulated combat mission to Lajes Field, Azores. The aircraft left Rapid City AFB from March 14th through 15th, flying via Milwaukee, Wisconsin, and Portsmouth, New Hampshire. The exercise was supported by four C-124s assigned to the 1st and 2nd Strategic Support Squadrons. The C-124s departed Rapid City AFB March 12th, carrying support personnel and essential mission support equipment and spare parts. One of the C-124s developed mechanical trouble and had to be left behind at Kindley AFB, Bermuda, but all its passengers and cargo could be accommodated in the three remaining C-124s, and all arrived safely at Lajes, Azores. The mission was flown March 17-20, 1953, with four aircraft leaving Lajes on each of three days. The mission produced a considerable amount of acceptable photography, both visual and radar, and provided some good electronic counter measure (ECM) experience, but it was marred by one of the worst air tragedies in the history of the 28th BW (more on this later in the chapter).[7]

Tail section of Rapid City AFB, 28th SRW RB-36H tail number 51-13721, with Brigadier General Richard Ellsworth on board, which crashed into a hill at Burgoyne's Cove, Newfoundland, on March 18, 1953, killing all on the aircraft. *Richard Stocker of St. John's, Newfoundland to Lt. Col. George A. Larson, USAF (Ret.)*

The mission of the 28th BG (H) was "to prepare to carry out the strategic bombardment mission of SAC," although it was designated as a VH reconnaissance unit. The Group won SAC's Second Bombing Competition for individual crew performance. The Group came in second in 1950. After the re-consolidation, the 28th Wing operated the RB-36 reconnaissance variant, mapping most of Europe's major cities by photography and radar. The 28th Wing also mapped populated regions of North Africa north of the Sahara and west of Egypt.

Six Turning and Four Burning

Crash site of the RB-36H in Newfoundland. Debris was scattered over a wide area. *Richard Stocker of St. John's, Newfoundland to Lt. Col. George A. Larson, USAF (Ret.)*

One interesting mission for the 28th Wing was to test North American Air Defenses. Coming in low from over the water, all transmitters, including bomb/navigation radar, would be turned off or on standby to avoid detection. At high altitude transmissions would be kept at a minimum, except to jam defense and communications frequencies at appropriate times. Flying out of Eielson, the RB-36s penetrated U.S. air space over a wide area toward Detroit, Michigan. The bombers were flying at 44,000-feet above a thick cloud cover. ADC launched fighters, but operated under a handicap due to the effective jamming of the ground based ADC radars and airborne radars, along with command frequencies. F-86Ds launched from Selfridge, Michigan, had to use afterburners in an attempt to climb to 44,000-feet, but were unable to climb to the RB-36s' altitude before reaching minimum fuel status. The RB-36s had to turn off their ECM jammers to allow the fighters to communicate with their base and to land.

The mission of the 28th included supporting SAC, Air Force, Department of Defense, and any other government agencies in need of the RB-36's outstanding capabilities, such as roominess, load lifting capacity, long range, high altitude, and adequate electrical power. In 1954, the Air Force was developing Distance Measuring Equipment to be used in conjunction with the directional capabilities of TACAN (Tactical Air Navigation). Pertinent questions were, would the equipment developed until then work at high altitude, and what was the range of the equipment? Air Force engineers at Wright-Patterson AFB wanted to fly on a RB-36, with three specialists tasked to fly up and down the U.S. East Coast. On the day of the mission everything went as planned at 40,000-feet, where the specialists could get azimuth measurements out to 240 miles.

On another mission, the 28th was tasked to test some high-resolution, side-looking radar at high altitude. One RB-36 was modified with special antennas, and the photo compartment was loaded up with matching equipment, including the capability of transmitting photo data directly to a Reconnaissance Technician. The higher the aircraft could fly above the surface terrain below, the better the resultant data. Wright Development Center wanted to verify the reconnaissance equipment's capabilities above 50,000-feet. Eglin AFB was selected as the test base since it was at sea level, offering the maximum true altitude possible. The crew had to wear high-altitude pressure suits for operations above 45,000-feet. The tests were completed, with the aircraft reaching 50,000-feet.

109

Six Turning and Four Burning

Operation Castle gave the 28th another important mission to perform, which was the nuclear testing program at Eniwetok Proving Ground, Marshall Islands, conducted by Army, Air Force, Navy, and AEC personnel from January to July 1954. They used RB-36H aircraft number 52-1386, one of the last ten RB-36Hs to roll out of the Convair production facility at Fort Worth, Texas. The aircraft was modified at Kirkland AFB with special radar, other electronics, and special photographic equipment, and all guns were removed. The aircraft was to be the controller aircraft and lead photo ship for *Operation Castle*. The crew was flown to Kirkland AFB in a C-47, which was still carrying on in the post-WWII support role for the Air Force. They were given training pertinent to the mission and took delivery of the aircraft. They subsequently proved that down time was reduced and operational efficiency was improved by keeping one flight crew and one maintenance crew involved in the operation of an aircraft over a long period of time. As controller aircraft they flew on all test shots, including the March 1, 1954, test of the thermonuclear device named *Shot Ivy* that destroyed the target island and contaminated several of the islands in the Bikini Atoll.

The highest priority of the 28th was to be trained to carry out its assigned mission in wartime. To this end, each combat ready crew was assigned its own target, with objectives to be accomplished staging sites, routes to and from the target area, and the recovery site. Unlike WWII, which generally involved large formations of bombers, each crew was to be almost totally independent of other crews. The crews spent a good deal of time studying all the latest information about their routes and targets. Since such information was classified Top Secret, it was never discussed outside the target study environment, and no crew knew any other crew's wartime assignment. The whole atmosphere was *need to know* only.

The vast expanses of remote areas the crew might fly over made it possible for a crew on a wartime mission to be forced down from any friendly flying facility, due either to battle damage, enemy ground activity, mechanical failure, or plain running out of fuel. SAC considered its most valuable asset to be its combat crews, so it wanted to keep them out of enemy hands, and to be able to return them to duty if they were forced down in Russian territory. To that end, SAC had a fleet of very long-range C-47s equipped with auxiliary fuel tanks in the main cargo cabin and JATO (jet assisted take-off) rockets mounted on the aft sides of the aircraft.

One of the six piston pusher engines which tore loose from the aircraft, minus the propellers.
Richard Stocker of St. John's, Newfoundland to Lt. Col. George A. Larson, USAF (Ret.)

The scenario was that such an airplane would be able to pick up downed crewmen from a crude, relatively unprepared strip (such as an open field, sand bar in a river, or road or flat area) in a remote area and return to U.S. controlled bases. This projected operation was one of the subjects covered in each crew's Advanced Survival Training program. Each SAC crew carried a long-range radio as part of its survival/Europe and Evasion (E&E) equipment. It was used at an appropriate time to contact the rescue center, giving encoded information as to location, etc. The crew would then prepare a rudimentary strip for the rescue aircraft to use, and at a scheduled time make contact with the rescue base, which would then provide a pick up time. The crew would stay away from the strip until the scheduled time, then prepare for a very quick pick up. The rescue crew would land, install JATO bottles, load the crew, and be airborne using the JATO in a very minimum amount of time, flying *nap of the earth* until safely out of enemy territory.[8]

There were many incidents and accidents with the RB-36s at Rapid City AFB. On an evening Wing Standboard Check Flight, after take-off the crew couldn't get the pressure to relieve in the gear system after the gear was up. After two or three cycles of the gear and several cross-checks of the system no determination could be made on the problem with the system. The crew leveled off at 10,000-feet and started a check of everything. The Left Scanner advised that the canoe door appeared to be hanging down and that there was some debris hanging out of the wheel well. The crew feathered number 3 engine and sent the Chief Gunner and Engineer to check it out. The left gear was up, the canoe door was jammed into the wing at the front end, cracked at the center hinge, and the read half was hanging and trailing out the well. The forward edge of the door had nicked the fuel line every time it cycled, and one of the door actuating arms was sheared off. The crew cleared the wing and dropped the gear. The right main and nose extended OK, but the left main hung up on the door. The crew locked the right and nose gears out of the system and tried to pry the left canoe door open. The crew cut holes in it with a fire axe. The crew positioned everything manually to make the gear mechanism indicate the door was open, and it hung up on the flimsy, beat up door every time.

After six hours of hard work and lots of over radio advice from maintenance personnel at Eighth Air Force, Headquarters SAC, and Convair engineers, the crew was asked if the pilots intended to order the remaining crew to bail out or ride the damaged aircraft back to the ground. The pilots indicated with ten hours of fuel remaining that work would continue on the gear. The pilot took the bomber over the top to create negative Gs and dropped the gear as it started down. The gear got hung up again. The pilot dived the bomber to 188 mph, pulled up hard, and dropped the gear. The gear came down and the door came off. There were thirteen holes of various sizes in the aircraft, from six-feet in diameter to three inches. The loose hydraulic lines beat the hell out of the leading edge of the left stabilizer and the fin. The crew put the ground lock on the left gear, the pilot got everyone into crash landing positions, and the plane landed without any problems.

A 28th crew was flying an RB-36 to Carswell AFB. The bomber was on its base leg, making a practice GCA to a planned full stop landing, when the aircraft began a right turn, losing 400-feet of altitude. Both pilots struggled to get the wings back to level. Using full left rudder, left rudder turn, and full left aileron inputs they were able to control it. The pilots retracted the landing gear and reduced power on engine numbers 1 and 2. They regained lost altitude and completed a cockpit check, but were unable to determine the cause of the problem. The Pilot/Aircraft Commander alerted the crew for possible bail out and told Carswell control tower they could not turn left. Using differential power, they were able to steer it onto the runway and come to a complete stop. Post flight inspection discovered that the left inboard aileron trim idler push-pull rod had failed, causing it to drop down and wedge itself into the leading edge of the left aileron trim when the aircraft was in a right bank. There was already a technical order affecting the failed part. From that point all B-36s that had already not complied with it had to do so before further flight.

In February 1953, a 28th crew suffered a major engine malfunction/propeller emergency. There were gear teeth around the hub of each propeller blade. These were engaged by toothed gears of the blade pitch changing mechanism. In this instance, the teeth around the hub of one propeller failed and the blade went into an uncommanded pitch angle, causing the propeller to stop and then turn backward. The sudden occurrence of this event resulted in a spiral descent of the aircraft. The Pilot, assisted by the Flight Engineers and selected crew members, got things under control, started the turbojets on the side of the failed propeller, and determined that they could fly the aircraft. The propeller failure occurred off the U.S. West Coast. With safe operation of the aircraft established, the Pilots flew back to Rapid City, landing safely.

The 28th lost five RB-36s during its operations of the Peacemakers.

111

Aircraft Type	Aircraft Tail Number	Crash Date	Crash Site Location
RB-36H/	51-13719	2-15-1953	Roswell, NM
RB-36H/	51-13721	3-18-1953	Labrador, Newfoundland
RB-36H/	51-13722	8-27-1954	Rapid City, SD
RB-36H/	52-1384	1-4-1956	Ellsworth AFB
RB-36H/	51-13720	3-4-1957	Denver, CO[9]

(14) 1.000 feet to 1,000-feet

(15) 800 feet to 800-feet

(16) 1,000 feet to 1,000-feet[9]

Just to the side of the peak where the aircraft crashed, there is a memorial to the members of the flight crew. One of the propellers has been mounted in concrete with a plaque listing those who perished. *Richard Stocker of St. John's, Newfoundland to Lt. Col. George A. Larson, USAF (Ret.)*

In March 1953, Brigadier General Richard E. Ellsworth led a flight of 28th SRW RB-36s to Lajes Airdrome, Azores, for a training exercise to be executed when taking off from there. Having flown south with a task force of eleven other RB-36Hs (a total of 12 deployed) on five days of intercontinental combat training, crew H15 took off for Rapid City AFB.[10]

Captain Jacob Pruett, Jr., Captain Clark, Brigadier General Richard Ellsworth, Major Frank Wright, and a crew of 19 took off in RB-36H serial number 51-13721, assigned to 28th SRW (H) from Lajes Airdrome at 11:00 p.m. (Azores local time). In company with a second RB-36H, the two aircraft were testing and attempting to sneak under the shield of U.S. East Coast ADC radar sites as part of a Cold War evaluation of the security of North America from attacks by Soviet LRA long-range bombers. Their flight plan took them across the Atlantic Ocean and over Newfoundland. The flight was expected to take 25 hours. The pre-flight weather briefing indicated that their flight path would take them south of a low-pressure zone. The counterclockwise rotation of the low would produce head winds that were forecast to average 117 knots from 300 degrees.[11]

Raleigh H. Watson, Jr., noted:

Arrived Rapid City AFB in April 1952, assigned to 4011th A&E Maintenance Squadron as an ECM mechanic. In August 1953, volunteered to go to Kessler AFB for two months to train on ECM operator. Returned to Rapid City, assigned to 7th BS. Assigned to Captain Grissom's RB-36H crew for training. Flew with this crew once or twice, until checked out. I was asked to fill in by the Wing/77th BS on Major Anthony Lovino on the Azores trip as he could not go. Made the trip on Lt. Col. Cumming's crew. It was on this exercise that General Ellsworth was killed. I recall that he came (General Ellsworth) down to see us off when we departed Lajes Airdrome. I understood from our Engineer, Major Dale Poeot that General Ellsworth was to return with us, but plans were changed in order that he might sit in for a sick pilot on the RB-36H that crashed.[12]

General Ellsworth and Major Wright were not current in take-offs and landings, so Captain Jacob Pruett, Jr., and Captain Orion Clark were probably at the controls during take-off. Major Wright then moved into the Pilot's seat on the left and General Ellsworth into the Copilot's seat on the right. Major Wright and General Ellsworth flew over the water portion of the flight about 1,000-feet off the water for best range performance. They monitored their altitude above the water with the radar altimeter as they flew through the darkness.

The Navigator intended to turn on the mapping radar an hour before the time he expected the RB-36H to reach land at Newfoundland. The Pilots planned to climb to an altitude that would carry the RB-36H safely over the mountains of Newfoundland while they were still 20 miles from land. Most of the flight was flown in overcast conditions that prevented the Navigator from using a sextant for celestial observation to determine the true position of the airplane.

The low-pressure zone moved south of its predicted position before the RB-36H reached its vicinity. The airplane passed north of the low. Instead of the anticipated headwinds, the airplane encountered tailwinds that averaged 12 knots from 197 degrees. Ocean Station Delta received a position update from the RB-36H at 0645 Zulu. The Navigator reported that the ground speed of the airplane was 130 knots. The position was in error by 138 nautical miles, and the true ground speed was close to 185 knots.

The RB-36H reached Newfoundland about 1 1/2 hours earlier than expected. The crew made no attempt to contact Newfoundland Air Defense when they were fifty miles off shore. The Navigator did not turn on the radar. The Pilots continued to fly at low altitude. In the last twenty minutes of the flight the ground speed averaged 202 knots. The visibility was less than one-eighth of a mile as the airplane flew straight and level through sleet, freezing drizzle, and fog.

At 0740 Zulu (5:10 a.m. Newfoundland time), thirty minutes after crossing the coastline the RB-36H struck an 896-foot tall ridge at a flight elevation of 800-feet. The six whirling propellers chopped off the tops of many pine trees before the left wing struck the ground. The left wing ripped off the airplane which spilled fuel, igniting into a huge fireball. The fuselage and right wing impacted 1,000-feet beyond the left wing. The entire crew was killed on impact. The bomber's wreckage was strewn for three quarters of a mile across the hillside.

USAF 1st Lt. Dick Richardson, assigned to Nut Cove, Newfoundland, heard the RB-36H approach; the sound of the engines stopped suddenly, replaced by a loud explosion. He indicated everything lit up real bright. He could see a fire burning on the hillside above. The Lieutenant woke other men, sending a search party up to the crash site through deep snow. They found no survivors.

113

Another RB-36H on a similar mission narrowly escaped a similar fate that night. Although still an hour from the estimated time of land fall, the Pilot spotted land below the airplane through breaks in the clouds. He immediately initiated a climb to safe altitude. It was determined that the airplane was 90 miles north of its intended flight path. Although the radar was operating, it was not properly tuned, so it failed to indicate the imminent landfall.

The Accident Investigation Board recommended that a forward looking radar should be deployed to provide warning of high terrain ahead of the airplane. Navigators were instructed to scan for land with the radar every six minutes, and Pilots were instructed to climb to a safe altitude whenever the estimated position of the airplane was within 200 miles of land.[13]

Dedication of Ellsworth Air Force Base. *Ellsworth Air Force Base, 28th Bomb Wing Historian*

Richard Stoker of St. John's, Newfoundland, visited the RB-36H crash site in 2005, writing an account for the *Atlantic Canada Aviation Museum* in Nova Scotia regarding the crash. He allowed inclusion of that matter in this account:

Today, the remains of the RB-36H are still very visible on top of the hillside in Burgoyne Cove, Newfoundland. Obviously, the cannons and ammunition have all been removed from the crash site, but pretty much everything else is still where it landed on March 18, 1953. The single biggest piece of wreckage is the immense tail plane, which is still standing upright in a short gully (see photograph above) just to the side of the peak where there is a memorial to the members of the flight crew, where one of the propellers has been mounted in concrete and there is a plaque listing those who perished.

It is somewhat surreal to approach the site, one has to undertake a fairly steep climb up what appears to be a dry river bed, although it is actually a trail cut a number of years ago by a local Air Cadet Group. Emerging from the trees at the top of the trail, and the first thing one sees, just lying in the brush is a jet engine. As you climb a little higher, the whole crash site is spread out in front of you, it is a little odd to see a number of picnic tables scattered about, and you soon realize how great it is

114

to take the weight of what by now are aching legs. It is sobering to realize what a tragedy occurred at this place. For many years, the crash site has been identified as being on Random Island just to the north of the actual crash site, and thanks to the steep climb up the hillside, not too many people have visited the site. Thankfully, there has been little vandalism or artifact theft.[14]

Members of the 28th SRW march behind the base's band as they head toward the flight line for the dedication ceremony. Notice the number of WWII-constructed wood buildings that remained on the base, even in 1953. Even with upgrades to the base's infrastructure, military personnel used WWII structures while supporting the mission of SAC's large RB-36Hs then currently operating at the base. *Ellsworth Air Force Base, 28th Bomb Wing Historian*

Richard Stoker had this to say about it:

The terrain on both sides of Smith Sound, including Random Island, rises in places steeply from sea level. Terrain heights in this area range from 250-feet, 350-feet and up to 800-feet plus. Any aircraft flying at 500-feet in bad weather and approaching the coast of this part of Newfoundland would be fortunate to avoid flying into high ground.

1. The wreckage pattern suggests the RB-36H first struck a sloping to the east and about 70-feet below the SW-NE ridge. Finding a pylon structure at the East Area, together with a jet engine and a single fractured compressor disc, together with a propeller spinner and a wing structure suggest an impact line at the level underside of the the main wing surface.

President Eisenhower flew into the base on Air Force One, a Lockheed C-121 Constellation named Columbine II. The aircraft was flown by the 1254th Air transport Group, based, at that time, at Washington National Airport. It had a tripe tail and was powered by four Wright R-3350 (749C-18BD1) Cyclone piston engines, with a combined 10,000 HP. It was replaced by Columbine III in 1954. The Presidential aircraft is surrounded by a security detail, and cars are lined up to take the President and dignitaries to the dedication. A large group of local and military dignitaries wait for the President to exit the aircraft. The President arrived a few days before the dedication to do some fishing in the Black Hills, following the fishing trend of a former President, Calvin Coolidge, who liked to fish in the Black Hills in Custer State Park streams. *Ellsworth Air Force Base, 28th Bomb Wing Historian*

Six Turning and Four Burning

2. The nose wheel strut and a fractured wheel rim in the debris at the CENTRE AREA suggests this unit unit was free to move and fall just further on from the initial impact point.

3. The separation and direction of the tail unit suggests aircraft breakup, as the long fuselage, travelled up and along the rising ground toward the valley and ridge top. The tail section may have separated at first impact, with the outer rocky edge and flipped.

4. The position of two piston engines on the east side of the ride, left of the valley, show complete separation from the engine mount structure and tumbling slightly upwards and forward. These may have been the two outer piston engines of the port wing.

5. A third piston engine, on the top left side of a small valley, still has its local structure attached and rests with the pistons facing forward.

6. The question raised is whether the crew saw rising ground in the last seconds before impact and attempted to climb through a small break or valley in the ridge line. In this respect, the front fuselage section went through the valley and most of the wing sections lie at the east end of the valley at the east side of the ridge.

7. Separations of the inner wing structure allowed two main wheel struts to follow the fuselage section through the small valley; these lie closely with the front crew area. One main wheel strut may have impacted through the crew canopy.

8. A partial fuselage, an opened wing structure, a piston engine and outer wing section lie below the west side of the ridge suggesting aircraft breakup or at on the EAST AREA and to VALLEY EAST AREAS. The speed and momentum of the aircraft at impact may have allowed the wing section to lift and "fly" over the ridge.[15]

RB-36Hs lined up on the flight line for the dedication ceremony, along with personnel assigned to the 28th SRW.
Ellsworth Air Force Base, 28th Bomb Wing Historian

South Dakota opposition against previous efforts to rename the military installation to Weaver AFB, then back to Rapid City AFB, ended with the death of Brigadier General Richard E. Ellsworth. A strong public and military sentiment grew in order to rename the base in memory of Brigadier General Richard E. Ellsworth. This was done on June 13, 1953. It was marked by the arrival of President Dwight D. Eisenhower to dedicate the changing of the base's name to Ellsworth Air Force Base.

RB-36H set up with a reviewing platform facing the flight line. *Ellsworth Air Force Base, 28th Bomb Wing Historian*

Colonel Ariel Nielson and President Eisenhower walked from Base Operations, under the wing of the RB-36H, to take their seats on the dignitary platform at the front of the aircraft's nose. *Ellsworth Air Force Base, 28th Bomb Wing Historian*

The dedication was a well-attended event, including: Major General Thomas S. Power, Deputy Commander SAC; John M. Montgomery, Commander Eighth Air Force; South Dakota Governor Sigurd Anderson; South Dakota U.S. Senators Francis Case and Carl Mundt; and U.S. Representatives Harold Loure and E.Y. Berry (today South Dakota only has one House of Representatives member).

President Eisenhower makes his remarks at the dedication of Ellsworth AFB. *Ellsworth Air Force Base, 28th Bomb Wing Historian*

Colonel Ariel Nielsen, Commander Rapid City AFB, remembered:

When the President arrived, he went straight up into the Black Hills to relax and do some fishing. I know there was a lot of stocking of the trout streams going on to insure that he had a good fishing time. We had a frantic couple of days before the dedication.

Six (four visible in photograph) 28th SRW RB-36Hs over water approaching land. The back of the photograph indicated the four aircraft were returning from a deployment to RAF Fairford, England, crossing the U.S. East Coast and heading back to Ellsworth AFB. *Ellsworth Air Force Base, 28th Bomb Wing Historian*

118

Six Turning and Four Burning

One thing we did was to back up an RB-36H, with the tail just at the exit from Base Operations to the flight line. One couldn't possibly go from Base Operations to the flight line without walking along side and underneath the airplane.

The photograph is not of 28th SRW RB-36Hs approaching Andersen AFB, Guam, but B-36Ds of the 92nd BW. The 92nd BW was the first B-36 unit to deploy to Guam on October 15, 1954. The formation was typical of long over water flights from the United States to the Pacific and Andersen AFB, Guam. *United States Air Force*

As the President and I proceeded out the door toward where the ceremony was set up, we walked under the wing. His eyes kept going up and up, and finally he said…"Colonel, that's quite an airplane." And it was.

Photo of a 92nd BW B-36D parked in front of Base Operations on Andersen AFB, Guam, on October 15, 1954. It was a historic event for the base and well attended by base dignitaries. *United States Air Force*

119

President Eisenhower's dedication followed:

We are here in tribute to a gallant and patriotic American – a man whose name will always be an honor to the members of his family and to this air base. It is my great honor to dedicate the base in honor of Brigadier General Richard E. Ellsworth.

As soon as President Eisenhower's part in the ceremony was complete he departed on Air Force One.[16]

28th SRW RB-36H landing at Andersen AFB, Guam. *Ellsworth Air Force Base, 28th Bomb Wing Historian*

In March 1953, a detachment of the 28th Wing consisting of aircrew and RB-36Hs from the 717th BS deployed to RAF Fairford, England, to continue a project to photo-map western Europe that was begun in September 1952 by the 77th BS. The personnel and planes of the 77th had been replaced in December 1952 by aircrew and aircraft from the 5th SRW. From March 15th through June, crews and aircraft of the 717th were rotated between the United Kingdom and the Zone of the Interior, with as many as six crews being in England at one time. The last mission report on this project was dated June 20, 1953. Throughout the second half of 1953 and the first nine months of 1954, members of the 717th BS, like other personnel in the Wing, were presumably engaged in training designed to permit the Wing to accomplish its assigned mission. Effective April 1, 1955, the 28th SRW was transferred from the Eighth to the Fifteenth Air Force.

Major William W. Deyerle, 247th SRS, Fairchild AFB, Washington, landed this rudderless RB-36H at Ellsworth AFB on July 11, 1955. The aircraft was in formation for a flyover during dedication ceremonies for the interim Air Force Academy at Lowry AFB, Colorado, when the eight by 32 foot rudder came loose. For his skill in landing the disabled aircraft safely at Ellsworth AFB, Major Deyerle was selected to become a member of SAC's Heads-up Flying Club. *Ellsworth Air Force Base, 28th Bomb Wing Historian*

Six Turning and Four Burning

In April 1955 the 717th BS, along with other components of the 28th SRW, deployed to Andersen AFB, Guam. When I was assigned to the 43rd Strategic Wing (1974-1976 and 1978-1981), behind my desk I had an enlarged photograph of SAC's first deployment of RB-36Ds to Andersen AFB.

CASTLE
AIR MUSEUM

Poster from a reunion at former Castle Air Force Base, Atwater, California, for B-36 aircrew and support personnel. *Courtesy: Castle Air Force Museum*

Six Turning and Four Burning

View of the left side of the RB-36H on display at Castle Air Museum, Atwater, California, at the location of the former SAC base, 93rd Bomb Wing, Castle Air Force Base. *Lt. Col. George A. Larson, USAF (Ret.)*

An advance party from the 28th SRW left Ellsworth AFB on April 1, 1955. It arrived in Guam four days later. Arrangements were made for taking over 5th SRW control and operation at Andersen AFB. Seven 28th SRW RB-36Hs took off on April 9th for Travis AFB, California, refueled, and flew nonstop to Andersen AFB, Guam.

Seven more RB-36Hs left Ellsworth AFB for Travis AFB on April 11th, with another seven on April 15th. During the Wing's stay on Guam a good will mission was flown to Bangkok, Thailand. Since the TDY at Andersen AFB was to be for 90 days it should have ended in July, but it was not until the end of August when the bombers returned to Ellsworth AFB.[17]

As military organizations upgraded man power and equipment from time to time to meet new national security requirements, Ellsworth AFB organizations changed as well. Headquarters SAC assigned the 28th SRW from the Eighth to the Fifteenth Air Force in a general order dated September 2, 1955, effective October 1, 1955. The 28th SRW (H) became the 28th BW (H).

Six Turning and Four Burning

Straight on view of the RB-36H taken from an elevated cherry picker basket to allow a full photograph on the large bomber.
Lt. Col. George A. Larson, USAF (Ret.)

In the latter part of 1956, the Wing began planning for the transition from the RB-36H to the Boeing B-52D Stratofortress. In March 1957, the 28th BW began flying its RB-36Hs to Davis-Monthan AFB, Arizona, for storage and eventual disposal (scrapped for metal). The final RB-36H 28th BW training mission was flown on May 7, 1957. The last Ellsworth AFB RB-36H (aircraft number 52-1386) took off for Davis-Monthan AFB on May 29, 1957. The First B-52D landed at Ellsworth on June 14, 1957, assigned to the 717th BS.[18]

After distinguished SAC service, four B-36s were retained and on display: (B-36J) Pima Air and Space Museum, Tucson, Arizona; (B-36J) Strategic Air Command Museum, Ashland, Nebraska; (B-36J) National Museum of the United States Air Force, Wright Patterson AFB, Dayton, Ohio; and (RB-36H) at Castle Air Museum, Atwater, California.

One of the 28th BW's RB-36s (serial number 51-13730) was flown from Ellsworth AFB to Chanute AFB, Illinois, in 1957 to be used as a ground instruction trainer. The aircraft was then retained at Chanute AFB Museum as a display aircraft. It was given the serial number of a B-36D (44-92065). The aircraft was one of six photo-reconnaissance versions that had all its gun turrets. Chanute wanted to get rid of the aircraft, so a group from the Castle Air Museum started collecting the funds to move the aircraft to California. A volunteer crew traveled to Chanute in August 1992.

Six Turning and Four Burning

CHAPTER FIVE

RUSHMORE AIR FORCE STATION

The Top Secret Nuclear Weapons Storage Facility

THE GHOST SQUADRON

LT-COL. LARSEN

SECURITY CLASSIFICATION OF THIS PAGE	

REPORT DOCUMENTATION PAGE	Form Approved OMB No. 0704-0188

1a. REPORT SECURITY CLASSIFICATION Unclassified	1b. RESTRICTIVE MARKINGS
2a. SECURITY CLASSIFICATION AUTHORITY	3. DISTRIBUTION/AVAILABILITY OF REPORT Approved for public release
2b. DECLASSIFICATION/DOWNGRADING SCHEDULE	

4. PERFORMING ORGANIZATION REPORT NUMBERS United States Air Force Air Mobility Command Cold War Series Report of Investigations Number 5	5. MONITORING ORGANIZATION REPORT NUMBER(S)

6a. NAME OF PERFORMING ORGANIZATION Geo-Marine, Inc.	6b. OFFICE SYMBOL (if applicable) CESWF-PL-RC	7a. NAME OF MONITORING ORGANIZATION US Army Corps of Engineers, Ft. Worth District
6c. ADDRESS (City, State, and Zip Code) 550 E. 15th Street / Plano, Texas / 75074		7b. ADDRESS (City, State, Zip Code) PO Box 17300 Fort Worth, Texas 76102-0300
8a. NAME OF FUNDING/SPONSORING ORGANIZATION US Army Corps of Engineers, Fort Worth District	8b. OFFICE SYMBOL (if applicable)	9. PROCUREMENT INSTRUMENT ID NUMBER DACA63-93-D-0014, Delivery Order No. 0089

8c. ADDRESS (City, State, Zip Code) PO Box 17300 Fort Worth, Texas 76102-0300	10. SOURCE OF FUNDING NUMBERS			
	PROGRAM ELEMENT NO.	PROJECT NO.	TASK NO.	WORK UNIT ACCESSION NO.

11. TITLE (Include Security Classification) McChord Air Force Base, Tacoma, Washington, Inventory of Cold War Properties

12. PERSONAL AUTHOR(S) Karen J. Weitze

13a. TYPE OF REPORT Final Report	13b. TIME COVERED FROM March '95 to Oct. '96	14. DATE OF REPORT (Year, Month, Day) October 1996	15. PAGE COUNT 85 + Appendices

16. SUPPLEMENTARY NOTATION

17. COSATI CODES			18. SUBJECT TERMS (Continue on reverse if necessary and identify by block number) Inventory of Cold War Properties located at McChord Air Force Base
FIELD	GROUP	SUB-GROUP	
05	06		

19. ABSTRACT (Continue on reverse if necessary and identify by block number)

The U.S. Air Force, Air Mobility Command, has conducted real property surveys and evaluations at selected installations throughout the continental United States, including McChord Air Force Base, to identify potentially significant Cold War buildings and structures. Identified resources are primarily associated with the tactical and strategic USAF network built up in North America during the 1949-1962 years. Specific property types discussed are radar enclaves; command and control facilities for gathering and disseminating information within defined air sectors; readiness and alert complexes for fighter (tactical) and bomber/tanker (strategic) aircraft; missile housings and assembly-test units; and weapons areas. All inventoried resources are less than 50 years in age and must meet the NRHP criteria of *exceptional significance*. Thus the methodology used in these assessments suggests that individual properties within such a group meet a high standard for NRHP integrity, and that interpreted significance is relevant only at a national level.

A total of 29 buildings and structures were inventoried at McChord AFB, including alert and readiness buildings of 1951-1962, AC&W command and control buildings of 1950-1953, an ADC nuclear weapons storage compound of 1957-1962, the SAGE buildings of 1955-1958, and TAC buildings of 1983-1985. Building 300 is interpreted as potentially eligible for the NRHP under criteria A and C, as well as criteria consideration G. It is a rare, little modified structure associated with a nearly continuous significant alert mission from the first years of the Cold War to its end and is one of the first standardized FIS alert hangars erected nationwide, additionally unusual is its expansion to a double-squadron capacity.

20. DISTRIBUTION/AVAILABILITY OF ABSTRACT ☐ UNCLASSIFIED/UNLIMITED ☒ SAME AS RPT. ☐ DTIC USERS	21. ABSTRACT SECURITY CLASSIFICATION Unclassified	
22a. NAME OF RESPONSIBLE INDIVIDUAL Joseph Murphey	22b. TELEPHONE (Include Area Code) 817-978-6386	22c. OFFICE SYMBOL CESWF-PL-RC

DD Form 1473, JUN 86   *Previous editions are obsolete*   SECURITY CLASSIFICATION OF THIS PAGE

I was authorized to have access to Rushmore Air Force Station (AFS) in June 1996. During the escorted visit to the former facility he was allowed to take photographs. Currently this area is referred to as the Ellsworth Air Force Base Special Weapons Storage Area. *Provided by author*

The information collected on Rushmore Air Force Station (AFS) was gathered on a June 1996 escorted visit to the former facility, today referred to as the Ellsworth Air Force Base Special Weapons Storage Area. It was a rare look at a former Cold War nuclear weapons storage facility. The author was able to take photographs and walk throughout the facility, going inside the various storage buildings and underground bunkers.

One of the lesser-known units on the base, and first assigned to Weaver AFB, was the 3081st Aviation Group. This unit was operational from 1952 to 1962, assigned to the 3081st Aviation Field Depot, South River Depot, Weaver AFB, South Dakota. South River Depot became Rushmore Air Force Station (AFS), first on Weaver AFB and finally on Ellsworth AFB as of June 13, 1953. The first contingent of 108 3081st Air Police arrived after completing thirteen weeks of basic training at the U.S. Army's Camp Gordon, Georgia, to maintain operational and physical security of the Top Secret Project *Rocky Creek*, the storage of nuclear weapons in early 1953.[1]

Air Police personnel moved into their new barracks on the top of a hill, north of the B-36 parking apron, inside a secured area. The administration area consisted of four barracks, mess hall, motor pool, and headquarters building, all inside a single chain link fence.

A security road is between two of the former Rushmore AFS, illuminated by floodlights on poles spaced around the facility. The fence was topped with barbed wire for added security. The water tower is visible in the right center of the photograph. *Lt. Col. George A. Larson, USAF (Ret.)*

The weapons storage area was secured by five fences. Between the outside and the second fence, a road used by Air Police to patrol the storage facility's perimeter was prepared with crushed rock. The second fence separated Air Police from the third electric fence, charged with 3,200 volts of electricity. The fourth fence was inside the weapons storage area, again with a road separating it from the fifth fence. A second, or interior Air Police patrol used this road to maintain close in security, just in case, however unlikely, the previous fences were actively breached by intruders.

Air Police patrols traveled the two perimeter roads twenty-four hours a day, seven days a week, alternating speed and patrol sequences so no one watching the armed guards could determine a repetitive guard schedule. The electric fence was charged by a 1,500 horsepower diesel-electric generator. A backup generator came on line when the primary generator failed.

125

View of the inside security road. Earth covered igloo type weapons storage bunkers are visible to the right of the fifth fence. *Lt. Col. George A. Larson, USAF (Ret.)*

Every thirty days, the primary diesel electric generator was turned off and personnel waited for the backup generator to start, picking up the electric load to charge the fence. The previous running generator was serviced and placed on standby status in case the operating generator failed. The entire weapons storage area was illuminated by floodlights, so there were no areas of shadow which could mask the entrance of infiltrators into the storage area.

South Dakota's severe winter weather sometimes required underground tunnels connecting barracks, mess halls, and support buildings so personnel did not have to go outside when not on duty. Whiteouts resulted when temperature and dew point produced a thick fog. Personnel could not see their hands in front of them, making it impossible to go safely from one building to the next. Perimeter patrols would put a man in front of the truck, slowly walking on the gravel path as the vehicle followed behind, shining powerful truck-mounted spotlights on the security fence.

Weapon's site security fencing was two and one half miles long. The first fence consisted of one continuous electric fence circuit. If a rabbit ran into the fence it set off an alarm and security patrols searched the area where the fence had been contacted. Security personnel located the spot where the animal contacted the fence and radioed to have power to the electric fence shut off. However, a powerful electric charge remained in the fence. Technicians with the security patrol used a *hot stick* (a copper loop at the end of a rod) to ground the fence's electric charge. Because of numerous animal strikes in the fence it was divided into sections, so if an animal hit the fence it was easier to locate and remove the dead animal without deactivating the entire electric fence. The entire facility had a four-hundred-yard-wide security zone marked with poles with red flags and signs warning entry was forbidden. No one was allowed to enter without being challenged and detained. One-half of the on-duty Air Police force conducted patrols while the other remained on recall. They were on duty seventy-two hours, carrying rifles and side arms. The use of deadly force was authorized.

View of the security entry position into the weapons storage area from outside the facility, looking out into the security area. In 1996, construction work was in progress inside the area. *Lt. Col. George A. Larson, USAF (Ret.)*

Air Police spent the first weeks in fatigues, finishing construction work inside the security area. They laid sod on top of the concrete atomic weapons storage bunkers and conducted a general cleanup of the construction debris. Air Police were specially trained, with security clearances run by the Federal Bureau of Investigation (FBI) and other security agencies. Guards were frequently evaluated in countering penetration attempts made by base or designated personnel. Selected Air Police personnel were pulled to work with base security in attempts to penetrate and get inside the security area. Air Police personnel viewed security procedures as serious business. Air Police had no way of knowing if a penetration was an exercise or real.

Colorado National Guard soldiers were brought to Ellsworth AFB to infiltrate Rushmore AFS. Air Police had no knowledge about the exercise. Guard troops deployed on the north side of the weapons storage facility began crawling up the hill. One of the perimeter patrols noticed a movement outside the security fence, issued a verbal warning challenge, and emptied their carbines into the area. The Guard troops stopped, and because of the rife fire, the perimeter guards were told over their radios this was an exercise. Fortunately no one was wounded by the gunfire, as the Guard troops rapidly retreated. Air Police were responsible for the security of the site, guarding the entrance gate, atomic weapon igloo bunkers, security for loading and unloading weapons from aircraft, and especially during the unloading of arriving nuclear weapons into the storage facility.

Site guard tower inside Rushmore AFS. *Lt. Col. George A. Larson, USAF (Ret.)*

127

Rushmore Air Force Station

Ellsworth AFB was served by a railroad side line to bring the weapons, and later Intercontinental Ballistic Missile (ICBM) components for the Titan I and Minuteman I and II. A switch engine moved the heavily guarded box car into the secure area. Technicians replaced Air Force personnel on the engine for the short trip into the storage area, where the weapons were unloaded. Air Police rode on top of the slowly moving box cars armed with machine guns and automatic weapons to keep unauthorized personnel away from the top secret cargo inside.

Railroad track inside the base to support Rushmore AFS. Not operational as this time. *Lt. Col. George A. Larson, USAF (Ret.)*

MK-6, the first atomic bomb built in quantity for SAC's nuclear arsenal. It was an upgraded version of the WWII *Fat Man* plutonium bomb dropped on Nagasaki, Japan, on August 9, 1945. On display at the National Museum of Nuclear Science and Technology, Albuquerque, New Mexico. *Lt. Col. George A. Larson, USAF (Ret.)*

Drawing of the MK-6. It's yield could be configured at 8, 26, 80, 154, or 160 kilotons. It armed SAC's bombers from June 1951 through 1955, and was retired from the nation's nuclear inventory in 1962. *Courtesy National Museum of Nuclear Science and Technology, Albuquerque, New Mexico*

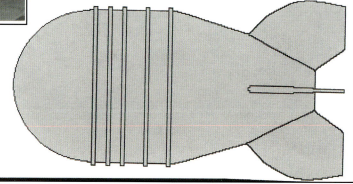

128

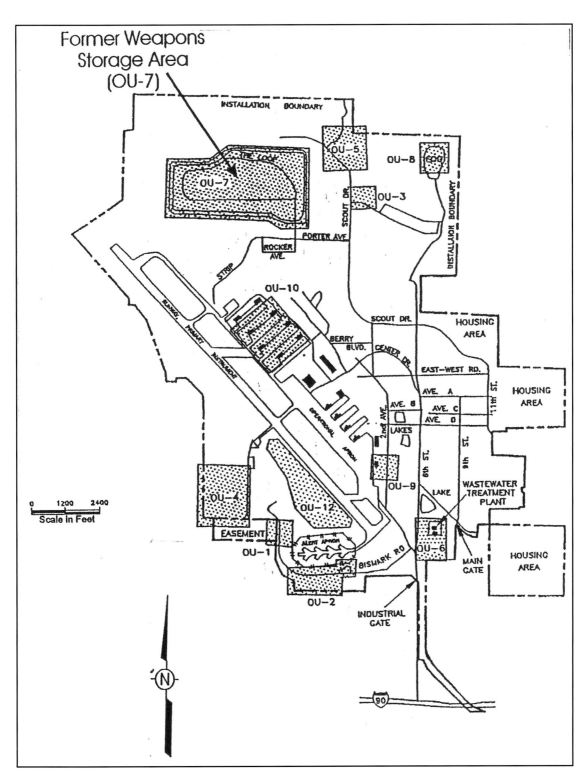

Map of the former Rushmore Air Force Station in relative position to Ellsworth Air Force Base.
Map: Ellsworth Air Force Base, 28th Bomb Wing Engineering Office

129

Rushmore Air Force Station

Air Police were not allowed to touch the weapons. They escorted weapons to the flight line and back to the igloo storage bunkers inside the security area. Equipment used at the weapons storage area came from Los Alamos, New Mexico (the classified U.S. atomic laboratory). Air Police never saw anyone wearing a radiation badge. During exercises, Air Police worked three-day shifts guarding the aircraft and weapons. When a RB-36H flew over the Air Police barracks with its six piston and four turbojet engines at maximum take-off power, the barracks' walls shook. The RB-36H was a huge aircraft, so Air Police guarding the aircraft on the flight line positioned one guard in front of the aircraft's nose, the second off the left wing, the third by the tail, and the fourth off the right wing. Two additional guards (one driving a Ford pickup truck with a second guard in the bed manning a .50-caliber machine gun) slowly circled the nuclear armed bomber (if so armed). This was repeated if other bombers were similarly loaded.[2]

Rushmore AFS had a construction cost of $1,175,408.00, with the classified facilities adding another $60,109.00.[3] The following discussion looks at the facilities inside the secure area.

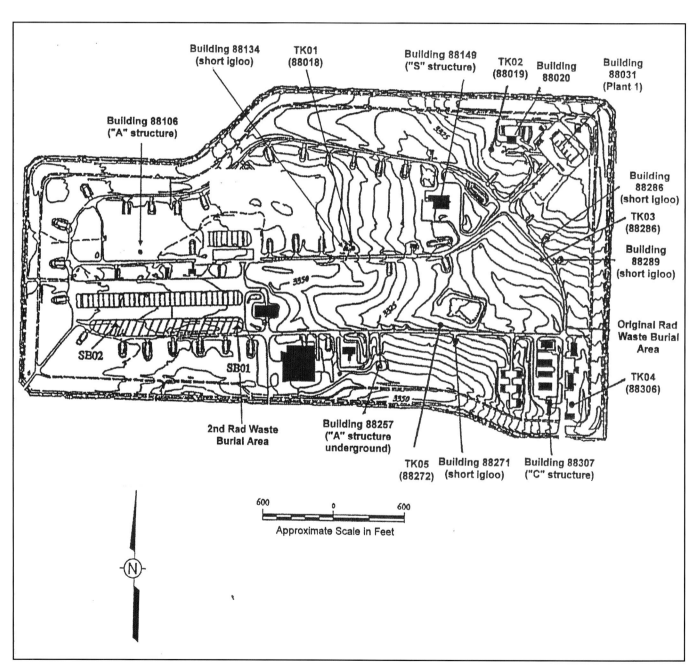

Map of Rushmore Air Force Station. *Map: Ellsworth Air Force Base, 28th Bomb Wing Engineering Office*

Rushmore Air Force Station

The Site "A" storage buildings were massive concrete structures. Each contained four small storage rooms ten-feet wide, thirteen-feet long, and nine-feet high with built in metal storage shelving. The first "A" storage building, built in 1952, could only store thirty capsules per room because of their bulky size. The second "A" storage building built in 1955 had smaller storage bins that were able to store seventy capsules per room.

Storage building "A." The concrete building was built above ground, and in 1950s Cold War mentality, it was fitted with false exterior window frames to disguise it as anything but a nuclear weapons storage building. Even on Ellsworth AFB, there was a concern that someone taking photographs or viewing from off base with a long distance telephoto lens would be able to identify the building's functions. *Lt. Col. George A. Larson, USAF (Ret.)*

Storage rooms were accessed through bank type vault doors fitted with dual combination locks, which required two officer control to guarantee security of the nuclear components inside the "A" storage building and storage room. *Lt. Col. George A. Larson, USAF (Ret.)*

Rushmore Air Force Station

Security procedures protected the economic value of the top secret atomic materials, preventing loss of technical information to possible espionage activities of foreign intelligence agents (primarily Soviet intelligence, GRU and KGB). This stored material represented a significant portion of the United States Gross National Product (GNP) in the 1950s. The author, while assigned to the 43rd SW on Guam, stored the Wing's Coded Switches in his office, secured by a similar two officer control vault combination door alarmed to prevent unauthorized access. The separate secured small vault was itself located inside a larger secured intelligence facility, again accessed by a vault door and alarmed. Security on a SAC base, even in the 1970s and 1980s, was not taken for granted.

Nuclear capsules stored inside the vaults could achieve nuclear criticality if positioned (stored) too close to each other. Criticality could generate a burst of neutrons, producing a health hazard to maintenance personnel. Capsules were built to prevent any chance of attaining criticality.

View inside one of the "A" storage building's vaults. The steel shelves were a simple but effective storage medium. Today, the shelves are used to store non-nuclear related items. *Lt. Col. George A. Larson, USAF (Ret.)*

The capsule was a metal cylinder fourteen-inches in diameter and twenty-five inches high. Steel braces were welded to the top, bottom, and sides to create a six-faced polyhedron of larger volume than the cylindrical variety, called bird cages. They were sealed with lead and wire seals similar to the seals used to prevent tampering on commercial and residential gas and electric meters. This assured the contents were not tampered with between authorized maintenance activities. Periodic security inspections matched the numbers of the seals with the inventory list. Seals were checked for tampering or attempts to access the material. This was the high-security level maintained on U.S. nuclear materials.

The "C" structure served as a nuclear materials inspection laboratory and maintenance building for the first generation of nuclear components stored at Ellsworth AFB beginning in the fall of 1952.[4]

132

The facility where the MK-6 atomic gravity free-fall bomb and MK-17 Thermonuclear (hydrogen) gravity free-fall bomb were stored required a high level of security. The large size of the MK-17 required a large storage bunker in which one of these weapons could be stored. It was a big deal to store one of these weapons, and its destructive capabilities required the weapon to be safe at all times, unable to reach criticality by accident or without the proper fusing and presidential release authorization.

Second "A" storage building; concrete covered with earth and sod. This was a large building.
Lt. Col. George A. Larson, USAF (Ret.)

The MK-6 and MK-17 were part of SAC's early atomic and thermonuclear deterrent arsenal while scientists worked on developing and reducing their size and weight, but increasing the next generation of nuclear weapon yields. The MK-6 was the first atomic weapon to be mass produced, entering the U.S. stockpile in 1951, and it was retired from the nuclear inventory in 1962. The MK-6 weighed 8,500 pounds, and was 128-inches long and 61-inches in diameter. It could be carried first by the B-29, then the B-50, and on early model B-36s.[5]

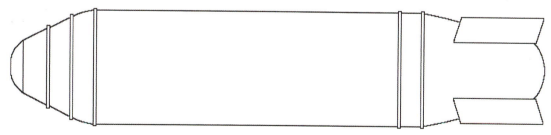

Drawing of MK-17 and data from the National Museum of Nuclear Science and Technology. *Courtesy National Museum of Nuclear Science and Technology*

133

MK-17 on Display at the National Museum of Nuclear Science and Technology. *Lt. Col. George A. Larson, USAF (Ret.)*

Bird Cage. *National Museum of Nuclear Science and Technology to Lt. Col. George A. Larson, USAF (Ret.)*

134

Rushmore Air Force Station

The MK-17 was a first generation thermonuclear weapon that could be only be carried by the B-36 in the early days of the Cold War between the United States and the Soviet Union. The weapon was crude by today's weapons standards and very large: 24.8-feet long and 61.4-inches in diameter and weighing 41,400 pounds, with a metal casing 3 1/2 inches thick. The weapon entered the U.S. stockpile in 1951 and was retired in 1962. The weapon's yield varied from 15 to 20 megatons.[6]

Every nuclear weapon used Polonium and Beryllium initiators to generate sufficient neutrons to produce and reach criticality, resulting in a nuclear detonation if released on a ground target. Polonium has a half life of approximately 138 days, requiring initiators to be replaced at scheduled intervals to retain a functional weapon. Polonium and Beryllium initiators were replaced by non-radioactive sealed neutron initiators in late 1954, eliminating any possibility of accidental release of Polonium waste during initiator replacement. Weapon capsules were disassembled on a scheduled interval to check the fissile material's integrity. The capsules were phased out in 1962, which ended nuclear component maintenance activities inside the "C" structure in the secure area. Thereafter, AEC maintenance activities did not require the handling of any exposed nuclear materials.

"C" structure. *Lt. Col. George A. Larson, USAF (Ret.)*

Weapons were put together in the two assembly and maintenance buildings. Each had heavy blast doors protected by thick earthen works designed to deflect an accidental explosion upward. The bunkered design is typical for facilities where large amounts of conventional chemical explosives are handled or stored. Chemical explosives compressed nuclear material into a critical mass in the early Mk-6 free-fall gravity atomic weapon, as on the Nagasaki bomb dropped on August 9, 1945.

Weapons assembly and maintenance building, with concrete walls, roof, heavy metal blast doors, and earth bunkered on each side, so if there was a conventional explosion, the force of the explosion was directed upwards. *Lt. Col. George A. Larson, USAF (Ret.)*

135

The "S" structure, or Surveillance building, was used to conduct quality and assurance inspections and test the operational readiness of weapons. It was a metal building designed to provide more space for quality assurance activities and keep work separated from other weapon maintenance activities.

Inside of the Surveillance building showing the heavy lift cranes mounted on steel beams on each side of the bay to lift weapons' components. *Lt. Col. George A. Larson, USAF (Ret.)*

The storage facility consisted of twenty-seven earth-covered concrete storage igloos designed to hold the assembled weapons, non-nuclear components, and casings. Nuclear material stored in these igloos was sealed inside weapons. Eleven super igloos were built in 1954 to hold the large MK-17s. Four short Type VI igloos stored Tritium booster cylinders for the MK-17s stockpiled between 1955 and 1957. A safety feature in the top of each igloo structure permitted the venting of escaping Tritium gas. Tritium gas has a half-life of 12.33 years.

One of eleven super igloos built in 1954 to hold the large MK-17s. The size of this concrete igloo is impressive. It had to be large enough to hold the MK-17 thermonuclear bombs. It had a large blast door that protected the weapons stored inside. It had a distinctive aerial photography signature. *Lt. Col. George A. Larson, USAF (Ret.)*

136

One of four short Type VI igloos that stored Tritium booster cylinders for the MK-17s stockpiled between 1955 and 1957. The top of the earth covered bunker had a tall ventilation stack out of which Tritium of the first generation of thermonuclear weapons was vented. This was not known to the surrounding community, and the amounts were considered small enough as to not present a hazardous environment (impact) to either on base personnel or civilians living outside the base's perimeter (by 1950s EPA and AEC standards). *Lt. Col. George A. Larson, USAF (Ret.)*

View inside one of the four short Type VI igloos showing the ventilation fan mounted on the wall outside the storage vault inside where Tritium initiators were stored. If any Tritium gas escaped from the initiators, it was vented up and outside the storage vault. *Lt. Col. George A. Larson, USAF (Ret.)*

137

Sandia National Laboratory technicians estimated that between 1955 and 1957, approximately six hundred Curries of Tritium per month vented from each of the four short igloos. Scientists determined Tritium venting was in small enough amounts to produce little Tritium contamination during release. However, monitoring continues inside the former Rushmore AFS. Thirteen Weapons Storage Areas (WSA) were built between 1948 and 1956. New facilities included storage buildings for nuclear component capsules, maintenance structures, assembly buildings, waste burial sites, and bunkers for weapon casings. Low-level radioactive wastes generated in the "C" structure were shipped to Sandia National Laboratory in New Mexico by secure courier between 1952 and 1954.

Limited low-level waste on-site burial was done on base consistent with accepted 1950s operational guidelines, which would not be permitted today. Low-level radioactive burial site number one, an area of approximately 5,000 square feet, was built in June 1952. A larger 10,000-square-foot land burial site was built in March 1955. Sandia National Laboratory determined small amounts of low-level radioactive waste are buried in one or two trenches in the WSA. No other low-level radioactive waste disposal areas have been identified.[7]

On June 8, 2006, the EPA summarized the partial deletion of the Ellsworth AFB site from the National Priorities list. OU-7 (Low-Level Radioactive Waste Burial Site) is located in the Munitions Storage Area (WSA), formerly identified as the Weapons Storage Area (WSA), at the northernmost end of Ellsworth AFB. The WSA covers approximately 65 acres. Radioactive wastes were generated at Ellsworth AFB between 1952 and 1962. During that time the WSA was under control of the Atomic Energy Commission (AEC). After 1962, control of the WSA was transferred to the Air Force. Contaminants in soil at OU-7 include VOCs (volatile organic compounds) and inorganic compounds. The concentrations of several inorganic compounds in soil and sediment exceeded background concentrations. Radionuclides detected in all media were within the normal background range due to natural variations in soil types and geological characteristics. The results of the risk assessment indicated that the risk due to exposure to contaminants in surface water and sediments was within the acceptable risk range. Therefore, it was determined that remedial action was not warranted for surface water or sediment.

There were eleven areas besides the low-level radioactive waste burial site which the EPA looked at. Overall, some ground water contaminants have moved beyond the boundaries of Ellsworth AFB to the east and south at low concentrations. The Air Force responded by installing cleanup systems and the construction of cleanup systems in the contaminated areas. Some private drinking well water was contaminated. The Air Force extended water lines from the Ellsworth AFB water system to these homes. Future expansion of Box Elder's water system will hook up to these homes. The ground water cleanup systems will be operating into the future, for an estimated time period of up to 30 years. This clean up is not without cost, and is estimated at $30,000,000.[8] Another cost of the Cold War.

View of Ellsworth Air Force Base high security area. *Lt. Col. George A. Larson, USAF (Ret.)*

138

Rushmore Air Force Station

The early atomic/thermonuclear weapons were so large that clearances from the bottom of the B-36 to the concrete parking apron did not allow direct uploading into the bomb bay. The B-36 had to be raised on hydraulic jacks, the bomb positioned underneath the open bomb bay, and the bomber slowly lowered down onto the concrete parking apron. The weapon was then hoisted up inside and secured at the top of the bomb bay.[9]

Huge ordnance vehicles were required to move the large thermonuclear bombs from the storage area to the flight line and return. The four-wheeled vehicle was tall enough so the operator could drive over the bomb and use hydraulic clamps or arms to lift the bomb and cradle off the ground. It would carry the bomb and its cradle to the flight line and position the bomb on the concrete flight line. The bomb would then be pushed underneath the jacked up B-36 (as described above) to provide the necessary ground clearance to position the bomb underneath the B-36.[10]

The RB-36Hs, if loaded with a MK-17, flew training missions with the weapon unarmed in a safe mode condition. At this time, SAC orders prevented bombers from flying with armed weapons within the Continental United States. A nuclear weapons technician flew as part of the crew, and if the aircraft received a nuclear launch order, decoded and verified, only then was the weapon armed. Later a coded switch was installed on the B-52, which after the same procedures armed the bombs, but also required a weapons technician.

Large truck-mounted snow blowers were used during winter to clear a path around the bombers on the flight line. Perimeter roads were kept clear so Air Police patrols could guard the storage facility. Air Police vehicles frequently became stuck in blowing and drifting snow. Security for Air Police assigned to Rushmore AFS was a twenty-four hour concern, even when they were not on the installation and Ellsworth AFB. Everyone was scared to open their mouths and talk about their jobs or Rushmore AFS. One Air Police enlisted man who went into Rapid City to drop off personal clothes at a dry cleaner could not find his security badge after leaving the dry cleaners. He called the Installation Security Officer, but a search was unable to locate the security badge. Consequently, everyone assigned to Rushmore AFS had to turn in the current security badge, the badge number was verified, and they signed for a new badge of different color and design.

One Saturday evening a group of Air Police went to a Rapid City bar. They walked in and sat down at a table, and suddenly they sat upright. They immediately recognized the bar tender who served them drinks as a First Lieutenant from the base, a counter-intelligence officer. Others would recognize the First Lieutenant at different Rapid City bars, perhaps wearing a cowboy hat, boots, and jeans, drinking beer and talking with airmen from the base. He and other security officers circulated throughout Rapid City bars to determine if anyone was talking about Rushmore AFS and its top secret mission. Security during the Cold War was constant and never taken lightly. United States political and military leaders feared the Soviet Union and its brand of Communism was out to take control of the free world, with its intelligence agents (many possible deep-sleeper U.S. residents planted by the GRU or KGB) prowling near top secret U.S. military installations.

Interior of bomb assembly building currently used at Ellsworth AFB. A 500-pound practice bomb is visible, used for training maintenance personnel. *Lt. Col. George A. Larson, USAF (Ret.)*

139

Rushmore Air Force Station

On March 4, 2011, airmen from the 34th and 37th BS raced against the clock to safely load bombs on a B1-B Lancer mock trainer. Comments from several airmen in the competition were noted in *The Patriot*:

Master Sgt James Price, 28th Maintenance Group Weapons Standardization Superintendent, recalled "It also helps build camaraderie. It also helps to get the weapons loaders to think about what they are doing, and to accomplish the load in the safest and most reliable way – but in a timely manner – to ensure we put bombs on target."

Technical Sgt. Taft Tubbs, 37th Aircraft Maintenance Unit Team Chief of the loading crew, noted "The Airmen and I down here today have a lot of experience with this, so I don't think the nervousness got to them." Senior Airmen Brandy Will, 37th AMU Weapons Load crew member, stated "Every time you load a bomb, you're building on what you already know. You learn something new every time you load."[11]

Modern weapons storage bunker inside the former Rushmore AFS. *Lt. Col. George A. Larson, USAF (Ret.)*

Security officers kept close tabs on anyone on leave from Rushmore AFS. Air Police, when they signed out, provided destination, route of travel, and expected arrival time. Security officers called the leave destination to verify the person signed out from the station was at the destination. A late arrival triggered a visit by an FBI agent, Federal Marshall, or local law enforcement officer. Air Police were constantly reminded of the security importance of calling the base, either for themselves or personnel assigned to Rushmore AFS, due to a possible late arrival to a leave destination and why. Law enforcement agencies would even trace the route of travel of a person who could not be located.

A few married men never had their wives on base while assigned to Rushmore AFS. Many Air Police spent their entire military career at Rushmore AFS. Tight security was even reflected on the birth certificates of children born to fathers who worked at the top secret military facility. Replacement personnel's orders indicated Rushmore Air Force Station, South Dakota, as their new duty station. However, this was not listed in any Air Force Military Installation Guide or at the Ellsworth security gate's guard entrance point. There also was no listing in the Ellsworth AFB civilian or military telephone directory/listing. One 3081st replacement arrived at the Ellsworth AFB main gate three years after Rushmore AFS began operations, showing his orders to Rushmore AFS, and the new security base guard at the gate did not allow him entry into the base. The gate guard said there was no such facility on the base. The gate guard called his on-duty shift sergeant, who was also new on the base, and said he had never heard of anything called Rushmore AFS. The on-duty shift sergeant called his commander, who came to the front gate and escorted the airmen to Rushmore AFS. This is how most people on Ellsworth learned about the operations on the hill at the north end of the base.

140

Rushmore Air Force Station

The current bunkered munitions storage area for supporting B-1B operations at Ellsworth AFB on the west side of the operational runway, visible in the center of the photograph, with the taxiway at the lower bottom. These are concrete bunkers, with doors on each side of a larger concrete- and earth-covered bunker. The area is separately secured and illuminated by flood lights on poles. The Black Hills, north of Interstate 90 and Rapid City, make a remarkable background. Note the operational flight line control tower to the right or north of the bunkers. *Lt. Col. George A. Larson, USAF (Ret.)*

From 1952 to 1962, Rushmore AFS handled large thermonuclear weapons, which were later replaced by the smaller and more powerful hydrogen weapons carried by the next generation of SAC bombers and warheads for Intercontinental Ballistic Missiles. Rushmore AFS Air Police did not come under the command authority of SAC. By 1962, after the Cuban Missile Crisis (sometimes referred to as the Missiles of October) of October 1960 and the transition to an all turbojet nuclear strike force, Rushmore AFS came under SAC control. For ten years Air Police performed a tough job at the secretive installation that even today not many people know about. Secrecy earned these men the nickname "Ghost Squadron of Ellsworth Air Force Base."[12] The transition of control of the weapons storage facility from AMC to SAC fully integrated all functions and operations into Ellsworth AFB.[13] The transition was possible because weapons design and stockpile changes for the new generation of nuclear weapons did not require maintenance of exposed nuclear material.[14]

View of the Ellsworth high-security area with two of the former Tritium initiator earth-covered igloo bunkers (identification feature is the tall vent on the top of each bunker). Other earth-covered bunkers can be seen in the photograph. The facility required a lot of power, and the numerous power distribution poles and lines are apparent. The security fences are visible and lights inside the area, along with current Air Force vehicles. The elevation is visible in this photo, indicating it was built on a hill at the north end of the base. Most of the buildings built in the 1950s are currently in use. *Lt. Col. George A. Larson, USAF (Ret.)*

141

CHAPTER SIX
NIKE AJAX AND NIKE HERCULES

Surface-to-Air Missile Defense of Ellsworth Air Force Base

Douglas Aircraft Company Nike-Ajax (MIM-3) on display on launch erector. The inert surface-to-air missile is on display at the South Dakota Air & Space Museum, outside Ellsworth Air Force Base, South Dakota. *Lt. Col. George A. Larson, USAF (Ret.)*

The history of Ellsworth Air Force Base is tied to the Cold War between the United States and the former Soviet Union, which resulted in a series of new weapons developments and deployments by both super powers. With the deployment of the B-36 to Ellsworth AFB and a perceived threat from the Soviet Air Force's long range bombers, a point defense system was needed.

Nike Ajax specifications and performance data. *Courtesy South Dakota Air and Space Museum*

Nike Ajax Specifications and Performance Data	
Weapon designation:	MIM-3
Prime contractor:	Western Electric
Airframe contractor:	Douglas Aircraft (Missile Division)
Power plant contractor:	Aerojet Corporation
Booster contractor:	Bell Aircraft
Overall length:	34 feet
w/o booster:	21 feet
Wingspan:	4 feet
Diameter:	1 foot
Firing weight:	2,455 pounds
w/o booster:	1,150 pounds
Maximum speed:	Mach 2.25
Speed at burnout:	1,669 mph
Maximum slant range:	25-to-30 miles
Initial maximum ceiling:	60,000 feet
upgrade:	72,000 feet
Booster:	Solid propellant
	59,000 pounds of thrust for 2.5 seconds
Sustainer:	Liquid fueled (IRNA, WDMH and JP-4)
	2,600 pounds of thrust for 21 seconds
Warhead:	Three HE fragmentation
Guidance:	Command guidance system

In the 1950s, U.S. national intelligence agencies believed Soviet LRA was seriously developing a force of four-engine, turbo-jet bombers, probably in the class of the Boeing B-52 Stratofortress, to carry atomic weapons to reach targets in North America and the Continental United States. The Nike-Ajax surface-to-air missile (SAM) system was designed as a weapon of point defense (last resort) for air defense around important U.S. military installations, which at that time were SAC's bomber bases. Later, this included SAC's first generation of soft ICBM installations (the Atlas series ICBMs that were not hardened in underground silos). The SAM sites were also to be placed in circles around important U.S. cities to provide protection from Soviet LRA bomber attack. The Nike-Ajax SAMs were to be used if USAF fighters and Air Defense Command interceptors on alert were unable to locate and destroy penetrating Soviet LRA bombers. The first Nike-Ajax site became operational at Fort Mead, Maryland, in 1957, and final deployment included 250 sites throughout the United States. It was the Cold War, and during this period of building civilian bomb shelters and school drills of drop and cover, there was a fear in the United States that it was only a matter of time for a nuclear war to break out between the United States and the Soviet Union.

Right side view of Nike-Ajax SAM and launch erector on display at the South Dakota Air and Space Museum. *Lt. Col. George A. Larson, USAF (Ret.)*

143

Though swiftly outdated as a weapon, this pioneer SAM was one of the great missiles in history. The first operational site at Fort Mead, Maryland, was tasked to defend Washington, D.C., from a Soviet LRA bomber attack. It became the world's first operational SAM system, before the famous and widely deployed Soviet SA-2 Guideline SAM, which in 1960 shot down a Lockheed U-2 flying over central Soviet Union. The Nike-Ajax was necessarily a large and cumbersome system by later standards, with large equipment installed in sites requiring thousands of tons of concrete and steel. It was determined that, in order to fight and win a nuclear exchange with the Soviet Union, military defense installations had to go underground, and be hardened and protected by layers of concrete reinforced with hardened steel rods. The system could later have been made mobile, but the original equipment was for the defense of the United States and its allies, with no deployment with a field army.

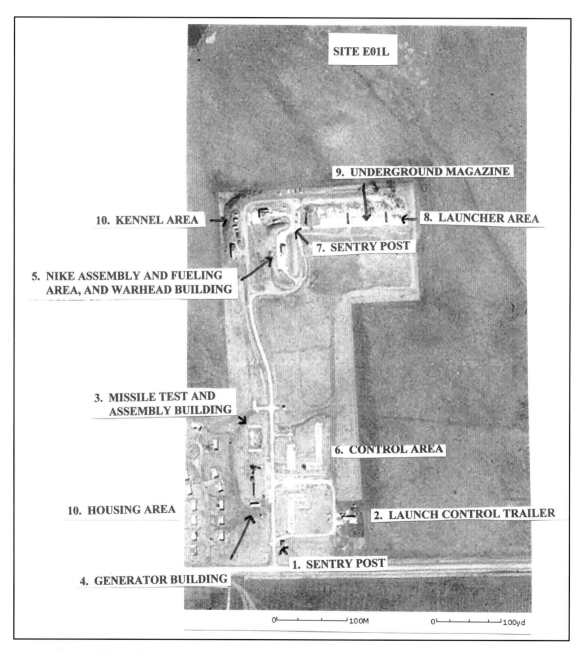

Photo of Nike-Ajax SAM site E-01L, located 5.2 miles north of Ellsworth Air Force Base. The major functions of the site are marked. *Sioux Falls, South Dakota: United States Geological Survey*

Nike Ajax and Nike Hercules

Most of the missile storage, checkout, and missile specialist accommodations were underground, though not sufficiently hardened to sustain and survive a near direct hit by a Soviet nuclear bomb. The guidance was derived exactly from the radar controlled anti-aircraft artillery (AAA) of WWII, including the M-9 fire control used for 90mm AAA. The prime contractors for this had been Western Electric and its chief operating subsidiary, Bell Telephone Laboratories. They were again selected to mastermind the U.S. Army's first SAM. The Nike-Ajax prime contractors were: Western Electric, missile airframe; Douglas Aircraft Company, power plant; Aero Jet; and for the booster, Bell Aircraft. Targets were acquired by acquisition radar, which in 1956 in the United States was just beginning to be tied into the military's Semi-Automatic Ground Environment (SAGE) system. The system used first generation enormous vacuum tube computers to assign every intruder into North American airspace to a specific airborne interceptor or SAM.

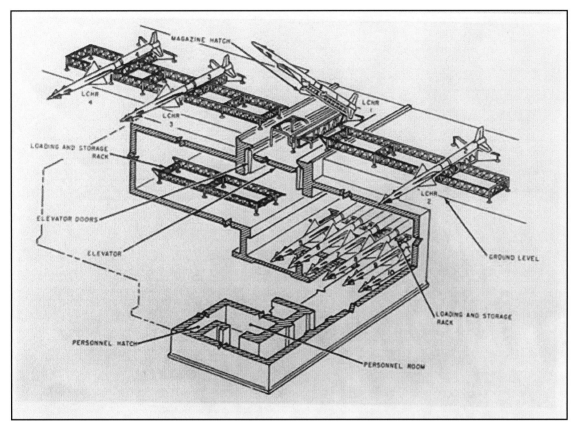

Line drawing of a Nike-Ajax underground missile magazine and facilities with four above ground launch erectors.
Drawing: Carlisle Barracks, Pennsylvania: United States Institute for Military History, United States Army War College

The acquisition radar passed the target information to a target-tracking radar (TTR), which continuously fed target data to a large computer with thousands of vacuum tubes referred to as therminic valves. The author, when he was a Second Lieutenant with the 544th Aerospace Reconnaissance Technical Wing (ARTW) at Headquarters SAC, Offutt AFB, used a second generation computer system. It was a huge computer. Data was stored in a room cooled to 60 degrees because of the heat generated by hundreds of reel-to-reel tape drives used to store and retrieve computer generated information. The computer had a four-foot diagonal screen. The reel-to-reel tape drive room was elevated three feet off the floor, which had removable steel panels to access the miles of computer cables. Airmen in the room wore fatigues and coats because the temperature was cold enough to warrant their use. Today, an HP laptop has more computing power and storage capacity than the early SAGE system. The large computer selected one or more missiles and commanded the launch, simultaneously driving the missile into the beam of a missile tracking radar. The computer thereafter drove the two radar beams into coincidence at a predicted future target point.

145

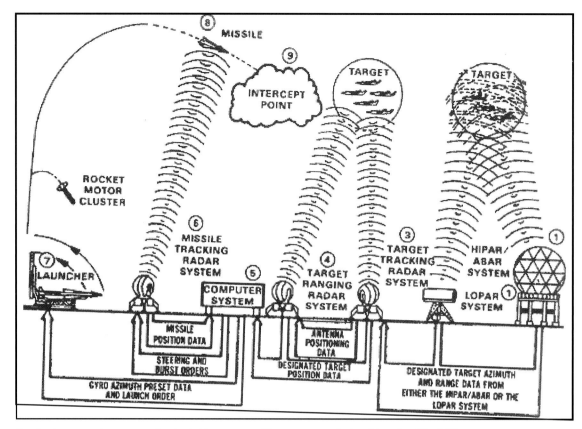

The drawing traces the sequence of radar locking onto an airborne target to the launching of a Nike SAM.
Drawing: Carlisle Barracks, Pennsylvania: United States Institute for Military History, United States Army War College.

Once the ground based radar beam was locked on, either to the target or the missile, at the predetermined point a detonation sequence was initiated. When the missile was just below the nose of the aircraft, the warhead detonation command was sent as a variation in the complex pulse coding of the beam transmissions.

The missile was a canard, with a large tandem boost motor. On early Nike 1 test vehicles, the boost was composed of multiple short motors with four giant cropped delta fins. Beginning in 1949 the whole missile became extremely slender, with the boost from a single Hercules charge giving 59,000 pounds of thrust for 2 1/2 seconds and having three modest fins. Bell Aircraft produced the acid/aniline sustainer, rated at 2,600 pounds of thrust. Burnout speed was Mach 2-3. Control was by cruciform for planes, and there were three warheads that could be mated to the missile: 12 pounds, 122 pounds, and 179 pounds. Each was wrapped in 1/4-inch optimum cubic fragments. Airframe manufacture and missile assembly was handled by Douglas Aircraft at an Army Ordnance Missile Plant at Charlotte, North Carolina.

Nike-Ajax SAM being raised from the below ground missile storage area to launch position. This was done during an open house, with civilians on the left watching the missile come up on the elevator to the surface.
Carlisle Barracks, Pennsylvania: United States Institute for Military History, Photo Archives, United States Army War College

146

By February 1958, approximately 16,000 missiles (firing rounds) had been delivered. The Army had 40 battalions, each comprising four batteries of either nine or twelve launchers. The Army designated the system M-1, and the DoD designated the Nike and Ajax the MIM-3 and MIM-3A, respectively. The name Nike is that of the Greek goddess of victory, while Ajax was a Trojan War hero. From 1957, Nike-Ajax installations were supplied to Belgium, Denmark, France, West Germany, Greece, Italy, the Netherlands, Norway, Taiwan, and Turkey. None shared in original system manufacture, though a few, notably Japan, became involved in supporting the existing installations. By the middle of 1978, approximately 7,000 Nike-Ajax SAMs had been fired in training, and a few sites remained on duty in Greece, Italy, and Japan.

Three Nike-Ajax SAMs are positioned on their respective launch erectors. Three U.S. Army missile site personnel have completed the manual operation to position the missiles on their launch erector and connect the electric cables for launch. They are headed toward the open personnel door in the lower right corner of the photograph. To the left is a decontamination shower if the toxic fuel leaked and had to be neutralized and cleaned off their clothes and skin in the shower. *Carlisle Barracks, Pennsylvania: United States Institute for Military History, Photo Archives, United States Army War College*

The standard Nike-Ajax configuration consisted of three areas. Area One was the Integrated Fire Control, or IFC. It was referred to as *Area C*. It was set on approximately six acres to provide spacing for radars and computers to track designated targets. Area Two was the missile magazine and launcher complexes. It was referred to as *Area L*. It was set on approximately 40 acres to accommodate one to three underground missile magazines. Each four-launcher complex maintained a staggered ready-to-fire or missile alert sequence: one on 15-minute alert, two on 30-minute alert, and one on 60-minute alert. Each magazine was equipped with an elevator to raise the missiles to the top of the magazine in preparation to be launched.

A Nike-Ajax site was converted to a Nike-Hercules site. This site consists of three missile magazines, each with one elevator, noted as the second launch position from the bottom (closest to the center of the photograph, and four launch positions). The entire area is fenced for security. The buildings at the lower center of the photograph are the missile assembly building, and to its right, the generator building. At the right of the photograph is the acid fueling and warhead assembly building. At the right of the three launch, complex site is the guard house, the controlling entrance into the facility. Below the photograph and not shown are the site's radars. *Carlisle Barracks, Pennsylvania: United States Institute for Military History, Photo Archives, United States Army War College*

Each magazine serviced four above-ground elevating/launcher rails. The magazine stored ten to twelve missiles equipped with launcher loading racks, crew shelter, launcher section control panel, ventilation equipment, test equipment, and hydraulic controls. Although the building was an underground structure, it was not hardened to withstand and continue operations after a near nuclear weapon impact and detonation. The resulting shock wave would destroy the magazine's missiles, with the above ground launchers and radars blown away by the resulting shock wave and intense heat. After a launch warning was received the Nike missiles were manually pushed one at a time onto the elevator, raised to the surface, and again manually pushed to the designated launcher. There the required electrical connections were made to the missile, and the missile was raised to an angle of 85 degrees in preparation to be fired.

Area Three was the administrative complex. It was co-located with the IFC to provide battery headquarters, barracks, mess hall, recreation hall, and motor pool. The entire complex was relatively compact. The launch complex was surrounded by a security fence topped with barbed wire. A few Nike sites consisted of four to six missile magazines, referred to as double sites to support two separate firing batteries, each with their own IFC radars.[1]

A site configuration of four batteries had a fire capability of sixteen missiles per battery, requiring 200 personnel. The 85-degree launch angle prevented the booster from falling back onto the launcher site and landing inside the empty, designated booster impact area. The control and launch areas were separated by 1,000 to 6,000 yards. Technical limitations with the guidance system (used at that time) required the control and launch areas to be separated by 3,000 yards. If possible, control area(s) were built on high ground to provide the necessary line of sight connection to the erected missile and had to be locked onto the missile prior to launch.

Four Nike-Ajax missiles in raised position at different angles for the staged photograph. These missiles were at the Nike Battery, San Francisco Defense Area near Fort Winfield Scott. *Carlisle Barracks, Pennsylvania: United States Institute for Military History, Photo Archives, United States Army War College*

148

Ellsworth AFB was an important SAC bomber, and later ICBM base. To provide air defense of the Ellsworth AFB, the U.S. Army created and established the Ellsworth AFB Defense Area (EDA) in 1950, constructing four missile sites later in 1957. The base was defended by four Nike-Ajax SAM sites, which were later reduced to one Nike-Hercules SAM site after deactivation of the Nike-Ajax sites. The EDA began operations during the Korean War (1950-1953) as the 531st Anti-Aircraft Artillery (AAA) Gun Battalion. It was originally equipped with 40mm AAA guns, which were later upgraded to M35 75mm Skysweeper AAA. The Battalion moved into SAM defense in 1957. It operated four Nike-Ajax SAM batteries. Later, its defense mission/operations were picked up by the 2/67th Artillery on September 1, 1958. Site E-01L was converted from Nike-Ajax to Nike-Hercules, and deactivated in August 1961. Site E-01L was located 5.2 miles north of Ellsworth AFB, 44 degrees 12 minutes 9 seconds north and 103 degrees 5 minutes 50 seconds west.

Site-E-20L is located 5.3 miles east-northeast of Ellsworth AFB at 44 degrees 9 minutes 23 seconds north and 103 degrees 10 seconds west. This was a typical three-magazine, 12-launcher Nike-Ajax site. In the photograph, the housing area is located in the upper right of the photograph. The two L-shaped buildings are administration and barracks, while the two buildings to the right are the mess hall and recreation buildings. The lower part of the photograph contains the three missile batteries in the long concrete area. The buildings at the top left of the missile battery are the missile assembly building, while to the right and below are the generator building. At the top left of the missile battery in the earthen bunker area is the acid fueling and warhead assembly building.
Sioux Falls, South Dakota: United States Geological Survey

149

The 28th Bomb Wing historian notes the history of the Army Battery as the following: The 20th Battalion, 67th Air Defense Artillery was constituted on January 22, 1926, in the Regular U.S. Army as Battery B, 67th Coast Artillery. It was activated on July 1, 1940, at Fort Bragg, North Carolina. It was recognized and redesignated on May 23, 1944, as Battery B, 67th Anti-Aircraft Artillery Gun Battalion, and was inactivated November 25, 1945, at Camp Shanlia, New York. Activated on January 31, 1949, at Fort Bliss, Texas. Redesignated on October 1, 1953, as Battery B, 67th Anti-Artillery Missile Battalion. Redesignated on November 1, 1957, as Battery B, Anti-Aircraft Missile Battalion. Reorganized and redesignated on September 1, 1958, as Headquarters and Headquarters Battery, 2nd Missile Battalion 67th Artillery (organic elements constituted on August 12, 1958, and activated on September 1, 1958). Battalion inactivated on August 25, 1961, at Ellsworth AFB, SD. Redesignated on September 1, 1971, as 2nd Battalion, 67th Air Defense Artillery. Assigned on September 13, 1972, to the First Infantry Division and activated at Fort Riley, Kansas.[2]

Site E-70L was located 4.9 miles west-southwest of Ellsworth AFB at 44 degrees 9 minutes 12 seconds north, 103 degrees 12 minutes 59 seconds west. As at the other three Nike-Ajax SAM sites, each consisted of four batteries, with a maximum of sixteen missiles per battery. They operated 24 hours a day, seven days a week. Their headquarters was at Ellsworth AFB. In 1958, Batteries E-20L, E-40L, and E-70L were removed as ADC active SAM sites protecting Ellsworth AFB only one year after activation. *Sioux Falls, South Dakota: United States Geological Survey*

E-40L was located 3.5 miles south-southwest of Ellsworth AFB at 44 degrees 6 minutes 14 seconds north, 103 degrees 5 minutes 54 seconds west. The housing area is immediately west of Radar Road, and was used for Air Force housing until the 1990s, then sold as private homes.

Four Nike-Ajax SAM on the training launcher at the Nike training site at Red Canyon, New Mexico. *Fort Campbell, Kentucky, U.S. Army Museum (Don F. Pratt Memorial Museum)*

150

Training for the Nike-Ajax SAM sites was continuous, including live fire exercises conducted at the Nike missile test range at Red Canyon, New Mexico. This is the only location in the United States where live or operational Nike missiles were fired. No missiles could be launched from active Nike SAM sites in the United States. These sites could only conduct dry fire training exercises. Missiles were removed from these sites as others were rotated in to maintain an inventory current with the missiles used in the training exercises at Red Canyon. Once each year, U.S. Continental and overseas Nike SAM units traveled to Red Canyon for training, firing three missiles per assigned battery. During training approximately 160 personnel assigned to the EDA attended one of the week-long training sessions, firing 80 missiles per week. Personnel from the EDA had the following results, according to the after action training report: "During the first week of training, out of six missiles launched, there were five kill hits and one hit. The second group, out of six missiles scored six kills." Note: the targets were unmanned aerial drones. Engagement rules specified that the Nike missile had to come within 105-feet of the drone for a kill and 150-feet for a hit.[3]

Nike-Hercules SAMs at Red Stone Arsenal, Huntsville, Alabama. *U.S. Army Aviation and Missile Command, Photo Archives, Historian Office*

151

At Ellsworth AFB, the remaining operational Nike-Ajax SAM site was upgraded to a Nike-Hercules facility. This was because the Nike system was so extensive and widely used, with over 3,000 launchers built, that it was extremely costly to either discard or modify. The system was cumbersome, complicated, and inefficient. A successor system had to be compatible with the existing radars, computers, auto-plotting boards, and major ground installations. The Nike system's original prime contractor, Western Electric, redeveloped the power and electric systems to meet the new specifications of what developed into the Nike-Hercules SAM. It dramatically increased the flight performance of the follow-on for the Nike-Ajax missile system. This allowed for an improved system to be installed at the former Nike-Ajax facility.

The Douglas Aircraft Company managed the U.S. Army Charlotte Ordnance Missile Plant at Charlotte, North Carolina. The new system became operational in January 1958. By June, the initial Nike-Hercules was deployed and operational at former Nike-Ajax SAM sites protecting New York City, Washington, D.C., and Chicago, Illinois.

Burnout speed was at Mach 3.35 in early production models, and increased to Mach 3.65 on follow-on variants. Except at extreme range, it was possible to intercept accurately at altitudes up to 150,000-feet. The boost motor was a four-barrel cluster built by Radford Arsenal and Borg Warner, incorporating four Hercules solid motors, each of Ajax performance. Thiokol's Longhorn Division at Marshall, Texas, cut its teeth on the high performance sustainer with a sold charge positioned on the center of gravity (CG) and a long tailpipe. AiResearch made the advanced auxiliary power unit (APU), which among other things drove the elevons on the trailing edges of the four extremely acutely swept delta wings. As with the Nike-Ajax, there were four small delta aerials to indicate the position of the guidance bay, behind which was the large conventional (fragmentation) warhead Type T-45. As a weapon of last resort, it could also be armed with a nuclear warhead in three variable or dial-able yields (low-2 Kilotons, medium-20 Kilotons, and high-30 to 40 Kilotons). There would have to have been an extreme threat to the security of the United States for the Presidential release of nuclear weapons for firing from stateside Nike SAM sites, though.

Nike Hercules specifications and performance data.
Courtesy South Dakota Air and Space Museum

Nike Hercules Specifications and Performance Data

Designation:	MIM-4
Overall length:	41 feet 6-inches
w/o booster:	27 feet
Wingspan:	6 feet 2-inches
Diameter:	2 feet 6-inches
Firing weight:	10,560 pounds
w/o booster:	5,250 pounds
Maximum speed:	Mach 3.5
Speed at burnout:	2,597 mph
Maximum slant range:	75 to 90 miles
Ground range:	When used in a tactical role it could deliver a tactical nuclear warhead out to a maximum range of 110 miles.

Note: The Republic of South Korea had this option with its deployed SAMs. The SAM could also be used in a coastal defense option.

Maximum ceiling:	100,000 feet
Booster:	Cluster of four Ajax solid propellant boosters
Sustainer motor:	Solid propellant
Warhead combinations:	Type T-45 HE fragmentation
	Initially Type W-7 nuclear until production of W31
	Type W-31 nuclear (standard)
	Yields
	Low-2 kilotons
	Medium-20 kilotons
	High-30 to 40 kilotons

Note: The nuclear warhead would allow one missile to attack and destroy closely spaced or formation flying hostile aircraft. However, it was doubtful if Soviet bombers would attack using conventional WWII bombing type formation.
Note: Warheads effective against small supersonic aircraft and standoff cruise missiles carried by penetrating Soviet bombers. The destruction of airborne cruise missiles was limited, as well as against tactical ground-to-ground missiles.

Guidance:	Alternate Battery Acquisition Radar (ABAR)
	High Power Acquisition Radar (HIPAR)

Note: The missile was continuously guided to target by the ground-based radars. The missile site was tied into the USAF's SAGE defense network, supported by the U.S. Army's Missile Master System allowing accurate acquisition of hostile targets.

152

At the peak of Hercules deployment from 1957 to 1960, Douglas Aircraft was operating not only the Charlotte Ordnance Missile Plant, but three other Nike Hercules facilities at Winston-Salem, Burlington, and Greensboro, North Carolina. Thiokol did not succeed with a self-consuming booster. General Electric provided the fusing for the nuclear warhead and much of the enhanced guidance capability, notably with Hipar (Hi-Power Acquisition Radar). This system, despite having a 43-foot aerial, could fit into three trailers, compared with 20 or 21 in the original system, and thus opened the way to a semi-mobile system. Hipar was used in early 1960, when a Hercules at WMSR destroyed an oncoming Corporal ballistic surface-to-surface missile. Later in 1960, a Nike-Hercules intercepted another Nike-Hercules at a combined closing speed of Mach 7 at an altitude of 100,000-feet. The Nike-Hercules was less effective against low-level penetrating targets attempting to sneak underneath ADC radars and missile systems, but did have a surface-to-surface capability. This was one of the reasons for sending the Nike-Hercules system to aid in the defense of the Republic of South Korea.

By 1960, the Nike-Hercules M6 and M6A1 were in operation with the U.S. Army in Taiwan, Okinawa, and in West Germany, and all the 73 battalions in the Continental U.S. had been converted from the Nike-Ajax to the Nike-Hercules. The U.S. Army's peak deployment was in 1963, when there were 134 Nike-Hercules batteries. The system, by then designated MIM-14A and B, was also in use in Belgium, Denmark, West Germany, Greece, Italy, Japan (where the Nike-Hercules missiles were produced under license by Japan's Mitsubishi Heavy Industries), the Netherlands, Norway, South Korea, Taiwan, and Turkey. Total production of the Nike-Hercules reached 25,500 missiles. In 1978, Mitsubishi Heavy Industries continued producing the non-nuclear Nike-Hercules missiles, disregarding the fact that it was considered to be an obsolete air defense system.

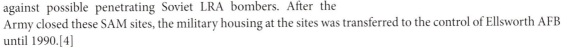

Scale model of a Nike-Hercules SAM site. Refer to the previous photograph of the Nike-Hercules operational site. The radar facility is visible in the scale model. *Carlisle Barracks, Pennsylvania: United States Institute for Military History, Photo Archives, United States Army War College*

The U.S. Army started replacing the Nike-Hercules SAM system with the Patriot SAM-D system beginning in 1975, and began phasing out the Nike-Hercules SAM system. In 1974, the final 48 Nike-Hercules SAM batteries were phased out. At Ellsworth AFB, site E-01L remained operational until 1961 as part of the reduction of air defenses in the Continental U.S. against possible penetrating Soviet LRA bombers. After the Army closed these SAM sites, the military housing at the sites was transferred to the control of Ellsworth AFB until 1990.[4]

Four Nike-Hercules in this staged photograph are at various angles of elevation at Fort Barry, San Francisco, California. *Carlisle Barracks, Pennsylvania: United States Institute for Military History, Photo Archives, United States Army War College*

By the end of the Cold War in 1989, only five Nike-Ajax SAM sites remained operational in Florida, providing defense against a possible Cuban Air Force penetration, and in Alaska, positioned along possible Soviet LRA bomber penetration routes from their arctic staging bases into the United States. The last Nike-Hercules SAM site was deactivated in 1974.[5]

153

CHAPTER SEVEN

TITAN I
INTERCONTINENTAL BALLISTIC MISSILE

Ellsworth AFB Enters the Missile Age

Atlas ICBM on display at the Strategic Air Command Museum. *Lt. Col. George A. Larson, USAF (Ret.)*

In the late 1950s, the Convair Division of General Dynamics was working on America's first intercontinental ballistic missile (ICBM), referred to as the Atlas. It was designed to deliver a thermonuclear warhead into targets inside the Soviet Union out to a range of 5,000 miles from its launch site. The USAF awarded a production contract to Convair in January 1955. It was a liquid-fueled missile, stored in a relatively soft, partially below ground installation. It served with SAC until April 1965. It also served as a nuclear ICBM response/deterrent until replaced by the larger and heavier Titan I ICBM.

In October 1960 – two years before the Cuban Missile Crisis – Ellsworth AFB entered the strategic missile age with the activation of the 850th Strategic Missile Squadron, initially assigned to the 28th Ballistic Missile Wing (BMW). For more than a year this squadron prepared for the employment of the Titan I, which arrived in 1962, shortly after the activation of the 44th Strategic Missile Wing (SMW) in January 1962. At that time, Headquarters SAC named the 44th SMW a host wing at Ellsworth AFB.[1]

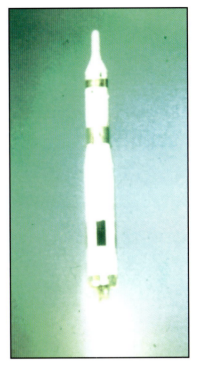

Titan I ICBM test launch.
Ellsworth Air Force Base, 28th
Bomb Wing Historian

SAC's next liquid-fueled ICBM was the Titan ICBM. Martin-Marietta began design work on the large and heavy warhead ICBM (designated the SM-68/LGM-25A) in 1955, at about the time the Atlas went into production. The Titan I was a two-stage ICBM with a framework constructed of an aluminum-copper alloy. Supported by fuel and oxidizer tanks, a 10-foot diameter booster stage, two engines providing a combined thrust of 300,000 pounds, and an eight foot diameter sustainer stage with 80,000 pounds of thrust, the ICBM had a range of 6,300 miles. This allowed the missile to deliver a four-megaton warhead able to destroy hardened targets in the Soviet Union. This was necessary because its accuracy, or CEP, was 4,600-feet.

Re-entry Vehicle (RV):	Mk 4 re-entry vehicle
Length:	14 feet
Weight:	3,800 pounds
Warhead:	W38-4 (four megaton yield)
RV contractor:	AVCO Corporation of Wilmington, Massachusetts
Airframe contractor:	Glen L. Martin Company of Denver, Colorado
Vernier engines:	(two) each with 1,000 pounds of thrust
Burn time:	50 seconds
Guidance:	Radio-inertial
Contractor:	Bell Telephone Laboratories of Allentown, Pa.
Guidance computer:	Ground based within launch complex
Contractor:	Remington Rand UNIVAC of St. Paul, Mn.
Maximum flight altitude:	541 miles above the earth's surface
Maximum flight time:	Thirty-three minutes
Range:	6,300 miles
Accuracy	4,600 feet, circular error probable (CEP)
Production cost:	(1962 dollars) $1,500,000.00
Time to build:	(initial) 75,000 hours
	(terminal) 19,000 hours
Rate:	One missile per week
Number produced:	158
Test launches:	64
Operational	54
Spares:	40

155

Titan I ICBM Specifications and Performance

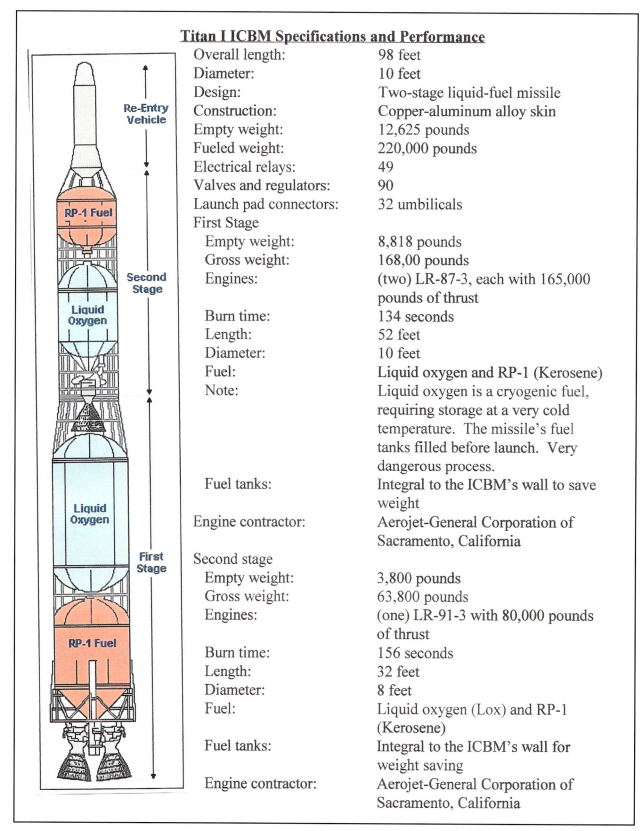

Overall length:	98 feet
Diameter:	10 feet
Design:	Two-stage liquid-fuel missile
Construction:	Copper-aluminum alloy skin
Empty weight:	12,625 pounds
Fueled weight:	220,000 pounds
Electrical relays:	49
Valves and regulators:	90
Launch pad connectors:	32 umbilicals
First Stage	
Empty weight:	8,818 pounds
Gross weight:	168,00 pounds
Engines:	(two) LR-87-3, each with 165,000 pounds of thrust
Burn time:	134 seconds
Length:	52 feet
Diameter:	10 feet
Fuel:	Liquid oxygen and RP-1 (Kerosene)
Note:	Liquid oxygen is a cryogenic fuel, requiring storage at a very cold temperature. The missile's fuel tanks filled before launch. Very dangerous process.
Fuel tanks:	Integral to the ICBM's wall to save weight
Engine contractor:	Aerojet-General Corporation of Sacramento, California
Second stage	
Empty weight:	3,800 pounds
Gross weight:	63,800 pounds
Engines:	(one) LR-91-3 with 80,000 pounds of thrust
Burn time:	156 seconds
Length:	32 feet
Diameter:	8 feet
Fuel:	Liquid oxygen (Lox) and RP-1 (Kerosene)
Fuel tanks:	Integral to the ICBM's wall for weight saving
Engine contractor:	Aerojet-General Corporation of Sacramento, California

Titan I ICBM specifications and performance. *Information from SAC.com*

156

Titan I ICBM first stage. It had an empty weight of 8,818 pounds, and a fueled weight of 168,000 pounds. The two main engines (LR-87-3) produced 327,000 pounds of thrust for 138 seconds. It had a diameter of 10 feet and a length of 52 feet. It was fueled by liquid oxygen (Lox)/RP-1 kerosene. The first stage engines were produced by Aerojet. The missile was photographed at the South Dakota Air and Space Museum's restoration facility on Ellsworth Air Force Base. The missile has never been displayed at the museum. *Lt. Col. George A. Larson, USAF (Ret.)*

View from the end of the first stage of the Titan I. Its fuel lines and engine detail are visible. As with all liquid fuel rockets, there was a lot of what engineers referred to as plumbing. The ignition chamber had to withstand a pressure of 580 PSI, which produced an initial thrust to weight ratio of 89 to 20. *Lt. Col. George A. Larson, USAF (Ret.)*

Titan I second stage. The Nike-Ajax SAM, along the missile's side, is currently on display at the South Dakota Air and Space Museum. It had an empty weight of 3,800 pounds and a loaded weight of 63,800 pounds. Its single LR-91-3 engine provided 80,000 pounds of thrust with a 210-second burn time. It had a diameter of 8 feet and a length of 32 feet. It was also fueled by liquid oxygen (Lox)/RP-1 kerosene. The rocket chamber had to withstand a pressure of 650 PSI, producing a thrust to weight ratio of 61 to 51. The white cone shape at the top of the missile held the warhead. *Lt. Col. George A. Larson, USAF (Ret.)*

Titan I ICBM

The final stage had two vernier engines with a total of 2,000 pounds of thrust for exoatmospheric operations prior to the glide phase or reentry of the warhead to the target.

The 98-foot long Titan I was stored inside a 115-foot deep and 40-foot diameter hardened concrete silo, with each missile complex containing three silos within a secured area. Later, the Minuteman I, II, and III ICBM systems had a hardened underground command/launch capsule physically separated from 10 underground silos, which were also geographically separated so one Soviet ICBM could not destroy the entire missile battery. The Titan I used the same liquid propellant as the Atlas. The Titan I could not be fired from inside its silo and had to be raised to the surface by an elevator, which it sat on. Once raised to the surface, on site personnel had to fuel the missile with oxidizer and oxygen. This was a dangerous procedure that was only eliminated with the development of solid-fuel technology used on the Minuteman ICBM. The Titan I used a radio-inertial guidance package that followed a radio beam during powered flight, providing guidance updates. It could have possibly been jammed by Soviet ECM or by near earth orbit disturbances from the detonation of a nuclear warhead along the path of incoming U.S. ICBMs, thus creating an intense burst of electromagnetic pulses, or EMPs.

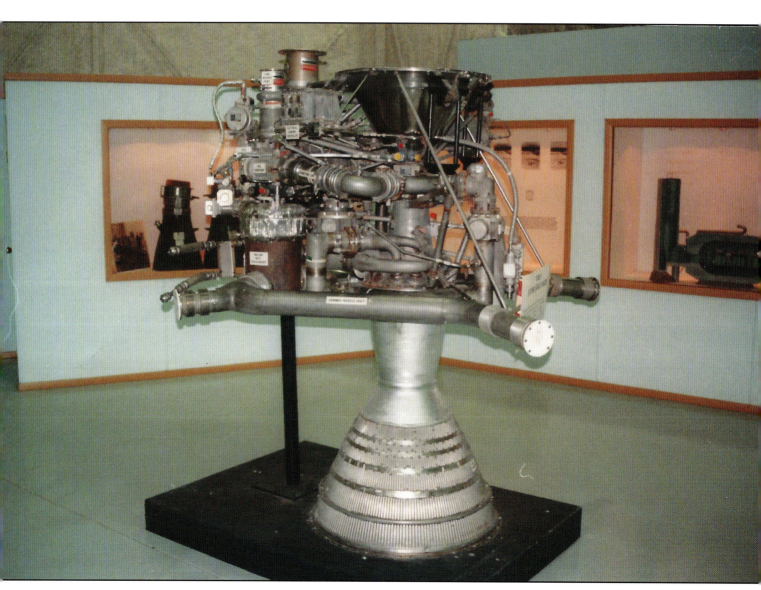

Second stage engine of the Titan I. Engine on display inside the museum building at the South Dakota Air and Space Museum. The date of the first flight for this type of rocket engine was August 10, 1962. It generated 80,000 pounds of thrust to propel the second stage and its nuclear warhead to a maximum speed of Mach 15. The rocket engine consumed 250 pounds of fuel and oxidizer per second. The numerous fuel lines that produced the 80,000 pounds of thrust was a complicated engineering structure and required rigorous quality control procedures and constant maintenance checks when the missile became operational. *Lt. Col. George A. Larson, USAF (Ret.)*

158

The first test flight of a Titan I was on February 6, 1959, and the first operational missile squadron was equipped with nine missiles on alert at Lowry AFB, Denver, Colorado, on April 18, 1962, under operational control of the 703rd SMW. The Titan I remained on alert with SAC from April 1962 to June 25, 1965. In addition to the two squadrons at Lowry AFB, other squadrons operated at: Ellsworth AFB, SD; Beale AFB, California; Larson AFB, Washington; and Mountain Home AFB, Idaho.[2]

Contractors for the Titan I project at Ellsworth AFB were led by Leavell-Scott & Associates, which represented a consortium of eight partners. On December 8, 1959, the consortium was awarded a contract to build three missile complexes, one each at New Underwood, Hermosa, and Sturgis. U.S. Army Corps of Engineers oversight initially came from the Omaha Engineer District. Ten months into construction, responsibility was transferred to the Corps of Engineers Ballistic Missile Construction Office at Los Angeles, California. The initial cost estimate for the contract was 47.2 million dollars. By March 1962, cost estimates escalated to 64 million dollars, a 31 percent increase. Additional costs resulted from 265 modifications requested by the Air Force to the original contract.

Finding skilled laborers proved to be a challenge. Eventually many workers were brought into the area, and as a consequence the construction project suffered high worker turnover rates. Labor management relations were amicable. At construction's peak, 2,500 laborers and skilled technicians worked on the three launch complexes. There were fifteen short labor work stoppages, most lasting less than one day. A labor strike at General Electric in October 1960 delayed receiving terminals, which pushed back construction and the completion date. There was a major technical difficulty at these first generation liquid-fueled ICBM sites, especially in the northern Black Hills construction site at Sturgis. The installation of the propellant loading system was expensive. Excessive ground water at Complex IC at Sturgis required an additional $500,000 to fix. Ground water penetrated the underground facility, especially the silos, which could damage the electrical equipment and could not be allowed to penetrate the liquid oxygen and oxidizer storage area. Pipes could corrode and lead to explosions. This was an expensive fix not included in the original contract. Also, difficulties between the Site Engineer and Site Activation Task Force (SATAF) arose after construction completion. Because SATAF refused to assume responsibility for site management until practically every item listed on the installation's punch list (items requiring repair or correction to reach full operational status) was in full working condition, contractors remained at the three sites for months, maintaining each site's equipment after installation, and again increasing final installation costs.

With Titan I sites under construction around Ellsworth AFB, the Air Force activated the 850th Missile Squadron December 1, 1960. About the same time, construction began on installations for the second generation ICBM, the Minuteman (to be discussed in the next chapter).[3]

Site excavation used an open cut to a depth of 163-feet. Construction of underground facilities was of reinforced concrete and structured steel line access tunnels. An unusual requirement was the blast proofing of elements incorporated into the work, with major mechanical and electrical elements shock-mounted to withstand nearby nuclear detonation, except a direct hit. The heavy construction phase moved a lot of earth and the completion of heavy steel and concrete structure deep underground.

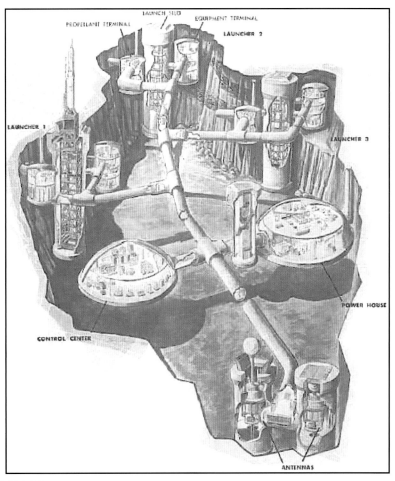

Drawing of an underground Titan I missile complex. The silo on the left has a missile raised to the surface on its elevator. Underground maintenance/access tubes connect the three silos to the command center and power house. Each silo had its own propellant storage area. *Drawing: SAC.com*

159

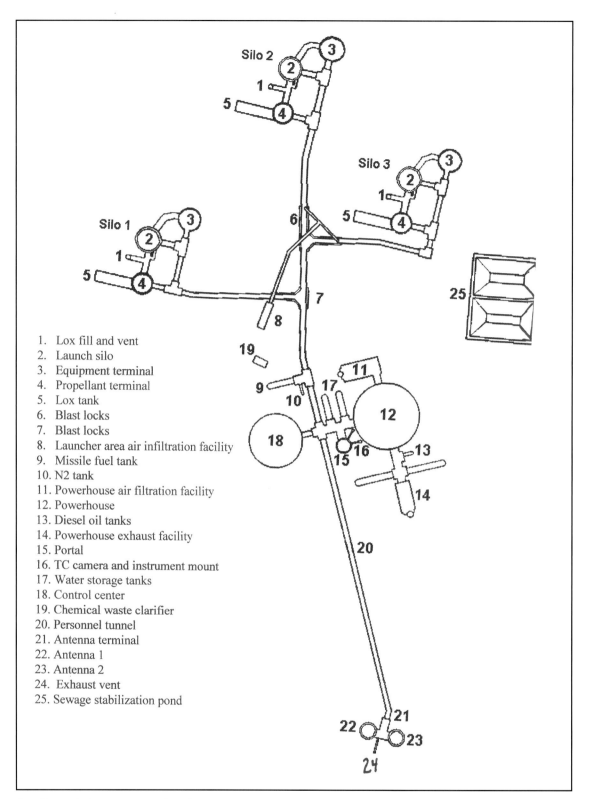

1. Lox fill and vent
2. Launch silo
3. Equipment terminal
4. Propellant terminal
5. Lox tank
6. Blast locks
7. Blast locks
8. Launcher area air infiltration facility
9. Missile fuel tank
10. N2 tank
11. Powerhouse air filtration facility
12. Powerhouse
13. Diesel oil tanks
14. Powerhouse exhaust facility
15. Portal
16. TC camera and instrument mount
17. Water storage tanks
18. Control center
19. Chemical waste clarifier
20. Personnel tunnel
21. Antenna terminal
22. Antenna 1
23. Antenna 2
24. Exhaust vent
25. Sewage stabilization pond

General layout of a Titan I ICBM site at Ellsworth AFB. *Drawing: SAC.com*

Each Titan I missile complex contained:

1. Three launch positions, each with one missile silo, supporting equipment terminal, propellant terminal, and propellant system.
2. One guidance facility with two antennas.
3. One hardened, underground control center.
4. Utility and service facilities, including an underground power facility for electrical generation to power the missile site, heating, ventilation, and air conditioning equipment.

Titan I ICBM

5. Interconnecting underground tunnels for utility distribution and personnel access.
6. Utilities, including road access, water storage and distribution, sanitary system, exterior lighting, entrance patrols, air intakes, and exhaust system.
7. Wells for independent water system operations.
8. Primary site access road and security fencing.[4]

The underground control center consisted of three levels housed in a concrete structure covered with a 1/4-inch thick, seam welded, steel radiation shield. The control center's floors were suspended on eight large springs to provide limited shock protection. Level number 1, the top command level, contained the on-alert crews' sleeping, eating, and rest/wash room facilities. Level number 2 was the control center. Level number 3 contained electrical equipment, power supplies, emergency electrical batteries, communications transmitters, air conditioning equipment, and the crews' escape hatch access.[5]

The Titan I Combat crew inside the underground command center had responsibilities to maintain and supervise all responsibilities of the site's operation. Their primary function was the alert operations of the complex: on-base pre-departure/assumption of alert briefings, Wing missile alert status, current weather, expected alert weather forecast, travel to the missile site, alert changeover, and assumption of 24-hour alert.[6]

The control center was staffed by two officers (required under control and security of nuclear weapons, mandatory for verification and authorization by the two officers who controlled the sealed launch codes for launch and release of nuclear weapons) and two enlisted personnel. The personnel stood a 24-hour alert shift, relieved by another alert missile crew. The Missile Combat Crew Commander (MCCC) was the Officer-in-Charge of the Missile Site, responsible for its safety, maintenance, and operational readiness. The Deputy Missile Combat Crew Commander (DMCC) was second in command, and responsible for site communications and monitoring the two enlisted maintenance personnel's location(s). The Ballistic Missile Analyst Technician (BMAT), usually referred to as the Missile System Analyst Technician (MSAT), was responsible for on-site maintenance of electronics, launch system electrical systems, missile airframe, and associated electronics. The Missile Facilities technician (MFT) was responsible for maintenance of the missile site facilities, heating and air conditioning systems, plumbing and water systems, utility air compressors, hydraulic systems, and inspection of the entire facility during each alert period. If on-site maintenance personnel could not correct the problem the crew called for a top side maintenance crew to repair the problem.

Double access gate to the former Titan II ICBM site, now the only Titan II ICBM museum in the United States at Green Valley, Arizona. The gate security was similar to that used at the larger Titan I three missile silo complexes, versus the one silo and command bunker at the Titan II ICBM sites. *Lt. Col. George A. Larson, USAF (Ret.)*

Titan I ICBM

When a launch message was received, both the MCCC and DMCC copied and jointly verified the message. They opened the red Emergency War Order (EWO) safe, removing the authentication package. The two officers independently decoded the message containing the authentication code (This authorized the release of nuclear weapons by the President of the United States), time of launch (pre-determined launch time to provide mutual support for ICBM and SLBM launches, preventing fratricide or the deterioration of nuclear weapons in the air from nuclear bursts on the ground), and the BVL unlock code (a six-digit code combination beginning the fuel release sequence).

MCCC's console on display inside the Titan II ICBM museum at Green Valley, Arizona. The MCCC's launch console contained the Commander's launch key switch. The DCCC's launch key switch was separated from the Commander's key so that one person could not insert both keys and turn them at the same time. The MCCC's responsibilities included: site commander, safety of personnel at the site, responsible for operational site decisions, enforcement/following all technical order manuals, monitoring launch control complex and facilities console (LCCFS), verification and authorization of nuclear missile and launch/execution order, and launching missiles with the DMCC. *Lt. Col. George A. Larson, USAF (Ret.)*

DMCC's console. The DMCC's responsibilities were to monitor communications, on-site location of personnel, Alternate Launch Officer's Console (ALOC), primary alert systems, VHF/UHF/SSB communications console, maintenance operations, and relay communications with Command Centers, on-site maintenance, and airborne Command & Control aircraft. The ALOC contains the DMCC's launch key Switch. The unit is on display inside the Titan II ICBM museum at Green Valley, Arizona. *Lt. Col. George A. Larson, USAF (Ret.)*

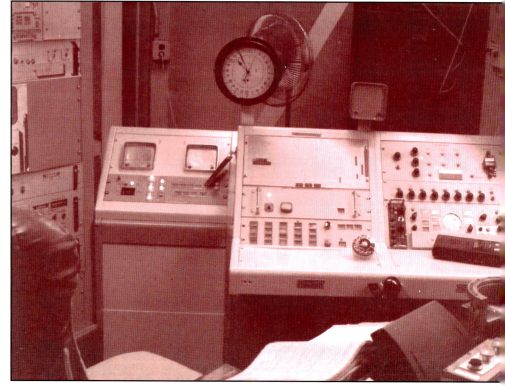

After receiving a verified launch order both officers inserted their launch keys simultaneously, then turned both at the same time, initiating the missile launch sequence and enabling the BVL to be unlocked, onboard missile batteries to be activated (taking approximately 28 seconds to reach the proper voltage), and activation of APS power (100 percent running on internal power).[7]

BMAT and MFT panels. The BMAT was responsible for missile site electronics; launch system electrical; missile airframe; electronics to include guidance system, warhead, and engines; supervise Control Monitor Group and missile guidance alignment-checkout group; and inspected all missile site systems once per day (referred to as Daily Shift Verification). The Missile Facilities Technician was responsible for site heating and air conditioning, power generation and control systems, plumbing, water systems, utility air compressors, and hydraulic systems, and inspected all facility systems once per day. The MFT also monitored the command center's Facility Power Control Board Power Distribution Control. *Lt. Col. George A. Larson, USAF (Ret.)*

The first operation Titan I ICBM was placed on alert on June 22, 1962. The 850th SMS controlled maintenance and operations of the missiles. The first Titan I ICBM was pulled out of its silo on January 7, 1965. The last Titan I was removed from its silo in February 1965, turning over its nuclear alert deterrence to the Minuteman I ICBM solid-fuel missile system.[8]

Interior of the access corridor to the Titan II inert ICBM on display at the Titan II Museum. The corridor is similar to the corridors constructed at the Titan I ICBM launch complex. *Lt. Col. George A. Larson, USAF (Ret.)*

163

Extreme care had to be taken when performing maintenance on the missile and its associated silo equipment. On December 5, 1964, a retrorocket located below a Titan I ICBM fired while two repairmen were working nearby, sending the Reentry Vehicle (RV) crashing down to the bottom of the silo. The Arming and Fuzing/Attitude Control containing the RV's batteries was torn loose on impact, removing all sources of power from the RV and causing considerable damage. The missile's safety devices operated properly and did not allow the warhead to become armed.[9]

On May 16, 1964, Secretary of Defense Robert McNamara directed an accelerated phase out of the Atlas and Titan I ICBMs. Titan I ICBMs were removed from their silos and transported to Mira Lomas AFS for storage in the spring of 1965. The Titan I ICBMs were temporarily stored for nine months. There was no need for the surplus Titan I ICBMs for use as launch vehicles or anti-missile targets. Consequently, in the spring of 1966 Aerospace Corporation advised against continued storage. The Atlas ICBMs were available for use as space launch vehicles in greater numbers than the Titan I ICBMs. Because the Atlas ICBMS removed from their launch facilities had already been adapted as space launch vehicles previously, it did not make economic sense to modify the eighty-five Titan I ICBMs as space launch vehicles. There are fifteen Titan I ICBMs on static display in the United States.[10]

Environmental suit required during fueling operations of the Titan I and II ICBMs. The oxidizer fumes were extremely toxic and required full body protection. Mannequin on display at the Titan II Museum at Green Valley, Arizona. *Lt. Col. George A. Larson, USAF (Ret.)*

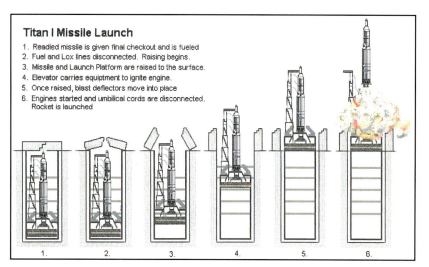

Titan I Missile Launch

1. Readied missile is given final checkout and is fueled
2. Fuel and Lox lines disconnected. Raising begins.
3. Missile and Launch Platform are raised to the surface.
4. Elevator carries equiptment to ignite engine.
5. Once raised, blast deflectors move into place
6. Engines started and umbilical cords are disconnected. Rocket is launched

Drawing of Titan I ICBM launch sequence (numbers 1 through 6). *SAC.com*

There are fifteen Titan I ICBMs on static display in the United States. One is on display at the National Museum of the United States Air Force (second from left in photo). *Lt. Col. George A. Larson, USAF (Ret.)*

164

Titan I ICBM

Ground photo sequence of a Titan I ICBM launch. *SAC.com*

Titan I ICBM Site Number 1, 850-A, located 31 miles from Ellsworth Air Force Base and 3 miles northwest of Wicksville, SD, at 44-08-12N and 102-37-03W. Silo 1 is at the top left of the photo, and clockwise are silos 2 and 3. *Sioux Falls, South Dakota: U.S. Geological Survey*

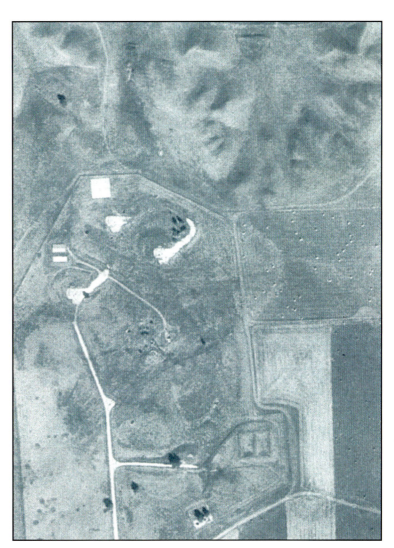

165

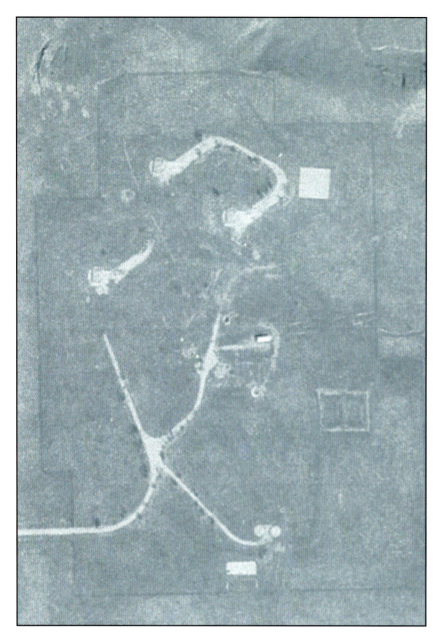

Titan I ICBM Site Number 2, 850-B, located 21 miles south-southeast of Ellsworth Air Force Base and five miles southeast of Hermosa, SD, at 43-43-38N and 103-08-47W. *Sioux Falls, South Dakota: U.S. Geological Survey*

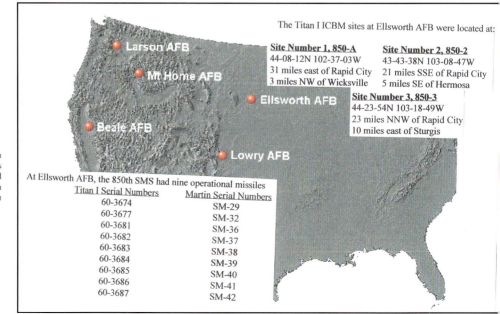

The Titan I ICBM sites at Ellsworth AFB were located at:

Site Number 1, 850-A
44-08-12N 102-37-03W
31 miles east of Rapid City
3 miles NW of Wicksville

Site Number 2, 850-2
43-43-38N 103-08-47W
21 miles SSE of Rapid City
5 miles SE of Hermosa

Site Number 3, 850-3
44-23-54N 103-18-49W
23 miles NNW of Rapid City
10 miles east of Sturgis

At Ellsworth AFB, the 850th SMS had nine operational missiles

Titan I Serial Numbers	Martin Serial Numbers
60-3674	SM-29
60-3677	SM-32
60-3681	SM-36
60-3682	SM-37
60-3683	SM-38
60-3684	SM-39
60-3685	SM-40
60-3686	SM-41
60-3687	SM-42

Map with annotation added on site coordinates and missiles assigned to the three launch complexes. *SAC.com*

Titan I ICBM Site Number 3, 850-C, located 23 miles north-northwest of Ellsworth Air Force Base and 10 miles east of Sturgis at 44-23-54N and 103-18-49W. *Sioux Falls, South Dakota: U.S. Geological Survey*

Titan I ICBM

The two-stage Titan I ICBM, together with the Atlas ICBM, comprised SAC's first generation of liquid-fueled missiles. Operation Titan Is contained an all-inertial guidance system to direct the nuclear warhead to the target. Liquid propellants for Titan I's Aerojet rocket engines were kerosene fuel and liquid oxygen. The HGM-25A, formerly known as the SM-86 (or B-68), was the first USAF ICBM to be placed in hardened underground silos for protection against enemy attack. However, they had to be lifted from their silos to the surface by elevator prior to launching. The Air Force launched its first test Titan I on February 6, 1959, and in April 1962, the SAC squadron of nine Titan Is was declared operational. Eventually squadrons of Titan Is were deployed at five different bases in the western United States. By 1965, however, Titan Is were being phased out in favor of Titan IIs, which offered greater range and payload, and launched from within their silos. Modified Titan IIs were used to launch Gemini astronauts into low Earth orbit for training and preparations for the Apollo missions to the moon and back. The larger and more versatile Titan II, developed from the Titan I, was one of the Air Force's heavy launch vehicles for military payloads launched into space, most importantly intelligence satellites.[11]

Titan I ICBM raised from its underground silo on its elevator launcher. *Ellsworth Air Force Base, 28th Bomb Wing Historian*

Titan I ICBM

CHAPTER EIGHT

MINUTEMAN I AND II INTERCONTINENTAL BALLISTIC MISSILE

Boeing Aircraft provided this model of the Minuteman II ICBM, which was displayed at Ellsworth AFB and is currently on display at the South Dakota Air & Space Museum. *Lt. Col. George A. Larson, USAF (Ret.)*

SAC wanted a solid-fueled ICBM that could be produced and deployment in large numbers to counter and balance the expanding number of Soviet ICBMs being deployed by the Soviet Strategic Rocket Forces (SRA). The Titan II storable liquid-fueled ICBM was to be retained until June 1987. Liquid fuel was dangerous, and high maintenance would end the operational life of the heavy ICBM, as its mission was picked up by the more accurate Minuteman ICBM. The small size of a solid-fuel missile made it ideal for a silo-launched ICBM spread over a large area to complicate Soviet targeting, and it had a near instant launch capability. A solid-fuel ICBM could be launched under receipt or warning of an incoming Soviet ICBM attack through satellite and ground based early warning systems in less than five minutes. In March 1958, the USAF reviewed seven different missile aerospace and aircraft design/production proposals for a three-stage, solid fueled, 68-foot long silo launched ICBM. In October 1958, the Boeing Aircraft Company's ICBM design proposal was selected by the Air Force. Boeing engineers created a subcontractor team of six companies to build the Minuteman I, followed by its upgraded variants, the Minuteman II and III.[1]

Minuteman II ICBM on display at the South Dakota Air & Space Museum. *Lt. Col. George A. Larson, USAF (Ret.)*

Minuteman I and II ICBM

LGM-30A Minuteman I (MMI) ICBM Specifications

Length	53 feet 8-inches
Diameter	5 feet 6-inches (first stage)
Weight	65,000 pounds at launch
Max speed	15,000 mph
Max ceiling	700 miles
Max range	6,300 miles
Propulsion	(three stages and post-boost phase)
First stage	Thiokol MSS solid-fueled rocket
Weight	210,000 pounds
Second stage	Aerojet General M56 Solid-fueled rocket
Weight	60,000 pounds
Third stage	Hercules M57 Solid-fueled rocket
Weight	35,000 pounds
Warhead	(One) W-59 Thermonuclear weapons
Yield	1.2 megatons (1.2 MT)
RV	MK-5 RV[2]

The Minuteman I's first powered flight was on February 1, 1961. It launched from Cape Canaveral, Florida, with the dummy warhead hitting its designated target 25 minutes later at a down range of 4,600 miles into the Atlantic Missile Test Range. The Minuteman I, approximately one-half the size of the Titan II, could destroy a hardened target with improved CEP.

One of many Launch Control Centers (LCC) under construction around Ellsworth AFB. The excavation was deep, and the structure hardened by steel and concrete. The workers are preparing the steel bars for the cylindrical shape of the underground LCC. The LCF's construction was standardized at all Minuteman ICBM complexes to reduce construction and maintenance costs, as well as construction time to allow the maximum number of ICBMS to be deployed in the shortest time to counter the Soviet ICBM threat, providing a credible and survivable nuclear deterrence that if necessary could ride on the first attack wave (strike) of Soviet ICBMs. *Ellsworth Air Force Base, 28th Bomb Wing Historian*

171

Silo construction started with earth scrapers digging a twelve-foot-deep trench, deepened to 32 feet by backhoes. A crane fitted with a clamshell excavation bucket dug the silo to a depth of 84 feet. A heavy crane lowered a steel cylinder 12 feet in diameter into the excavation. Reinforced concrete was poured around the steel cylinder, and allowed to harden, then it was back-filled with dirt, covered by more hardened concrete, and fitted with a ballistically activated door mounted on three railroad type travel/guide rails that allowed access into the silo and opened for a launch. *Lt. Col. George A. Larson, USAF (Ret.)*

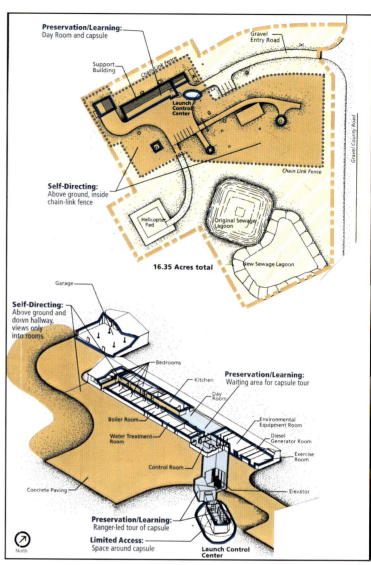

Line drawing of D-1 LCF/LCC. *Drawing: Minuteman Missile National Historic Site, United States Department of the Interior, National Park Service, February 2002*

Entrance to LCF D-1, exit 131, east of Wall, SD. The above ground building was the launch facility, and held the crew elevator inside for access to the LCC below. All support crews for the 10 ICBMs in the Delta launch complex worked out of the building, as well as security personnel and any maintenance crew personnel that had to remain overnight when working on one of the missiles, guarded by security police. In periods of bad weather, especially during the winter months, many missile personnel stayed in the building. They were able to eat, take a shower, and sleep in one of the bunk beds set up inside. Each silo is connected to its LCF by an underground trench several miles long containing shielded cables. Communications backup was provided by hardened VHF radio receivers, an above ground mast antenna receiver, and alternate buried cables from other LCFs, providing redundant command and control. Each LCF was an above ground building, with the LCC 31 feet below. *Lt. Col. George A. Larson, USAF (Ret.)*

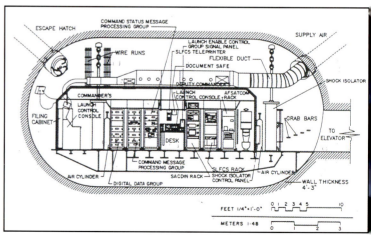

Engineer drawing of underground LCC. *Drawing: Minuteman Missile National Historic Site, United States Department of the Interior, National Park Service, February 2002*

Minuteman I and II ICBM

A large excavation was dug by earth scrappers for the external shell, a 29-foot diameter, 54-foot long structure built out of reinforced concrete four feet thick poured around a steel framework. Inside a one-quarter inch hardened shell was added. Suspended inside the plate steel shell was a box-like acoustical enclosure containing two Launch Control Officer (LCO) consoles, communications, missile monitoring equipment, and crew accommodations.

Access/entrance into the LCC is protected by an eight ton blast proof steel and concrete door. The blast door at D-1 LCC was painted in a knockoff of a Domino's Pizza box, with the top half of the box retained, with the addition of Minuteman ICBM and Minuteman II lettering. Above the box is the statement *World-Wide Delivery In 30 Minutes Or Less* and below *Or Your Next One Is Free*. Interestingly, SAC inspectors never asked that the blast door be repainted to its original gray color. It was retained at the Minuteman Missile Historic Site. *Lt. Col. George A. Larson, USAF (Ret.)*

LCC Deputy LCO Commander's position. *Lt. Col. George A. Larson, USAF (Ret.)*

The LCC had two consoles, each with a swiveling high-backed aircraft type chair equipped with a seat belt and shoulder harness. The Deputy LCO Commander's chair slid along a double rail mounted directly to the LCC's steel floor to permit monitoring of the equipment. The LCO Commander's location was positioned at the far end of the LCC, opposite the entrance, with instruments to monitor the operational and security status of the flight's ten missiles, as well as those from other flights, to provide redundant launch and control capabilities. The Deputy LCO Commander's console area contained the radio, telephone, decoding equipment, and other communications equipment.

LCC LCO Commander's position. *Lt. Col. George A. Larson, USAF (Ret.)*

View of the bank of electronics the Deputy LCO Commander had to monitor. The aircraft chair with the steel rails allowed the Deputy LCO to slide along the floor and monitor all the equipment. The blast door is behind the chair. This photo was taken on the dedication of the Minuteman Missile Historic Site on September 16, 2002. *Lt. Col. George A. Larson, USAF (Ret.)*

Each of the LCC's two consoles had a spring-loaded, key-operated missile launch switch. The two keys were secured inside a red, double-locked (LCO Commander's and Deputy LCO Commander's padlock, visible in the photograph to the left, the only red box on the wall) steel box located above the Deputy LCO Commander's console. An audio alarm alerted the two LCOs to an incoming missile launch warning message. After verifying the coded launch message, both officers must agree on the message's formatted content, and they would unlock their combination locks on the red box, removing the missile launch keys. Each officer took a key to his console, inserting it into the launch switch, then continue the proper launch sequence. Both officers strapped themselves into their chairs and completed the final missile launch sequence. At the proper time both would turn their keys simultaneously, launching the flight's ten missiles under their immediate control. Because the two LCO consoles containing the respective launch switches were 12-feet apart, this provided two-officer control, so that one officer could not independently launch the flight's missiles. Before the flight's missiles could be launched, the proper Nuclear Weapons Release Authorization Message had to be authenticated and verified by another LCC or airborne command post (the airborne command post's operations

will be discussed in the next chapter). This added another layer of positive nuclear control to prevent an unauthorized missile launch. Once launched, a Minuteman ICBM, in approximately 30 minutes, would deliver a warhead to its assigned target. Under START, ICBMs are limited to one nuclear warhead. Current upgrades give the Minuteman III near zero-zero impact accuracy at intercontinental range.

Eventually operational procedures were modified so that one of the LCOs could rest during the 24-hour shift. A bunk was installed with a curtain to close out lights from the LCC. He would rise to his position at the end of the rest period, at the sound of an alarm, or at the request of the on-duty and awake LCO. *Lt. Col. George A. Larson, USAF (Ret.)*

Even though the LCC receives its day-to-day electrical and air conditioning from the LCF above, it could be completely sealed off, operating independently on emergency power, air conditioning, and oxygen equipment. This isolation was only initiated if an incoming Soviet missile attack was detected.

D-1 LCF. *Lt. Col. George A. Larson, USAF (Ret.)*

Currently, U.S. early warning detection systems not only monitor a possible Russian ICBM launch, but also from the People's Republic of China and North Korea. Other possible threats are constantly reviewed and evaluated. The isolation of the LCC allowed operations to be continued in a nuclear, post-attack environment. The whole ground of the LCF is secured by a chain link fence monitored by video cameras, with entry through a remote-controlled, chain-link vehicle-size sliding gate. The support building provides lodging and cooking facilities, security control, and environmental and electrical equipment for the LCC.

Security team's Peacekeeper vehicle on display at Ellsworth AFB's former Minuteman Training Facility. Currently still located on Ellsworth AFB under control of the South Dakota Air & Space Museum, and is open to museum visitors who pay for a guided tour of the base by bus. *Lt. Col. George A. Larson, USAF (Ret.)*

The Minuteman III ICBM is still operational today. As of January 2013, there are LGM-30 MMIIIs on alert at F.E. Warren AFB, Wyoming, and Minot AFB, North Dakota. These missiles will remain in the Air Force inventory until 2030 at a total of 500 MMIII ICBMs, pending development and replacement of new SSM.

The LCT is manned by two security controllers, (two) two-person armed response teams (security police), a cook, and a facility manager. The two LCOs in the LCC remain on duty for 24 hours before being relieved by two LCOs. Support personnel work a three-day shift. The LCF support area contains a vehicle storage building for two security vehicles and a front end loader for snow removal at the LCC, the flight's ten silos, two hardened high-frequency (HF) antennas, one ultra-high frequency (UHF) antenna, and one underground, survivable low-frequency (LF) communication system antenna to provide continuous communications with the surviving National Command authorities during a nuclear attack. A helicopter landing pad is located outside the security fence for emergency support.

175

Bunk beds inside D-1 LCF for support personnel and other missile personnel who had to remain overnight at the facility. *Lt. Col. George A. Larson, USAF (Ret.)*

The Air Force began building its first Minuteman ICBM field March 16, 1961, at Malmstrom AFB, Great Falls, Montana. The first ten silos and the flight's LCF and underground LCC became operational on October 27, 1962, at the height of the Cuban Missile Crisis. In October 1962, the Soviet Union and the United States neared the brink of thermonuclear war. Fortunately, President John Kennedy and Soviet Premier Nikita Khrushchev found a way for both super powers to back away from a nuclear confrontation. By 1967 SAC fielded and controlled 1,000 Minuteman IIs, reduced from the original planned deployment of 1,500. Minuteman I ICBMs were deployed at Malmstrom AFB, Montana; Ellsworth AFB, South Dakota; Minot and Grand Forks AFB, North Dakota; Whiteman AFB, Missouri; and Francis E. Warren AFB, Wyoming.[3]

44th Missile Wing shield of *Aggressor Beware* on display at the South Dakota Air & Space Museum. *Lt. Col. George A. Larson, USAF (Ret.)*

Squadron shield from Ellsworth AFB Minuteman II ICBMs on display at the South Dakota Air & Space Museum. *Lt. Col. George A. Larson, USAF (Ret.)*

Shield removed from Ellsworth AFB on Space Command. On display at the South Dakota Air & Space Museum. *Lt. Col. George A. Larson, USAF (Ret.)*

SAC shield removed from 28th BW Headquarters, now on display at the South Dakota Air & Space Museum. *Lt. Col. George A. Larson, USAF (Ret.)*

Minuteman ICBM launch complex. The site was physically secured by a wire fence and electronic security. The silo blast door is visible because of the rails which allowed the blast door to open for maintenance or prior to an operational launch. The entire area was lighted at night and protected by layers of entry doors requiring combinations and intrusion alarms. *Ellsworth Air Force Base, 28th Bomb Wing Historian*

Minuteman I and II ICBM

The 44th Missile Wing traces its roots back to WWII, when the unit was the 44th BG, nicknamed the "Flying Eight Balls." The 44th BG was first activated in January 1941, and was trained on and equipped with the B-24 aircraft. The group became fully operational in February 1942. During the 44th BG's tour in England the group bombed submarine installations, industrial sites, airfields, harbors, and shipyards in France and Germany. The unit earned the Distinguished Unit Citation after its participation in the attack on the naval installation at Kiel, Germany, on May 14, 1943. In late June 1943, a large detachment of the 44th BG was moved to North Africa to take part in the invasion of Sicily. The detachment also participated in the low-level raids on the oil fields at Ploesti, Romania, on August 1, 1943. For this raid the unit received its second Distinguished Unit Citation for its participation, and the Commander, Col. Leon Johnson, was awarded the Medal of Honor. In October 1943 the Group was reformed in England, where it remained until the end of the war. During the last year of the war the unit participated in the D-Day assault, the Battle of the Bulge, and assaults across the Rhine River. The Group returned to the United States in August 1945 and was stationed at Sioux Falls Army Air Field, South Dakota. From 1945 until January 1962 the Group went through several inactivations and reactivations.

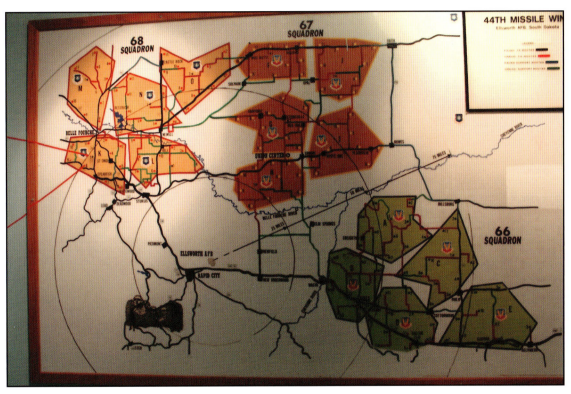

Map of Minuteman I and II ICBM flights at Ellsworth AFB on display at the South Dakota Air & Space Museum. *Lt. Col. George A. Larson, USAF (Ret.)*

Minuteman ICBM launch complex. The site was physically secured by a wire fence and electronic security. The silo blast door is visible because of the rails which allowed the blast door to open for maintenance or prior to an operational launch. The entire area was lighted at night and protected by layers of entry doors requiring combinations and intrusion alarms. *Ellsworth Air Force Base, 28th Bomb Wing Historian*

44th Missile Operations crew in capsule at LCF at D-01. *Ellsworth Air Force Base, 28th Bomb Wing Historian*

177

Minuteman I and II ICBM

Minuteman ICBM Transporter-erector backing up to a missile silo. *Ellsworth Air Force Base, 28ᵗʰ Bomb Wing Historian*

In 1962, the unit was reactivated at Ellsworth AFB and redesignated the 44th Strategic Missile Wing (SMW) as part of the 821st Strategic Aerospace Division (SAD), also located at Ellsworth. The Wing received its first operational Titan I ICBM on June 22, 1962. The 850th SMS controlled maintenance and operations of the missiles. During 1962 three new missile squadrons (the 66th, 67th, and 68th) were activated to support the Minuteman Is around Ellsworth AFB. A 44th Missile Maintenance Squadron (MMS) was established at the same time. Each strategic missile squadron supported five flights of Minuteman missiles with 50 missiles per squadron. A total of 150 launch facilities were constructed to house the missiles. The sites were scattered over 13,500 square miles of western South Dakota. The first Minuteman missile was positioned near Wall, South Dakota (Flight Delta), in April 1963. All Minuteman I missiles were in place by the end of 1963. The first Titan missile was removed from its silo on January 7, 1965. The last Titan was removed in February 1965, turning over its mission of deterrence to the Minuteman I system.[4]

Each SMS controlled five flights:

66th SMS	
A-01	19.9 miles S Howes
B-01	7.5 miles NNW Wall
C-01	10.1 miles N Philip
D-01	6.7 miles SSW Cottonwood
E-01	6.3 miles NNE Kadoka
67th SMS	
F-01	61 miles NNE Ellsworth AFB
G-01	11.3 miles N Union Center
H-01	10 miles SW Union Center
I-01	5.7 miles E white Owl
J-01	13.8 miles SE Maurine
68th SMS	
K-01	5.6 miles N Spearfish
L-01	6.2 miles SSE Vale
M-01	17.7 miles NNW Belle
N-01	6.7 miles N Newell
O-01	38.5 miles W Opal[5]

The following information was authored by Major Aida E. Roug-Compton and MSgt Paul Gremse in *A Day in the Life of a Missileer* for the history of the 44th MW, and was provided by the 28th BW Historian. It provided a look at the duties performed by missileers assigned to Ellsworth AFB.

Minuteman silo blast door at the Ellsworth AFB training facility. The three rails on which the heavy concrete and steel blast door opened and closed are visible. Note the angle of the blast door's side facing the three rails. If debris from a near Soviet ICBM was thrown onto the rails and door, the door would act as a plow, pushing the debris to either side and allowing the silo to be cleared and the missile inside launched. *Ellsworth Air Force Base, 28th Bomb Wing Historian.*

While here at Ellsworth, their sixteen-hour days could begin as early as 3:30 a.m. They often worked under the harshest of winter conditions. Security was at a premium—the stakes were high. Isolated and alone, the two officers of the Operations crew were entrusted with keys of unleashing awesome power. An involved array of teams was used to provide the security for and maintain the missile systems. Under the 44th MSS; the Flight Security Controllers (FSC), and the Alert Response Teams (ARTs). Under the 45th MSS: Wing Security Control (WSC), and the Key & Codes Control Center (KCCC). Under the watchful eye of the WSC were the Fire Teams (FTs), the Camper Alert Teams (CATs), and the Security Escort Teams (SETs). All of them gathered for the purpose of providing for our nation's defense.

Every morning without fail, it would begin in the Pride Hangar. There were only a few hours of preparation before everyone would depart for their various assignments. Teams and crews' start times were often staggered between 3:30 and 6:00 a.m., in order to diminish the crowd of people waiting for briefings, supplies, vehicles, etc. The farthest missile sites and launch control centers needed more travel time—often two or three hours—so teams assigned to these had the earliest start times.

Always a flurry of activity to get ready. Missile maintenance personnel could only work sixteen hours before they were required to stop for an eight-hour break. If they could not return to base and debrief by the end of a sixteen-hour timeline, they would be forced to remain overnight or *RON*, in the LCF. Missile maintenance personnel waiting for an alarm to reset could jeopardize breaking the timeline, so security police were often dispatched from the nearest LCF to relieve them. Scheduled RON's were often planned to accommodate extensive maintenance, but no one liked to spend a night away from home if they could help it, so people stayed busy.

179

The insertion of a Minuteman ICBM into a silo was a delicate and precise process. Once the silo blast door was open, the transporter-erector backed to a specific spot in front of the silo, locked into place, the rear doors opened, and hydraulics raised the transporter-erector to vertical. The missile inside was secured by steel cables attached at the bottom of the missile. Technicians guided the slowly lowered missile into the silo. Once secured in place at the bottom of the silo, the cables were retracted and the transporter-erector lowered to horizontal and pulled away from the silo. *Ellsworth Air Force Base, 28th Bomb Wing Historian*

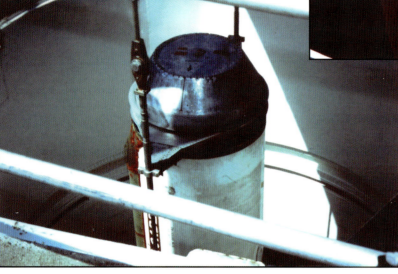

Minuteman ICBM completely lowered into its silo. The steel suspension cables are still attached. The warhead is not attached at this time. *Ellsworth Air Force Base, 28th Bomb Wing Historian*

Security in the missile field was straightforward and effective. It revolved around the concept of *Split Knowledge*. The idea was to ensure no one person possessed enough information or gained access to a missile. Security Police controlled the *A-Side* information, and maintenance and capsule crew personnel controlled the *B-Side*. Security was monitored around the clock by security police on base and within the missile complex. The two Missile Security Squadrons, the 44th MSS, and the 45th MSS, were both subordinate to the 44th Security Group.

The Operations crews met together in the crew pre-departure alert briefing room. During their briefings, they would learn of any scheduled maintenance at the launch facilities within their squadron or flight for that day. They were issued an M-15, 38-special revolved (later the M-9, 9mm), ammunition, and the coded combinations for the B-plug lock at the missile site. Their alert would last 24-hours in the LCC. Once there, they were prepared to enable and execute launch orders by the President of the United States. Their duties included reporting off-alert conditions to Wing Job Control, investigating anomalies using computer queries, and assisting the maintenance effort by identifying problems, On days they were not scheduled for alert, Operations crew were frequently scheduled for recurring training and evaluations. Many of these officers were Second and First Lieutenants, and Captains, and often performed additional duties for their Squadron Commanders. When fresh crews arrived, the grateful returning crews would drive the Suburbans back to base.

The maintenance team had their work cut out for them. Their dispatches were among the most involved, and everyone was unique. The dispatch could be for periodic, routine, or high-priority maintenance, each involving the coordination of many agencies and squadrons. And while a day did not go by that didn't have its share of glitches and snafus, an acronyym for *situation normal, all fouled up,* the Wing usually enjoyed an alert rate of 95 percent, thanks to the dedication and skill of the maintenance teams and supporting personnel.

Minuteman I and II ICBM

Minuteman ICBM with its RV (nuclear warhead) attached. The RV was lifted by a crane onto the top of the missile. At this time all electrical and communications connections (umbilical cords) were attached. Note the cables at the left of the missile coming through the silo's wall. *Ellsworth Air Force Base, 28th Bomb Wing Historian*

At the daily scheduling meeting the maintenance squadrons would assign specific teams for specific jobs. As many jobs as possible were scheduled at a site that was going to be opened, and every job had a standard time span. This took into account, briefings, vehicle checkout, pre-departure loading, travel time to-and-from the site, maintenance and inspection time, vehicle washing and gas-up, equipment downloading, and debriefing. Every detail had to be hammered out. During winter, each team needed a winter survival kit, with emergency blankets and first aid equipment, in case a team was left stranded out in the field during a sudden blizzard. The missile sites did not have latrines. Port-a-potties were available for dispatch.

Inside the D-01 LCF recreational area. The author toured the LCF as part of the VIPs on its dedication as the National Minuteman Historic Site on September 27, 2002. The author's wife is in the back left of the photograph wearing glasses. The woman in the pants suit in the center of the photograph is Fran P. Mainella, National Park Service Director (in 2002). The LCF is intact, containing all Air Force issued furniture and equipment as furnished during active operations. *Lt. Col. George A. Larson, USAF (Ret.)*

Dispatches began with the team chief checking in and getting briefed. He or she would learn about weather and road conditions, site maintenance to be performed and what supplies were needed. A computer printout was provided, and the tram chief noted any additional maintenance. The team chief also picked up the site key and combination codes from the KCCC. Sometimes he or she also had to pick up a weapon, even though all launch facilities had a supply of 12-gauge shotguns.

Minuteman II ICBM on display at the South Dakota Air & Space Museum. The missile previously was displayed at Ellsworth AFB outside the Pride Hangar. *Lt. Col. George A. Larson, USAF (Ret.)*

The rest of the team checked vehicles and loaded equipment. The equipment had to be strapped down to prevent damage, Much of it was sensitive to jarring or extreme temperature. Before leaving the base, the team would pick up one or two security guards. Then maybe a quick stop at the Shoppette or Mini-Mart to get a breakfast snack, coffee or a soft drink.

Minuteman I and II ICBM

Flight lunches were picked up, or you brought your own. As they departed base, they would contact Job Control and Traffic Control Centers by radio. Lights on, seat belts on, and a control number from Traffic Control, and you were on your way—already up to two hours into your timeline.

Display set up at the South Dakota Air & Space Museum featruing an LCC capsule for the Deputy LCO Commander. This gives a close up view of the many electronic controls requiring monitoring during a 24-hour alert tour inside the LCC. The red box with two combination locks inside stored the nuclear launch *Cookies* used to verify and authenticate a presidential nuclear release and launch order. Everything inside the LCC was completed by following checklists. Note the open manual on the desk. The two rails on the floor of the capsule allowed the Deputy LCO Commander to slide his chair to either side while monitoring his equipment. *Lt. Col. George A. Larson, USAF (Ret.)*

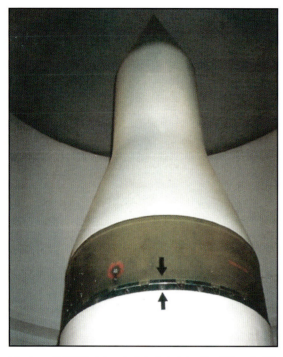

Minuteman II ICBM training missile in the training silo at Ellsworth AFB. *Ellsworth Air Force Base, 28th Bomb Wing Historian*

Periodic maintenance was priority. Routine maintenance might be postponed, but periodic maintenance attended to critical systems, systems needing periodic attention to ensure continued alert status. Any off-alert condition was considered high-priority and teams were immediately dispatched. Data Control and Scheduling Control would print out computer sheets with every discrepancy pending against that site. Scheduling Control would meet daily and weekly with involved agencies to schedule all maintenance. These included Job Control, Material Control (Supply), Vehicle & Equipment Control, Safety, Security Police, and representatives from the maintenance squadrons. Technical Engineering Branch was often consulted in narrowing down the fix for a problem. Quality Control & Safety were often dispatched with teams to oversee maintenance practices.

The missile sites were usually located on a rancher's large acreage. With so few roads and crossings, maintenance teams used a map for reference, especially at night, and could not do without one in a blizzard. They approached the site on a short gravel road, the launch facility fenced in. Flight Security Control, stationed at the LCF, was radioed for permission to enter. Once granted, the maintenance team unlocked the padlock to the chained gate. Once they entered, and passed one of the large blue antennae, the outer-zone security alarms would sound in the LCC. The maintenance team had to quickly get to the soft-support building, so that alarms could be shut off. While the team chief got the ladder stored beneath the soft-support building, and attached it to its base, the team members parked their vehicles, facing them outward from quick egress. In winter, snow could hide the ladder and parking spaces, slowing the entry. They only had five minutes to unlock the building and pick up the handset locking them to the LCC. After identifying themselves with a secure personalized code, they would receive the B-plug combination. Combinations were printed on bright pink pages that were burned to ashes after use. In the meantime, the security guard got the secure combinations for the A-circuit lock from the Flight Control NCO.

182

Minuteman II ICBM silo site L-3 near Vale, South Dakota. The gate was closed to keep out unauthorized entry while cleanup of site destruction was in progress. The site's lights are stacked to the right of the gate. To the left of the gate is the destroyed silo upper works, visible by the concrete rubble on the site's surface. After the warhead and missile were removed from the silo, as much equipment as possible was salvaged for repair parts at remaining operational ICBM installations or on other Air Force bases, or for scrap. The silo head works, once destroyed, was left untouched for 90 days to allow for overhead Soviet reconnaissance systems to verify site destruction. *Lt. Col. George A. Larson, USAF (Ret.)*

Once the maintenance team and the security guard had the combinations, they would begin site entry. The A-circuit was troublesome in winter, prone to freezing, and if the A-circuit combination locks were not correctly changed by the last security guard, the locks would not open. If that happened, it was called a lockout. In order for the maintenance team to get in, they had to have the lockout corrected. A special team was called from the LCC to drill open the locks and put new ones in their place. Next, you had to get in the *clamshell* lid. Once opened, it revealed the B-plug. A set of poles and chains were set up around the B-plug hole as a safety measure to prevent falling. The maintenance team chief would go down a short ladder and spin the B-plug combination to get a switch to lower the plug. B-plug lockouts were extremely serious, but fortunately a rare occurrence. The plug slowly moved down, drawing the 30-foot extension ladder with it. Once all the way down, with its ladder fully extended, the maintenance team could start lowering equipment into the launcher.

A technician would strap on a body harness, hook the strap to an attachment point to prevent falling. Up to 60 pounds of equipment would be placed into a large canvas bag, and lowered down by a rope. A special winch, attached to the clamshell lid, was used to lower anything bulky or heavier. At the bottom, someone would collect and move the equipment, constantly staying within view. The two-man policy applied inside the launcher, and no one was allowed to remain unobserved.

Some equipment required extensive assembly and sometimes-heavy moving equipment had to be used. It took time to assemble the work cage that fit inside the launch tube. A large tractor-trailer, nicknamed the *dog-bone*, was used to slowly inch the launcher door open, if they had to remove or replace any sections of the missile. During inclement weather, special hoods were fitted around the opening to keep everything dry.

Combat Targeting Teams worked with the Missile Maintenance Teams and Missile Handling Teams (MMTs), each time a Re-entry Vehicle (RV), Missile Guidance Section (MGS), or launch section had to be removed, repaired, or replaced. Each team consisted of five personnel, lead by an NCO. Opening the launcher door to gain access required that they disconnect the explosive device. It was designed to blast the door open during an actual launch. The launcher door was massive, 3-feet thick—made of solid concrete. It had to be inched open along its rails to reveal the missile. In the meantime, a work cage large enough for two people was lowered in sections down the B-plug opening. An opening in the launch tube, appropriately called the diving board, was dropped open, with safety poles and chains erected. The work cage had to be assembled on the diving board, and slipped into a groove that circled the launch tube. Winches and rail wheels allowed it to move completely around, up, and down the missile. For additional safety, workers wore hard hats, safety harnesses, and wrist lanyards attached to every tool and piece of equipment. Even the dropping of a small bolt into the launch tube was enough to damage the thin-skinned missile. Different sections of the missile were lifted out with crane hoists. MMTs handled the RV and MGS with their own vans, while the MHTs lifted the missile fuel stages out in one piece using the transporter-erector.

183

Electro Mechanical Teams (EMTs) were usually composed of three members, and they would dispatch to both the launch facilities and the launch control facilities. There they would remove and replace large electronics racks or smaller drawers with various indications. These required special handling and precise work. Often cold and inclement weather meant they had to take special care in moving their equipment.

Backing out of a site often took just as long as entering did. The equipment was raised up and loaded as carefully as it had been loaded at the base. If the launch tube had been entered, the teams would have to check for any loose or foreign debris at the bottom of the tube before leaving. A detailed site inspection was required before departing. Every discrepancy that could be cleared was documented, no matter how trivial it seemed. Items such as burnt out light bulbs, missing site maps, and minor corrosion were often noted. When finished, the teams informed the LCC they were done, and the teams waited for alert status to return.

Security of the site was achieved in phases. First, inner zone security was achieved, with the B-plug raised and locked, and the A-circuit locked and scrambled. This would take 15- to 30-minutes. Only when all personnel and vehicles were outside the gate could outer-zone security be established. Only then would they receive permission to depart for the base.

Back on Ellsworth, they still had plenty of work left. Every vehicle had to be gassed up and washed completely, unless it was below 20 degrees. Weapons and ammunition had to be turned in, and the security guard dropped off. Keys and codes were turned in. All equipment and parts had to be down loaded, and if any were broken, turned into Maintenance Processing Facility (MPF), where it had to be documented. The team chief went to debriefing and documented all site discrepancies that required parts and ordered them on the spot. Everything had to be accomplished before the end of the sixteen-hour timeline. Often it was very close. It was nice to have the next day off, but often they were scheduled for training or had some administrative requirement to accomplish. Most maintenance and operations teams had about 12 dispatches per month.

Close up view at site L-3 and the destroyed upper silo headworks. *Lt. Col. George A. Larson, USAF (Ret.)*

Missileers worked sixteen-hour days, or on 24-hour alerts. They had to overcome distance, weather and isolation. The strict security measures and intense coordination required men and women committed to the highest standards. Perhaps the Missileers of the 44th and 45th MSS were somewhat invisible in our own country; their duty performed out of sight, an occasional prairie dog, sheep, or stray cow for company. Perhaps their presence was only fully realized in the nightmares of those enemies gauging our nation's awesome destructive power, power protected and maintained by the Missileer.[6]

Blast door open at silo L-10 near Vale, South Dakota. Demolition personnel were then able to set explosive charges inside the silo to destroy it and remain undisturbed for 90 days for Soviet reconnaissance verification. *Lt. Col. George A. Larson, USAF (Ret.)*

Ellsworth AFB was scheduled as host ICBM base for a series of as-realistic-as-possible missile launch tests without an actual response to a nuclear attack. These tests were approved by the Secretary of Defense in November 1964 as *Project Long Life*, which called for a short-range operation base launch of three modified Minuteman IB ICBMs to provide a realistic test for this system. The Minuteman IB was longer (55 feet 11 inches long), providing more propellant, and was fitted with an improved second stage engine that was able to lift the heavier MK-11, W-56 thermonuclear warhead with a yield of two megatons. The guidance package held two target designations, allowing target flexibility while in flight. Each missile in the test for *Project Long Life* would only contain sufficient propellant for seven seconds of powered flight, and would be fitted with inert upper stages and re-entry vehicles. These were referred to as boilerplate missile sections and warhead (RV package). On March 1, 1965, *Project Long Life* was conducted. This was the first of three scheduled test launches of the Minuteman I ICBM. The modified ICBM landed two miles down range from the operational silo from where the missile was launched. The test proved that SAC's combat missile crews could launch a Minuteman ICBM from its silo. The test was also a well-publicized and documented series of launches to show Soviet political leadership SAC's nuclear deterrence capability.

The 44th SMW played an important function in establishing the Airborne Launch Control System (ALCS) in the late 1960s (to be covered in the next chapter). On June 30, 1971, the 44th SMW became host unit at Ellsworth AFB. With the activation of the 821st Strategic Aerospace Division (SAD) it was inactivated. The Wing was reassigned under the 4th Division, headquartered at F.E. Warren AFB, Wyoming. The Wing was later assigned as part of the 57th Air Division at Minot AFB, North Dakota. In October 1971 the transition to the Minuteman II ICBM began, kicking off a forty million dollar project at Ellsworth AFB. A large portion of the project's dollars was spent in Rapid City. Over three hundred workers were hired for the modernization project, referred to as *Force Modernization*. The expenditure of these funds was on top of the estimated (1971) forty million dollars injected into the area economy by Ellsworth AFB each year. Conversion to the LGM-30F Minuteman II was completed by March 1973.[7]

The last Minuteman II ICBM was blown up on September 13, 1996. After each silo was blown up and the ninety-day verification period completed, the 30-foot diameter silo headworks was mechanically demolished using large excavators. The concrete rubble was dumped into the silo, and any recovered steel piled to one side for recovery and hauling to a scrap recovery facility. *Photo: Lt. Col. George A. Larson, USAF (Ret.)*

LGM-30F MINUTEMAN II ICBM (MMII) SPECIFICATIONS AND PERFORMANCE

Length	57 feet 6-inches
Diameter	6-feet
Propulsion elements	
First stage	Thiokol TU-122 solid-fuel
Thrust	210,000 pounds
Second stage	Aerojet SR 19-AJ-1 solid-fuel
Thrust	60,300 pounds
Third Stage	Hercules 57A1 solid-fuel
Thrust	34,400 pounds
Weight	73,000 pounds
Range	8,000 miles
Speed	15,000 MPH
Warhead	(1) W-56 Thermonuclear
Yield	1.2 megatons
Throw weight	1,625 pounds
Re-entry vehicle	(1) MK-11C[8]

Minuteman I and II ICBM

With these new missiles in place, Ellsworth AFB was selected to host *Giant Pace test 74-1*, the first Simulated Electronic Launch-Minuteman (SELM) exercise. During this test, 11 SELM-configured Minuteman II ICBMs underwent successful simulated launch on command from both underground launch control centers and the ALCS. During February 1991, the Secretary of Defense announced that the Air Force would begin retirement of older weapon systems in response to the end of the Cold War and a declining defense budget. The deactivation of the Minuteman II ICBM system was announced on April 15, 1991. The schedule for Ellsworth AFB included a drawdown of one squadron per year, beginning with the 67th SMS, followed by the 66th SMS, and finally the 66th SMS. On September 1st, under the *Objective Wing* concept developed by the Air Force, the Wing was renamed the 44th Missile Wing.

LCF L-10 near Vale, South Dakota. The LCF building was not destroyed. It was offered for purchase to the original landowner, and if the original land owner was not the same or not interested, then it went for sale to an interested buyer. *Photo: Lt. Col. George A. Larson, USAF (Ret.)*

The ICBM squadrons were reassigned to the newly established 44th Operations Group, along with the lineage, honors, and history of the 44th BG. On September 28, 1991, in response to President George Bush's directive to stand down, Minuteman II personnel of the 44th OG worked around the clock to dissipate launch codes and pin safety control switches at 15 launch control facilities. Removal of the first Minuteman II missiles assigned to the 44th OG occurred at G-02, near Red Owl, SD, on December 3, 1991.

Silo L-10 after it was destroyed. *Lt. Col. George A. Larson, USAF (Ret.)*

Site L-02 during the final phase of destruction, which consisted of the site being backfilled, compacted, and rough-graded. *Lt. Col. George A. Larson, USAF (Ret.)*

187

Minuteman I and II ICBM

On April 6, 1992, the first LCC shut down. On June 1, 1992, the 44th SMW was relieved of its Emergency War Order (EWO), and its primary focus was deactivation of the Minuteman II ICBM weapon system. This day also marked the end of SAC and the beginning of Air Combat Command (ACC). The 67th MS was inactivated on August 15, 1992, and the 66th MS was inactivated on September 1, 1993. On July 1, 1993, the 44th MW changed hands from ACC to Air Force Space Command, along with the remaining operative ICBM wings. Deactivation of the entire 44th MW formally took place July 4, 1994.[9]

On September 27, 1991, President George H.W. Bush ordered Minuteman II ICBMs removed from nuclear alert, with their silos and LCCs below their respective LCF destroyed. The deactivation and destruction process used on the Ellsworth AFB's Minuteman II ICBM force was followed at other Minuteman II ICBM installations.

Site L-02 during the final phase of destruction, which consisted of the site being backfilled, compacted, and rough-graded. *Lt. Col. George A. Larson, USAF (Ret.)*

Site L-06 after gravel was spread over the former operational missile silo launch facility. *Lt. Col. George A. Larson, USAF (Ret.)*

The destruction process on Ellsworth's Minuteman II ICBM sites started on December 3, 1991, when the first missile was pulled from its silo (G-02, located north of Red Owl, South Dakota). The destruction of the remaining 148 missiles, with one held from destruction for a possible future museum, followed. Each silo was destroyed by a demolition charge down to a depth of 20-feet 6-inches. This was quite a spectacle. The author witnessed three such detonations at Lima sites near Vale, South Dakota.

188

Delta Nine launch site from the gravel access road to the missile site. The gate is open and an Air Force pickup truck is inside the site, cleaning up so it could be turned into a National Park site. The photograph shows the commercial power supporting the missile launch site, with power pole and transformers and poles inside the fence area with lights. The communication's antenna is visible behind and left of the Air Force truck. The majority of the site is below ground. D-09 is an ideal Minuteman II ICBM silo site for viewing due to its location immediately south of Interstate 90 at Exit Number 127. *Lt. Col. George A. Larson, USAF (Ret.)*

The original D-09 silo was uncovered, then enclosed with a steel and glass cover, allowing visitors to the former launch facility to look down inside the silo. The silo contains an inert Minuteman II ICBM, verified by the Soviet government as being non-operational and for display only, and is unable to be launched or armed with an operational nuclear warhead. The steel railing around the viewing glass prevents accidental falls into the glass and a balance framework for visitors looking down into the silo and taking photographs of the missile inside. The steel beam in front of the viewing area is the contact/lockdown point for the Minuteman transporter/erector when lowering or removing a missile. *Lt. Col. George A. Larson, USAF (Ret.)*

The last Minuteman II ICBM silo was blown up on September 13, 1996. After each silo was blown up and the ninety-day verification period completed, the 30-foot diameter silo headworks was mechanically demolished using large excavators. The concrete rubble was dumped into the silo and any recovered steel piled to one side for hauling to a scrap recovery facility.

Minuteman I and II ICBM

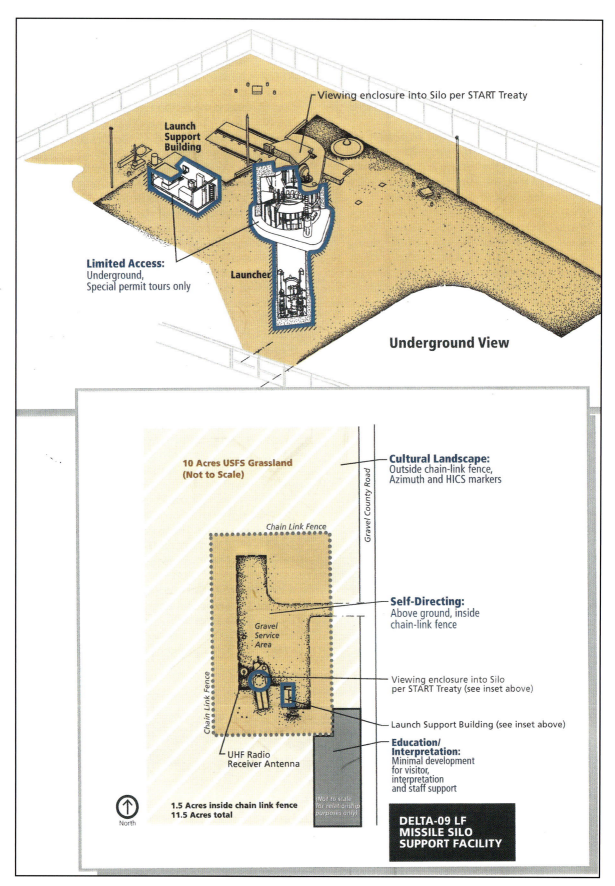

Viewing enclosure into Silo per START Treaty

Launch Support Building

Limited Access:
Underground,
Special permit tours only

Launcher

Underground View

10 Acres USFS Grassland
(Not to Scale)

Gravel County Road

Chain Link Fence

Chain Link Fence

Gravel Service Area

UHF Radio
Receiver Antenna

↑
North

1.5 Acres inside chain link fence
11.5 Acres total

(Not to scale for relationship purposes only)

Cultural Landscape:
Outside chain-link fence,
Azimuth and HICS markers

Self-Directing:
Above ground, inside
chain-link fence

Viewing enclosure into Silo
per START Treaty (see inset above)

Launch Support Building (see inset above)

Education/ Interpretation:
Minimal development
for visitor,
interpretation
and staff support

DELTA-09 LF MISSILE SILO SUPPORT FACILITY

Artist rendering of a LCF and underground LCC. *Drawing: Ellsworth Air Force Base, 28th Bomb Wing Historian*

During demolition, an observation cone was dug 20 feet deep and 70 feet wide over the silo. This was in compliance with START. All United States compliance with START and Soviet reconnaissance verifications were completed by January 1997. This was a complicated and time-consuming process.

After the site was graded, gravel was trucked in by dump trucks, dumped over the graded dirt, spread out, and leveled by bulldozer or road grader, with the security fence retained. The underground LCC was also destroyed and backfilled, then sealed with a concrete floor over the elevator shaft inside the LCF. As noted previously, the LCF was left intact and offered for sale. The sewage lagoon was back filled and graded. Silo D-09 and LCF/LCC D-01 was not destroyed, pending transition into a National Parks facility as the Minuteman Missile National Historic Site authorized under START.[10]

Inert Minuteman II ICBM on display inside the D-09 silo. Visitors can look into the silo through the viewing glass at the top all the way to the bottom of the silo. *Lt. Col. George A. Larson, USAF (Ret.)*

First public tour of the dedicated Minuteman Missile National Historic Site's LCC capsule (from left to right): Lieutenant General Robert Hinson, Vice Commander, Air Force Space Command; Victoria E. Larson (author's wife); Fran P. Mainella, Director, National Park Service; Ronald L. Orr, Principal Deputy Assistance Secretary of the Air Force for Installations, Environment and Logistics; and to the right edge of the photograph, giving the briefing was Tim Pavek, 28th Bomb Wing, Civil Engineer Squadron. Author took the photograph between the capsule's door and Deputy LCO's position. *Lt. Col. George A. Larson, USAF (Ret.)*

Hardened UHF communications antenna at the D-09 interpretive site. *Lt. Col. George A. Larson, USAF (Ret.)*

Minuteman II ICBM (second from right) in the Missile Gallery of the National Museum of the United States Air Force. *Lt. Col. George A. Larson, USAF (Ret.)*

Ellsworth AFB Minuteman II ICBM sites were among SAC's oldest. They were also the least altered from the original Minuteman I ICBM configuration, with most of their technology dating to the time period of the Cuban Missile Crisis. At the end of the Cold War, an opportunity was presented to preserve one silo and LCC. Through an interagency agreement, the National Park Service and the Air Force decided to preserve two representative Minuteman sites at Ellsworth AFB (Delta One LCC and Delta Nine Launch Facility) until their long term preservation could be evaluated.

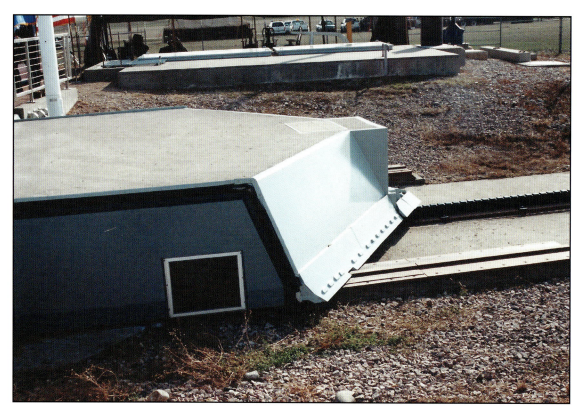

The heavy blast door was moved on its rails to open the silo. The silo was then loaded with a training missile and covered by a glass viewing enclosure. The concrete pad at the top of the photograph is the site's underground launch support building, with access at the left side through a locked door. *Lt. Col. George A. Larson, USAF (Ret.)*

Photograph taken at the dedication ceremony at D-09 on September 27, 2002. The Air Force formally turned over control of the missile silo and its controlling LCF/LCC to the NPS. A set of ceremonial keys was passed from the Air Force to the NPS. From left to right: Fran P. Mainella, Director, National Park Service; Craig Manson, Assistant Secretary for Fish and Wildlife and Parks, Department of the Interior; Colonel James Kowalski, Commander, 28th Bomb Wing; William R. Supernuagh (holding the ceremonial silo facility keys), Superintendent, Badlands National Park; Lieutenant General Robert Hinson, Vice Commander, Air Force Space Command; and Ronald L. Orr, Principal Deputy Assistance Secretary of the Air Force for Installations, Environment and Logistics. *Lt. Col. George A. Larson, USAF (Ret.)*

192

In December 1993, the National Park Service (NPS) began a special resource study of Delta One and Delta Nine. The Minuteman Special Resource Study Team, which included representatives from the National Park Service, Air Force, and the United States Air Force Museum (today renamed the National Museum of the United States Air Force), South Dakota Historical Society, and the Ellsworth Heritage Foundation spent much of 1994 evaluating possible preservation of Delta One and Delta Nine to make them available to the public as historic sites. The result was the Special Resource Study completed in 1995.

Patch of the Minuteman Missile National Historic Site presented to the author during the dedication ceremony of the site. *Lt, Col. George A. Larson, USAF (Ret.)*

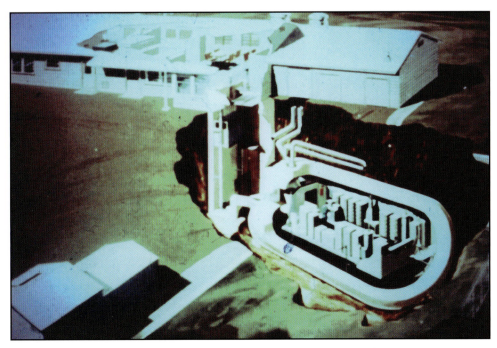

Presentation drawing/layout of the Delta Nine Launch Facility.
Drawing: Minuteman Missile National Historic Site, United States Department of the Interior, National Park Service

South Dakota U.S. Senator (at that time) Tom Daschle (Democrat) and Tim Johnson (Democrat) jointly introduced a bill to establish the Minuteman Missile Historic Site in 1998. The author provided one of his published magazine articles on Ellsworth's Minuteman ICBM sites to show their historic significance, which was provided to Senator Daschle for distribution to Congressional members. Congress began hearings and taking testimony in 1998. The bill failed in 1998, but was reintroduced in 1999 to the 106th Congress. Representatives from the NPS and Air Force testified in favor of establishing the historic site. On November 29, 1999, the U.S. House of Representatives and Senate passed legislation which resulted in the establishment of the Minuteman Missile National Historic Site. The law describes the purpose of the Minuteman Missile National Historic Site as:

> To preserve, protect and interpret for the benefit and enjoyment of present and future generations the structures associated with the Minuteman II missile defense system.
> 1. To interpret the historical role of the Minuteman II missile defense system
>   a. as a key component of America's strategic commitment to preserve world peace, and
>   b. in the broader context of the Cold War; and
> 2. to complement the interpretive programs relating to the Minuteman II missile defense system offered by the South Dakota Air & Space Museum at Ellsworth Air Force Base.

After the Federal government officially endorsed the creation of the historic site, the NPS and USAF began preparation for the opening of the LC and LCF to the public. Once the legislation was passed, the specific plans to convert D-09 to a static display were formally presented to the START Compliance Review Group in Washington, D.C., for approval. Moreover, the new law gave the NPS funds to produce a General Management Plan (GMP) for the site. A GMP attempts to establish a clear management philosophy and provide direction for interpretive themes, resource preservation, and visitor use.

The NPS began work on the GMP for the site in the spring of 2001, and hosted a series of public meetings to gather input from interested groups and individuals on their vision for the new historic site. The GMP team included representatives from the USAF, NPS, South Dakota Air & Space Museum, Badlands National Park, NPS Midwest Regional Office, NPS Denver Service Center, and NPS Harper's Ferry Center. The GMP was available for public comment in the summer of 2004 and finalized in 2005.

While the NPS occupied itself with the GMP for the Minuteman Missile Historic Site, the USAF worked on the conversion of D-09 into a static display complex to comply with START. This conversion was necessary prior to the transfer of ownership to the NPS. The USAF worked to acquire an unarmed training missile, refurbishing and painting it prior to shipment to D-09. Procuring the display missile proved to be no small task for the USAF staff working on the site with the NPS. Intense competition for training missiles and/or static display missiles existed, as military museums desired them for their own Minuteman II ICBM exhibits. Additionally, many of the high-level officers who once staffed the USAF's six operational missile wings had moved on after wings were inctivated. These Colonels and Generals had supported the establishment of a monument to the Cold War, and their transfer or retirement reduced high-level support for the new historic site and made obtaining a inactivated Minuteman II ICBM more difficult.

Once Air Force staff at Ellsworth AFB located component parts of a training missile at Hill AFB, Utah, technicians there refurbished the missile. On June 12, 2001, an Air Force team lowered the missile into D-09's silo using a transporter-erector. A missile crew from the 90th Logistic Group of F.E. Warren AFB, many of whom had been based at Ellsworth AFB, assisted in the installation of the training missile. Local media were invited to the missile's installation, increasing the profile of the Minuteman Missile National Historic Site.

Construction of a viewing enclosure for the missile silo began shortly thereafter. The design of the enclosure represented a joint effort between the NPS and Air Force, meeting START and interpretive needs for the display to the general public.[11]

On September 27, 2002, exactly ten years after President George Bush stood down the Minuteman II ICBMs from alert, the USAF formally transferred D-09 and D-01 to the NPS. At 10 a.m., a 28th BW B-1B Lancer flew over the crowd of 500 at the dedication ceremony. The author was one of the VIP guests at the dedication.

Badlands National Park Superintendant William Supernaugh opened the formal dedication ceremony, introducing Fran P. Mainella, NPS Director, who commented:

These two new Park Services Sites will help teach children about the Cold War era. They need to know the terms Berlin Wall, fallout shelter and Mutual Assured Destruction or MAD. This is a tribute to American Morals while others, are hiding their weapons, America is turning hers into a national park. This shows who believes in peace.

Lieutenant General Robert Hinson, Vice Commander, Air Force Space Command, said:

The missiles were so effective that they eventually put themselves out of business. We celebrate that victory here today. I recommended when stationed at Ellsworth AFB, seven years ago, that two of the missile facilities should be preserved to help tell the Cold War story. Young people have a hard time understanding the fear and tension of a time when many thought nuclear war would breakout.

Ronald L. Orr, Principal Deputy Assistant Secretary of the Air Force for Installations, Environment and Logistics noted "The Minuteman II ICBMs succeeded in their mission of deterrence because they were reliable and could not be destroyed."

Craig Manson, Assistant Secretary for Fish and Wildlife and Parts, Department of the Interior mentioned:

This is a particularly personal day because I was an Air Force Missile Launch Officer in the mid 1970s, assigned to Delta Flight, LCC Delta-Nine. If you had told me 26-years ago, I would have considered it pure science fiction. In those middle years of the Cold War, we believed this was a permanent condition, something that would have made this transfer impossible.

Colonel James Kowalski, Commander, 28th BW, said "The missiles' effectiveness helped win the Cold War. This was a war we had to win without any direct combat. It was the war we could not afford to lose."

Because the 44th SMW had previously been inactivated, the Air Force tradition of passing the unit's flag from one commander to another could not be performed. Colonel Kowalski passed (handed) a set of ceremonial keys to D-09 and D-01 to Badlands National Parks Superintendent Supernaught. This concluded the formal dedication ceremony at the missile silo. At this time, a selected group of VIPs and distinguished guests (including the author and his wife) were invited on a tour of the LF and underground LCC at D-01.

Once at the LCF, Tim Pavek, 28th BW Civil Engineer Squadron, escorted a small group of VIPs and distinguished visitors into the elevator in the LCF for the short ride down to the entrance of the LCC (see photograph). Once inside the LCC, he explained the functions of the LCOs who pulled a 24-hour combat duty shift, as well as their equipment. He added his personal experiences of being a LCO. This was also an information tour for Fran Mainella on how best to present the new national park to the American public.[12]

Similar to the United States, the Soviet Union built and maintained an extensive underground command and control bunker network to detect and monitor a launch of U.S. ICBMs and/or SLBMs. The author served in this capacity as an Intelligence Watch Officer with the Defense Intelligence Agency at the Alternate National Military Command Center (ANMCC) in the southern mountains of Pennsylvania, north of Camp David, Maryland. The primary Soviet ICBM early warning facility was Serpukhov-15, located outside of Moscow, with secure communications to the Soviet Politburo. With the technology available in 1983, a near war-starting mistake was made at the warning center at 12:30 a.m. (Moscow time) on September 26, 1983. In its orbit path over the central U.S., a Soviet Oko early warning satellite (Cosmos 1382) indicated and transmitted information on a U.S. Minuteman launch from Malmstrom AFB, Montana.

Lt. Col. Stanislaus Petrov, a software engineer, was the duty officer at Serpukhov-15. He was not scheduled to work, but was assigned as a replacement. When the launch alarms rang, Petrov did not accept the information to be a valid U.S. launch, as it consisted of only five Minuteman ICBMs. Petrov noted:

I decided the launch indicators were false. I had been told that a US nuclear attack would be massive, an overwhelming attack to knockout Soviet offensive and defense systems. Our ground-

195

based radar did not verify the launch but needed a few more minutes to let the ICBM indicated as launched to break its line of sight radar beams. These radars were monitored at a separate command bunker and would be the necessary independent verification. If the US was to attack the Soviet Union, it would not launch only five ICBMs. I reported the launch indications as a false warning. We did not retaliate. Soviet strategic nuclear posture of launching its ICBMs on warning of a retaliatory nuclear response implied a risk, possibly responding to a false early warning indication of an incoming U.S. missile attack. Our early warning system verification from ground-based radars reduced decision time to approximately 30-minutes, in reality 10 to 20 minutes in command decision time. I was investigated by the Soviet General Staff over the incident, determined not culpable nor punished. However, I was determined to no longer to be trustworthy. I was transferred from Serpukhov-15 to a less sensitive assignment. I was eventually forced to resign from active duty onto a military pension. [13]

The following analysis of the false launch of the Soviet Oko satellite was provided by Association of Air Force Missileers. The non-launch of five U.S. Minuteman ICBMs from Malmstrom AFB was classified as a false event. It may have resulted from a series of natural phenomena. Cosmos 1382's viewing angle in its elliptical high-earth orbit interpreted sunlight reflecting on five Malmstrom silos as the correct readings to indicate exhaust plumes of missile launches. The satellite's infrared sensors, operating near the time of the autumn equinox, where sunlight could lineup with certain Malmstrom ICBM silos and was intensified by the high cloud cover at the time, thereby reflecting more light than under normal conditions, might have produced conditions for an indication of a false event.[14]

Stanislaus stated:

> While here at the Minuteman Missile National Historic Site I am excited to visit a former U.S. Minuteman II missile silo and underground launch control bunker. I never imagined of being able to visit one of my former enemy's most secure sites.[15]

Even with the success of U.S. ICBMs at maintaining a successful nuclear deterrence, there is another legacy of these Cold Warriors seldom mentioned. The U.S. Army Corps of Engineers identified sites throughout the United States requiring environment cleanups. This list includes Ellsworth AFB's former Titan I ICBM and Minuteman ICBM facilities/operational sites. The Corps of Engineers identified landfills and burn pits with concentrations of Trichloroethylene, or TCE. Exposure to high concentrations of TCEs can result in nervous system problems, liver and lung damage, an abnormal heart rate, coma, and death, according to the U.S. Department of Health and Human Services Agency for Toxic Substances and Disease Registry. As previously commented during the discussion of Rushmore Air Force Station, the Air Force at Ellsworth AFB has undertaken extensive measures to clean up, protect, and monitor any ground water contamination.[16]

CHAPTER NINE

THE 4TH AIRBORNE COMMAND AND CONTROL SQUADRON

4th ACCS Boeing EC-135A

Shield of SAC's 4th ACCC on display at the South Dakota Air & Space Museum. *Lt. Col. George A. Larson, USAF (Ret.)*

The 4th Airborne Command and Control Squadron (4th ACCCS) was one of the most specialized squadrons in the former Strategic Air Command (SAC). The Squadron was established to act as the Western Auxiliary Command Post (WACCP), and on occasion was the SAC Airborne Command Post. It provided a survivable airborne launch capability for the Minuteman ICBM force. This allowed the Minuteman I and follow-on Minuteman II and III ICBMs to be launched from the surviving silos if their LCC was destroyed or damaged, or if they were unable to electronically communicate a verified and authorized Presidential nuclear launch order. The WACCP served as an integral part of the SAC Post Attack Command Control System (PACCS).[1]

The PACCS force converted to KC-135s. On March 25, 1965, SAC's PACCS force was reorganized. The 4063rd and the 4364th PACCS, located at Lockbourne AFB, Ohio, and Mountain Home AFB, Idaho, respectively, were discontinued. Their radio-relay missions were absorbed by EC-135As that were assigned to air refueling squadrons at Lockbourne AFB, Ohio, and Ellsworth AFB, South Dakota. Thereafter, the PACCS force became an all KC-135 system, with EC-135s performing the Looking Glass and Auxiliary Airborne Command Posts (AACP) missions.[2]

Aerial photograph of Ellsworth AFB in 1965. Two of the three WWII runways are now sites of the new hangars and support facilities for the 28th BW. The main runway has been extended to first support B-36 operations, then B-52 and KC-135 squadrons. A new large aircraft parking apron is visible at the left end of the parallel taxiway. At the end of the runway is the concrete aircraft apron referred to as a "Christmas Tree." This was for the nuclear alert force of B-52s and supporting KC-135s. This allowed the nuclear alert force to taxi directly onto the runway and takeoff within 15 minutes, and was set up by SAC for a credible nuclear deterrence. *Ellsworth Air Force Base, 28th Bomb Wing Historian*

In 1965, the 28th BW gained the new EC-135A PACCS aircraft to perform airborne launch control functions for the Air Force's Minuteman ICBM wings, and later one Peacekeeper ICBM squadron (which remained on nuclear alert for a short period until inactivated, leaving only the Minuteman III ICBMs operational and on alert).[3]

EC-135A aircraft number 61-0278, on display at the South Dakota Air & Space Museum. This aircraft looked like a standard SAC KC-135A on the exterior, but inside it was extensively modified and carried a large crew to serve in the capacity of an airborne battle staff and command post. It was operational in the Cold War, when the United States wanted to be able to fight and win a nuclear confrontation against the Soviet Union. *Lt. Col. George A. Larson, USAF (Ret.)*

198

The EC-135A is a variant of the Boeing KC-135A Stratotanker. It was modified with internal workstations for SAC battle staff personnel and communications personnel, along with external antennas for increased communications. For long-range aerial endurance, EC-135As were equipped for aerial refueling with a forward/ top fuselage mounted fuel receiver to mate with the KC-135A's flying boom. Its maximum flight endurance is approximately 72 hours, which is limited by engine lubricating oil, which eventually would shut down the engines. EC-135As retained their aerial refueling boom. If for some reason it could not take on fuel through the receiver, its boom could be connected to a tanker or any other aircraft with a receiver receptacle to pull in fuel through the flying boom. This allowed the EC-135A to pull fuel from any compatible airborne aircraft to remain flying, thus staying secure from nuclear attack and performing its battle staff and communications functions. To increase endurance and flight capabilities, its original J57 engines were replaced by more powerful and fuel efficient JT3D/TF33 turbofan engines. Six KC-135As were converted to EC-135As.

EC-135A on takeoff from Ellsworth AFB. The side view shows that externally it looked identical to a KC-135 Stratotanker used for aerial refueling operations. *Lt. Col. George A. Larson, USAF (Ret.)*

4th ACCS EC-135A with its landing gear down during a series of touch and go landings at Ellsworth AFB. *Ellsworth Air Force Base, 28th Bomb Wing Historian*

The 4th ACCS was one of the largest tactical squadrons in the Air Force, with more than 185 officers, non-commissioned officers, and airmen assigned. It was created to fulfill the needs of the Joint Chiefs of Staff (JCS) and Commander-in-Chief SAC for a survivable command and control element during the trans- and post-attack period, and as an alternate means of launching land-based ICBMs. It was one of the few tactical squadrons that conducted a formal Air Force school, the Airborne Launch Control System Operational Readiness Training School. In addition to alert commitments, the squadron flew the higher headquarters of Minuteman and Peacekeeper missiles from the Western Test Range launch silos at Vandenburg AFB, California.[5] From April 1, 1970, to September 1, 1991, the 4th ACCS was attached to the 28th BW. It performed airborne launch functions for the 44th SMW and other Minuteman ICBM wings.[6]

EC-135A Specifications and Performance Data

Wingspan	130-feet
Length	128-feet
Tail height	42-feet
Empty weight	106,300 pounds
Combat weight	300,000 pounds
Powerplant	(4) JT3D/TF33 turbofan engines with 18,000 pounds of thrust per engine
Max speed	600 mph
Range	5,000 miles (unrefueled)
Flight Crew	Aircraft commander
	Copilot
	Crew instructor
	Navigator
	Boom operator
Operations	[7] Communications personnel
Battle Staff	[15 personnel-as indicated below]
	[1] Communications Control Officer
	[2] Operational Planning Officers
	[1] Airborne Launch Control System Officer
	[1] Communications Control Officer
	[2] Operation Planning Officers
	[2] Intelligence Officers
	[1] Weather Officer
	[1] Battle Staff Commander
	[1] Logistics Officer
	[1] Airborne Emergency Action Officer
	[1] Engineering Officer
	[1] Airborne Launch Control System Officer
	[1] Spare staff position[4]

Part of this system for SAC was the Airborne Command Post referred to as "Looking Glass." Its nickname came from its mission, as it was a mirror of the ground-based system. At least one airborne command post was in the air at all times, and its highly trained crew and staff ensured there was a viable means to direct bombers and missiles from the air should ground-based command centers become inoperable in a post-nuclear attack environment. It guaranteed U.S. strategic forces would act only in the precise manner dictated by the U.S. President.

Pilot or copilot helmet from an EC-135A of the 4th ACCS on display at the South Dakota Air & Space Museum. *Lt. Col. George A. Larson, USAF (Ret.)*

200

EC-135A landing at Ellsworth Air Force Base. *Ellsworth Air Force Base, 28th Bomb Wing Historian*

The EC-135As began their duty with SAC on February 3, 1961. From then on a Looking Glass was always airborne, 24 hours a day for over twenty-nine years. Author's note: as an Intelligence Officer assigned to Headquarters SAC (1971-1974 and 1989-1992), I directly supported Looking Glass, providing intelligence/ situation briefings to the SAC General Officer on board the aircraft, especially on worldwide intelligence updates. This was a highly classified operation, and vital to the security of SAC's nuclear deterrent forces. On July 24, 1990, Looking Glass ceased continuous airborne alert operations, but remained on ground or airborne on alert 24 hours a day. Crews accumulated more than 281,000 accident-free flying hours. The last EC-135A was retired from service on October 1, 1998.[7]

The capabilities of the EC-135A PACCS as a survivable platform capable of directing retaliatory strikes against the Soviet Union was powerfully demonstrated on April 17, 1967, when an airborne battle staff successfully launched a Minuteman II ICBM from Vandenburg AFB, California, after receiving the necessary launch signal. Airborne command and control squadrons took on the PACCS functions: the 38th SRS, 55th SRW at Offutt AFB, Omaha, Nebraska; 305th AREFS and 305th AREFW at Grissom AFB, New York; and 28th BW transferred their EC-135s and crews to the 2nd ACCS, 55th SRW at Offutt AFB, Nebraska, and 4th ACCS, 28th BW.[8]

Scale model of a EC-135A assigned to the 4th ACCS. Note the communications antennas on top of the aircraft from the top of the fuselage to the middle of the tail. The refueling receptacle is visible behind the cockpit on top of the fuselage. Model on display at the South Dakota Air & Space Museum. *Lt. Col. George A. Larson, USAF (Ret.)*

201

On April 1, 1970, SAC reorganized its PACCS and moved some of its EC-135s out of Westover AFB, California. In this reorganization, all EC-135s were assigned to the 2nd, 3rd, and 4th ACCSs, which were activated at Offutt AFB, Nebraska, Grissom AFB, Indiana, and Ellsworth AFB. The basic function of PACCS remained unchanged. Looking Glass continued to remain airborne at all times in the vicinity of Offutt AFB, and relay aircraft remained on round-the-clock ground alert.[9]

EC-135A on display at the South Dakota Air & Space Museum. *Lt. Col. George A. Larson, USAF (Ret.)*

The 4th ACCS

Chapter Ten
Boeing B-52 Stratofortress

Ellsworth Begins the Turbojet Era (B-52D, G and H)

B-52H on display during Ellsworth AFB's Dakota Thunder 2000 Air Show and Open House. This B-52H was assigned to the 5th BW at Minot AFB, North Dakota. *Lt. Col. George A. Larson, USAF (Ret.)*

B-52D on the flight line at Ellsworth AFB. Control tower is visible in the left of the photograph, behind the aircraft and above the cockpit. *Ellsworth Air Force Base, 28th Bomb Wing Historian*

In early 1961, as part of evolving U.S. national defense strategies to counter possible Soviet nuclear aggression and expansion of Communism into other parts of the world, the 77th BS began participating in airborne indoctrination sorties code named *Chrome Dome*. The program had its origins in 1958 under the program called *Head Start*, which was an Airborne Alert Indoctrination Program. The *Head Start* program placed a B-52 armed with nuclear weapons in flight, circling Thule, Greenland. If the Ballistic Missile Early Warning System (BMWES) radars were taken out by a Soviet preemptive attack, SAC's nuclear-armed bombers would be unleashed. *Chrome Dome* airborne B-52s were armed with four B-28 (or larger B-36 or B-53) thermonuclear weapons. The *Chrome Dome* orbit kept SAC's nuclear-armed B-52s outside Soviet air space, ready to attack. The 77th BS, in carrying out the Wing's mission, played a major role in this operation, which maintained a percentage of SAC's bomber force airborne at all times to provide an immediate nuclear response to a Soviet Nuclear attack, LRA bombers, or SRF ICBMs.[1]

B-28 thermonuclear weapon. The Mark 28 was SAC's primary Cold War weapon for many years. The weapon's production began in January 1958. However, a problem was identified with the weapon's safety mechanism, preventing the weapon from being activated by coded response after receipt of a Presidential launch order. The concern was also that the weapon might arm itself when the electrical connections were connected to the aircraft. It took six months to correct before mass production started. Production continued through May 1966. The MK-28 was a highly dependable, multi-purpose tactical and strategic thermonuclear bomb carried by virtually all SAC bombers. It was in service for 33 years (1958-1991). It was 22 inches in diameter, 170 inches long, and weighed from 1,700 to 2,320 pounds, with a variable yield from 70 KT to 1.1 MT. It was able to be set for either an air burst or ground contact detonation. Weapon is on display at the National Museum of Nuclear Science and Technology. *Lt. Col. George A. Larson, USAF (Ret.)*

MK-53 thermonuclear weapon. The Mark 53 was carried by SAC's B-47, B-52, and B-58 bombers. Two variants were produced between August 1962 and June 1965 (clean and dirty). The *Y2* clean fissile material was all HEU, no plutonium. It required three parachutes to retard the heavy weapon's descent to target, allowing the aircraft time to escape the detonation area and blast waves. SAC began retiring the early weapon variants in July 1967, but they remained in permanent stockpile until early in 1997. As of 2009, fifty were in the U.S. nuclear stockpile. The weapon was 50 inches in diameter, 150 inches long, weighed 8,850 to 8,900 pounds, and had a yield of 9 MT. It was able to detonate after a retarded fall for air burst and contact detonation. A total of 350 were produced. Weapon is on display at the National Museum of Nuclear Science and Technology. *Lt. Col. George A. Larson, USAF (Ret.)*

MK-36 Thermonuclear weapon. The Mark 36 was a two-stage thermonuclear weapon. Between April 1956 and 1958, a total of 940 were produced. Modifications were made to 275, converting them to MK-21s, designated *Y1*. This was a particularly nasty weapon, with 95 percent Li-6 and a 4 to 5 MT yield in an air burst, contact, or lay down detonation. They were considered *dirty* bombs, but were never deployed. They were converted back to MK-36s. The MK-36 was widely carried by SAC aircraft, representing one-half the U.S. mega tonnage at the time of its retirement in August 1991. The clean version was designated *Y2*. Because of its powerful warhead, when released from a SAC bomber it required two parachutes to slow the bomb, giving the bomber sufficient time to clear the target (detonation) area. It had a 58 inch diameter, was 150 inches long, weighed 17,500 to 17,700 pounds, and had a 9 to 10 MT yield. Weapon on display at the National Museum of Nuclear Science and Technology. *Lt. Col. George A. Larson, USAF (Ret.)*

M-61 thermonuclear weapon. The B-52 sometimes carried the Mark 61. The weapon was produced between October 1966 and the early 1990s. It is a multipurpose tactical and strategic bomb. Its basic design has been adapted to many other weapon systems. It had four different yield configurations, and was slowed by a single parachute. It enjoyed the longest production run of any U.S. nuclear weapon and was the oldest design in service as of 2009. A total of 3,150 were produced, but the early models were retired in the 1970s, later models in the 1980s. As of 1997 there were 1,350 in service. It is impossible to get exact numbers of nuclear weapon types currently scheduled to be retained in the U.S. thermonuclear stockpile after the START update in 2011, which significantly reduced U.S. and Russian thermonuclear weapons. Weapon on display at the National Museum of Nuclear Science and Technology. *Lt. Col. George A. Larson, USAF (Ret.)*

Boeing B-52 Stratofortress

KC-135A, 28th Air Refueling Squadron, 28th BW. *Ellsworth Air Force Base, 28th Bomb Wing Historian*

In 1964, the 77th BS had a strength of 16 B-52Ds:

Serial Number	Boeing Aircraft Designation
55-0066	B-52D-20-BW
55-0107	B-52D-65-BO
55-0675	B-52D-25-BW
56-0615	B-52D-80-BO
56-0657	B-52D-25-BW
56-0658	B-52D-25-BW
56-0660	B-52D-35-BW
56-0674	B-52D-35-BW
56-0676	B-52D-35-BW
56-0680	B-52D-35-BW
56-0682	B-52D-40-BW
56-0683	B-52D-40-BW
56-0691	B-52D-40-BW
56-0693	B-52D-40-BW
56-0694	B-52D-40-BW
56-0697	B-52D-40-BW[2]

BO: Boeing Seattle BW: Boeing Wichita

On Monday morning, September 10, 1962, a KC-135A (aircraft number 60-0352) assigned to the 28th Air Refueling Squadron took off from Ellsworth AFB for a two and one-half hour flight to Fairchild AFB, Washington. Aboard the aircraft, in addition to the crew, were 38 airmen, one civilian, and one U.S. Army enlisted soldier. The airmen on the KC-135A were B-52D combat crew personnel and maintenance personnel from Fairchild AFB who had recently completed runway repairs at Ellsworth AFB. At 11:00 a.m., the KC-135's pilot contacted Fairchild's control tower, stating the aircraft was descending through 14,000-feet toward the runway. At 230 mph the tanker crashed through the area's thick forest, plowing a path 750-feet long and 75-feet wide. The tanker smashed into the mountain, exploding on contact and catching fire and burning fiercely. All on the tanker were killed. It appeared the pilot did not level off at the proper landing approach altitude of 5,000-feet, thus smacking into the northeast side of the mountain at an elevation of 4,400-feet.[3]

KC-135A on the Ellsworth AFB flight line *Ellsworth Air Force Base, 28th Bomb Wing Historian*

B-52D on alert with crew members running from the alert facility to the parked bombers on the alert pad, referred to as a "Christmas Tree." *Ellsworth Air Force Base, 28th Bomb Wing Historian*

B-52D returning from a bombing mission in SEA, landing at U-Tapao Royal Navy Air Force Base, Thailand. A KC-135A is waiting off the end of the runway to takeoff to support airborne operations, refueling fighters or bombers. A U.S. guard is at the lower left corner of the photograph, in front of a sand bag bunker. There were no attacks against Air Force aircraft on the ground in Thailand as at air bases in South Vietnam. *Andersen Air Force Base, Guam, 43rd Strategic Wing, Intelligence Division*

Boeing B-52 Stratofortress

KC-135A SPECIFICATIONS AND PERFORMANCE

Wingspan	130 feet 10-inches
Length	136 feet 3-inches
Tail height	41 feet 8-inches
Empty weight	106,300 pounds
Max weight	316,000 pounds
Powerplant	(4) 13,750 pound thrust Pratt & Whitney J57-P-59W turbojet engines
Max speed	600 mph
Service ceiling	45,000 feet
Range	10,000 miles
Crew	Pilot
	Copilot
	Navigator
	Boom operator

In 1964, the 28th BW became involved in a tactical fighter aircraft deployment, as the 28th ARS supported Tactical Air Command (TAC) fighters on their first non-stop deployment from the U.S. west coast to Europe. More than one hundred aerial refuelings were required before the fighters reached their destination. This was a demonstration of TAC's ability to transfer fighters from the U.S. to NATO for the defense of Western Europe, aimed directly at the Soviet political and military leadership after the Cuban Missile Crisis. This was a flexing of conventional, not nuclear air power, in the defense of Western Europe.

In March 1966, the 28th BW went to war for the first time in Southeast Asia (SEA). The wing's B-52Ds deployed to SEA eventually completed the Big Belly modification. Up until this time, SAC had deployed its B-52Fs to SEA. The B-52F could carry a total of fifty-one 750-pound bombs, broken down to 27 in the bomb bay and 12 each on external pylons under the wings between the fuselage and inboard engines. In December 1965, the Big Belly modification began allowing the B-52D's bomb bay capacity to expand from 27 to 84 500-pound bombs, or an increase from 27 to 42 750-pound bombs. On the two external pylons bomb capacity remained at 24 500-pound or 12 750-pound bombs. This gave the B-52D the ability to deliver a devastating bomb load on ground targets.

77th Bomb Squadron Deployments to Southeast Asia
March 9, 1966 through September 21,1966
January 15, 1968 to July 19, 1968
September 9, 1969 to March 18, 1970

The Wing's B-52Ds flew their first bombing mission to North Vietnam on April 11, 1966, against the southern area of North Vietnam containing the geographic and strategic supply routes (Mugia, Bankarai, and Ban Raving Passes) that fed the famous Ho Chi Minh Trail. It was nicknamed *Tally Ho*. The 28th BW's B-52Ds flew the 1000th and 5000th B-52 sorties against Communist targets in SEA. The Wing's extensive involvement in SEA aerial combat operations, flying B-52Ds and KC-135As, continued for the next nine years.

Boeing B-52 Stratofortress

B-52D on the ramp at Andersen AFB, Guam, being loaded with 500 pounders for a bombing mission to SEA. The bombs have already been loaded, one at a time, onto the two wing pylons under each wing. Two weapon carts hold 500-pound bombs on racks ready to be uploaded into the bomber's bomb bay. The bomb clips were pre-loaded, decreasing loading time for the large numbers of bombers at Andersen AFB. *Andersen Air Force Base, Guam, 43rd Strategic Wing, Intelligence Division*

In 1972, Captain John D. Mize, 77th BS became the first SAC crewman to receive the Air Force Cross for heroism for his efforts to save the lives of his crewmen after a Russian built SA-2 Guideline surface to air missile (SAM) attack during the Linebacker II mission over Hanoi. During its involvement in SEA, the Wing flew thousands of ARC Light Combat and Linebacker missions, including participation in Linebacker II. Major John Mize, USAF (Ret.), recalled:

Photo from a B-52D bomb bay dropping a load of 500-pound bombs onto a target in South Vietnam. *Andersen Air Force Base, Guam, 43rd Strategic Wing, Intelligence Division*

Boeing B-52 Stratofortress

On December 20, 1972, I flew my 292nd Combat mission in a B-25D out of U-Tapao Royal Navy Air Force Base, Thailand. I was Brick 02. We were the last ones into the area around Hanoi to attack the Petroleum Products Storage Area. After bomb release, we counted seven SA-2 Guideline missiles inbound.

Post strike aerial reconnaissance of the Giap Hhi Railroad Yards. B-52s caused heavy destruction of warehouses and the associated railroad marshaling yard. *Andersen Air Force Base, Guam, Eighth Air Force, Command Historian Office*

My electronics warfare officer's scope was going in and out, bet he could hear the approaching missiles. One SAM detonated nearby but did not bring the aircraft down. I checked in with the crew and there were no casualties. We landed safely at U-Tapao at which point we found a hole in one side of the fuselage and a matching one on the opposite side. We determined that the EWO had moved his head forward just prior to the SAM's detonation, so that when the shrapnel cut through the fuselage as he was looking into his scope. If he had shifted position, the shrapnel would have taken his head off. This was a good day for my crew but a bad day for SAC because of the losses it had sustained.

North Vietnamese SAMs brought down six SAC bombers (two B-52Ds and four B-52Gs). The author was a Captain, Intelligence Division (544th ARTW), Airfields Team, assigned to the command center as a briefer on Linebacker II. He reported the losses of these six bombers, damage to others, and a bomb damage assessment of targets hit.[5]

Photo book produced by Eighth Air Force Headquarters after Linebacker II. *Courtesy: author*

B-52D crew prior to a training flight for the 28th BW. *Ellsworth Air Force Base, 28th Bomb Wing Historian*

From December 18 through 29, 1972, referred to as the Eleven Day Air War over North Vietnam, B-52s and their associated support aircraft forced their way into the world's most heavily concentrated belt of air defenses. The North Vietnamese fired over 1,000 SA-2s, shooting down fifteen SAC B-52s. From December 28th through the 29th, North Vietnamese defenses had been neutralized and no more B-52s were lost. The cost to SAC was significant: out of 93 B-52 crew members in the shot down Stratofortresses, 26 were recovered by air-sea-rescue teams, while 33 bailed out over North Vietnam and were captured and held as Prisoners of War (POWs) until released after North Vietnam signed the Paris Peace Accords. Another 29 crew members were reported as Missing in Action (MIA), with four dying during a B-52 crash landing at U-Tapao, Thailand.[6]

Not all the Wing's B-52Ds were assigned to SEA all the time. On April 3, 1970, a 28th BW B-52D caught fire and crashed during a landing at Ellsworth AFB, skidding into a brick storage building containing 25,000 gallons of aviation jet fuel. Base fire fighters fought the flames to save the nine crewmen on the B-52. Only one suffered broken limbs, with three Ellsworth fire fighters injured. One of the bomber's eight turbojet engines kept running for 40 minutes until running out of fuel.[7]

B-52Gs armed with the nuclear warhead tipped Hound Dog missile. *Ellsworth Air Force Base, 28th Bomb Wing Historian*

Ground Attack Missile-77 (GAM-77), later redesignated Air-to-Ground Missile-28 (AGM-28) Hound Dog standoff, air to surface attack missile. On display at the National Museum of the United States Air Force. *Lt. Col. George A. Larson, USAF (Ret.)*

B-52G landing at Ellsworth AFB. *Ellsworth Air Force Base, 28th Bomb Wing Historian*

Boeing B-52 Stratofortress

The Hound Dog was operational with SAC from 1961 to 1976. The AGM was produced by North American Aviation. Design work began in the 1950s to increase the B-52's capability to attack heavily defended Soviet targets by destroying anti-aircraft defenses ahead and to either side of the bomber's penetration route to its assigned strategic targets inside the Soviet Union. SAC B-52Gs and B-52Hs were modified to carry two Hound Dog missiles on under-wing pylons between their fuselage and inboard engine nacelles. The B-52 could use the Hound Dog engines for added power on takeoff or while airborne, and refuel them from the aircraft's fuel tanks before launching. The Hound Dog had a sleek aerodynamic attractive configuration with small canard foreplanes, a delta wing, and a small tail fin with rudder. A turbojet engine was mounted in a pod under the aft fuselage. The Hound Dog had an inertial navigation system that was updated from the B-52 before launch. Some variants were equipped with anti-radar and terrain-following systems.

HOUND DOG AGM-77 SPECIFICATIONS AND PERFORMANCE

Wingspan	12 feet
Length	42 feet 6-inch
Diameter	28-inches
Weight	10,140 pounds
Propulsion	(1) 7,500 pound thrust Pratt & Whitney J52-6 turbojet engine
Max speed	Mach 2.1
Range:	
Low-level	500 miles (treetop)
High-level	700 miles (55,000 feet)
Warhead	W-28
Yield	1 MT
Number built	593

The 28th BW had replaced its B-52Ds with B-52Gs by June 1971, with fourteen assigned to the base. Its aircrew continued to fly B-52Ds in SEA. The Wing operated B-52Gs from 1971 to 1977. The Wing received the B-52Gs from the 60th BS, 72nd BW.

The aircraft carries a SEA camouflage color scheme. *Ellsworth Air Force Base, 28th Bomb Wing Historian*

KC-135A flying south of and within FAA limitations and guidelines over Mount Rushmore, west of Keystone, South Dakota. The aerial photograph was taken from an accompanying KC-135A assigned to the 28th BW. *Ellsworth Air Force Base, 28th Bomb Wing Historian*

Boeing Aircraft designed and produced the B-52G to reduce the aircraft's weight. It had an eight-foot shorter tail, increased chord, the self-sealing bladder type fuel cells were removed and replaced with a wet wing for fuel storage that increased capacity, and the tail gunner's position was removed from the tail as in the B-52D and B-52F models and moved into the forward crew compartment. The aircraft had additional modifications to allow improved low-level, all terrain, and night operations to penetrate heavily defended Soviet targets.

211

B-52G Specifications and Performance	B-52Gs assigned to 28th Bomb Wing	Aircraft Number	Boeing Designation
Wingspan	185 feet		
Tail height	40 feet 8-inches	57-6468	B-52G-75-BW
Length	160 feet 11-inches	57-6469	B-52G-75-BW
Powerplant	(8) 11,200 pound thrust (dry) Pratt &	57-6470	B-52G-75-BW
	Whitney J57-D-43WB turbojet engines	57-6471	B-52G-75-BW
	13,750 pound thrust with water injection	57-6472	B-52G-75-BW
Max speed	630 mph at 20,000 feet	57-6473	B-52G-75-BW
Combat speed	650 mph	57-6474	B-52G-75-BW
Cruising speed	526 mph	57-6475	B-52G-75-BW
Service ceiling	47,000 feet	57-6516	B-52G-90-BW
Combat ceiling	50,000 feet	58-0178	B-52G-95-BW
Unrefueled range	8,800 miles	58-0184	B-52G-95-BW
Empty weight	185,000 pounds	58-0211	B-52G-100-BW
Max weight	488,000 pounds	58-0221	B-52G-105-BW
Conventional	(12) 750 pound bombs on external pylons	58-0232	B-52G-105-BW
Armament	(84) 500 pound bombs internally	58-0253	B-52G-115-BW
	Able to be GPS/INS guided	59-2578	B-52G-125-BW
	(12) Conventional Air Launched Cruise	59-2580	B-52G-125-BW
	Missiles (CLAMS)	59-2590	B-52G-130-BW
Nuclear	(4) Gravity B-61 thermonuclear bombs		
Armament	(12) AGM-86B ALCM		
Electronics	Inertial Navigation Unit (INU)		
	Data Transfer System (DIS) cartridges		
	Upgraded ALQ-172 set		

- Combat Network Communications Technology (CONECT) improvement provides a modern cockpit information avionics architecture, color displays and enhanced situational awareness, network-centric war-fighting capability, fully integrated line-of-sight (LCS) and beyond-line-of-sight (BLOS) data link capabilities and mission/weapon reprogramming capability.
- UHF/EHF satellite communications
- Global Position System (GPS)
- ARC-210 radio with Have Quick II anti-jam feature
- KY-100, which provides secure voice and data transmission
- Combat Track II (CTII) radio, permitting an interim secure BLOS reachback connectivity allowing aircraft to receive targeting data from the Combined Air
- Operations Center over
- CTII and then update mission data in the Offensive Avionics System.
- An Electro-Optical (EO) Viewing System that uses forward-looking infrared (FLIR) and high-resolution low-light-level television (LLLTV) sensors to augment the targeting, battle assessment, flight safety, and terrain-avoidance systems, improving combat and low-level flight capability
- Integrated self-targeting and battle damage assessment (BDA) capability
- Upgraded radar systems
- MIL-STD-176D interface supports advanced precision weapons capabilities
- ECM suite uses a combination of electronic detection, jamming, and infrared (IR) countermeasures to protect against hostile air defense systems
- Pilots provided with night-vision goggles for low-level night operations

Crew:
- Pilot/Aircraft Commander
- Pilot
- Radar navigator
- Navigator
- Electronic Warfare Officer
- Gunner (this position has been deleted from the B-52H crew)

In 1984, the 28th BW was tasked to conduct sea reconnaissance, surveillance, and conventional operations from bases overseas. This was part of SAC's Strategic Projection Force and maintained readiness for global conventional operations with specially modified B-52s. The author, from 1978-1981, was assigned to the 43rd SW on Andersen AFB, Guam, which was tasked by SAC to develop these procedures, including building conventional combat mission folders, training crews in photograph and ship recognition, and other classified missions in the western Pacific. The 28th BW tested its capability to operate from forward overseas bases during its first deployment to Egypt during *Bright Star 85*. The exercise blended U.S. and Egyptian Air Forces into a combined operations package for the defense of the U.S.' Middle East allies.[8]

KC-135R landing at Ellsworth AFB. *Ellsworth Air Force Base, 28th Bomb Wing Historian*

KC-135R flying north to south, paralleling the runway and taxiway. A B-52G is at the lower left section of the photograph. *Ellsworth Air Force Base, 28th Bomb Wing Historian*

SAC's KC-135A was re-engined with new and more powerful 21,634-pound thrust CFM International CFM-56 turbofans. The airframe was rebuilt to allow increased fuel load and extend the operational hours of the aircraft due to increasing aerial commitments by the U.S. military, which was critical due to the delay in acquiring a new Air Force tanker and fighting three wars (as of 2011) in Iraq, Afghanistan, and Libya (as part

Boeing B-52 Stratofortress

of a NATO coalition). The new engines reduced aircraft fuel consumption, allowing two KC-135Rs to refuel as many aircraft as three KC-135As. This provided an air refueling multiple which proved to be fortunate given the delay in the contract for the new tanker (the Boeing KC-767), which was finally formalized in 2011.

B-52H on Ellsworth AFB's flight line during Dakota Thunder 2000 air show and open house. This B-52H is assigned to the 5th BW at Minot AFB, North Dakota. *Lt. Col. George A. Larson, USAF (Ret.)*

BOEING KC-135R SPECIFICATIONS AND PERFORMANCE

Wingspan	130 feet 10-inches
Tail height	41 feet 8-inches
Length	136 feet 3-inches
Powerplant	(4) 21,634 pound thrust CFM International CFM-56 turbofan engines
Speed	530 mph at 30,000 feet
Max ceiling	50,000 feet
Combat range	1,500 miles with 150,000 gallons of Fuel available for refueling operations
Crew:	Pilot
	Copilot
	Navigator
	Flight engineer
	Boom operator
Ferry range	11,000 miles
Takeoff weight	322,500 pounds
Fuel transfer	200,000 pound
Cargo capacity	83,000 pounds[9]

Boeing B-52 Stratofortress

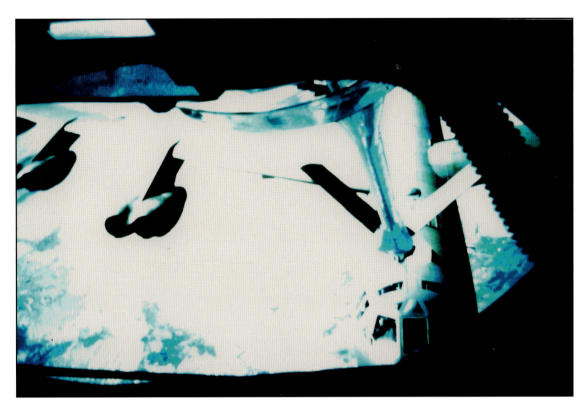

B-52G being refueled by a KC-135R. *Ellsworth Air Force Base, 28th Bomb Wing Historian*

B-52H SPECIFICATIONS AND PERFORMANCE

Wingspan	185 feet
Length	156 feet
Tail height	40 feet 7-inches
Empty weight	172,740 pounds
Max weight	488,000 pounds
Powerplant	(8) 17,000 pound thrust T33-P-3 turbofan engines
Armament	See previous B-52G specifications
Tail gun	M61 cannon (later removed)
Max speed	630 mph
Service ceiling	46,200 feet
Range	8,350 miles (unrefueled)
Crew	Same as B-52G
	Tail gunner's position later eliminated

This is a good time to cover SAC's ground-based nuclear alert procedures for the B-52 fleet. This is not specific to Ellsworth AFB, but covers general SAC operations, providing an excellent view of nuclear alert during the Cold War.

Crew members running to a nuclear cocked B-52G. *Ellsworth Air Force Base, 28th Bomb Wing Historian*

Four B-52Gs on alert. At the sound of the Klaxon, the crews raced to the nuclear cocked bombers, started engines, and headed toward the end of the runway. With a launch order, they performed what SAC called a Minimum Interval Take Off or MITO, followed by the tankers to clear the base before the detonation of Soviet ICBM warheads. The Klaxon often sounded for the alert crews, requiring them to taxi to the hold point and the end of the runway and then, after holding, receive orders to return to the alert pad. At this time maintenance personnel had to service the aircraft, which included refueling and correction of any items noticed by the crews while taxiing to the runway. *Ellsworth Air Force Base, 28th Bomb Wing Historian*

SAC was given the responsibility of creating a credible nuclear deterrent force to counter, at that time, a growing and apparent will to use nuclear force to possibly attack the United States. In 1957, SAC created a 55 air base nuclear bomber dispersal pattern, complete with security and alert facilities for bomber crews, nuclear cocked bombers, and supporting tankers on a designated pad. In 1957, SAC was in transition with the B-36 as it fielded the B-47 and B-52. SAC war planners positioned these aircraft at one end of the runway to respond to the alert klaxon.

B-52H just prior to wheels touching onto the runway. The pilots will keep the aircraft temporarily on the ground, then, increasing engine power, rapidly attain takeoff speed before climbing up for another go around for a repeat landing and takeoff. The earth-covered structures in the background are munitions/weapons bunkers. *Lt. Col. George A. Larson, USAF (Ret.)*

The cocked B-52 crews trained to get airborne before bombs delivered by Soviet LRA bombers or ICBM warheads could detonate over the base, thus destroying the ground-based alert bomber and tanker force. In the early years of the Cold War, due to the CEP limitations of the first generation of ICBMs, nuclear warheads were classified as area weapons, and were not comparable to the current technological pinpoint weapons of nine feet or less (exact accuracy is highly classified). At the start of the Cold War, because of the deployment of the B-36 and based on the threat of incoming Soviet LRA bombers, response time was approximately 60 minutes for bomber crews to take off and clear the perimeter of the base.

Later, SAC reduced its alert response window to 15 minutes in 1961, based on the estimated arrival time of launched Soviet ICBMs over SAC bomber bases and missile complexes. This was an evolutionary process with upgrades and constant changes, especially after the Soviet Navy deployed nuclear powered submarines equipped with SLBMs off the U.S. west and east coasts, dramatically reducing warning times to approximately five minutes. This did not give the President more than a few minutes after receiving indications of a Soviet SLBM attack to decide on the appropriate nuclear response and authorize the release of nuclear weapons. At the same time, it was doubtful the President could fly from the White House (if he was at that location) to Andrews AFB and board Air Force One to take off and clear the Washington, D.C., area, away from incoming nuclear SLBMs. Of course, if the international situation was deteriorating, the President would either be airborne in Air Force One or at one of the underground command bunkers in the United States.

B-52H crew practicing touch and go landings at Ellsworth AFB in March 1997. The B-52H was assigned to the 5th BW at Minot AFB. This was a common scene during the B-52's presence at Ellsworth. The author, along with his granddaughter Calisandra, sat on one of the concrete highway dividers used to close off the taxiway from the access road. Photographs were allowed, and crews often waved during taxiing to the end of the runway for takeoff. However, the terrorist attacks of September 11, 2001, ended this, and photographs on base are prohibited unless unless one receives prior written permission. *Lt. Col. George A. Larson, USAF (Ret.)*

216

Christmas Tree alert area at the south end and to the west of Ellsworth AFB's main runway. It was separately secured. The photograph shows three of the four 28th BW B-52Gs on alert and the corresponding five tankers (one of which was an on alert spare). At the sound of a Klaxon all were launched. The extra tanker was to be available if one of the primary tankers had to abort or could not off load the required fuel, or fill in for another bomb wing's tanker which suffered a similar occurrence. To the right rear of the alert pad is a security guard tower, which monitored internal security. The entire area was lighted at night. No unauthorized entry was permitted, and each aircraft was protected by security guards who checked access to the aircraft. *Ellsworth Air Force Base, 28th Bomb Wing Historian*

It took time for SAC to build supporting crew alert facilities. Initially, pre-existing on-base facilities were used. If required, temporary alert crew quarters were set up to reduce response time and travel distance to cocked nuclear-armed bombers. Existing parking aprons were used if none existed at the end of the runways until separate alert facilities and aircraft parking alert aprons could be constructed.

B-52H on the Ellsworth AFB Christmas tree. The power cart is located at the rear of the aircraft, along with a maintenance scaffolding in case work was required to launch the alert aircraft. In the foreground another B-52H is parked. The sign in the center is connected to others with a rope. The security sign indicates the area around the aircraft is two-officer policy for entry into a nuclear cocked aircraft and was guarded by security police. *Ellsworth Air Force Base, 28th Bomb Wing Historian*

Once initial alert facilities were set up, SAC had to fully integrate a permanent structure in 1954. As SAC's alert force grew, there was an increased requirement for an alert crew and support personnel barracks. Changes were made daily, requiring constant evaluation and program officers following construction timetables.

SAC headquarters created an initial alert basing plan, which consisted of a colored base layout chart indicating distances between parked alert aircraft, alert crew quarters, mess hall, and routes on base from buildings to parked alert bombers marked in red on a wall map in the Base Command Center. When the alert crews were not in the alert facility they had Air Force crew cab trucks assigned, with a flip chart mounted on the dash to indicate direct routes from authorized buildings back to the alert facility/alert bombers. If recalled by the alert Klaxon, the crews activated their sirens and flashing lights to clear traffic to the alert area.

SAC went through two alert configurations:

1. In 1956, alert aircraft were parked on stubs at 90-degree angles to a 45-degree alert taxiway, with earth barricades between the parked aircraft at each alert stub.

2. By the end of 1957, SAC opted for a bomber alert herringbone configuration. The alert taxiway was at 45 degrees, along with each alert aircraft stub. Crews nicknamed the alert parking stubs *Christmas Trees*. This name became accepted SAC terminology. SAC constructed this configuration at Mountain Home, Robbins, Wright-Patterson, and Wurtsmith AFBs.

KC-135R scale model on display at the South Dakota Air & Space Museum. *Lt. Col. George A. Larson, USAF (Ret.)*

The author supported four B-52Ds on alert at Andersen AFB. These B-52s were parked at a 45-degree angle to the taxiway near the west end of the runway, secured by guards, and were separated from the general flight line. There were steel aircraft revetments where the bombers were parked. This was a holdover from bombing operations during the Vietnam War to prevent complete destruction of the bombers by fratricide if one bomber exploded due to a bomb detonation during rearming after a completed bombing raid. No changes were made on Andersen AFB to the Vietnam-era bomber flight line.

B-52H at Dakota Thunder 2000 at Ellsworth AFB, assigned to the 5th BW at Minot AFB. Notice at the rear of the aircraft that the M61 cannons have been removed and the aircraft no longer carries an enlisted gunner. *Lt. Col. George A. Larson, USAF (Ret.)*

The majority of SAC's alert configuration was *Christmas Trees*. SAC engineers determined that when terrain at the ends of the runways did not allow the construction of a *Christmas Tree* without the extensive purchase of needed land or large amounts of fill material, alert aircraft were parked on pre-existing rectangular apron areas near the end of the runway, or they would expand an existing parallel taxiway into a small rectangular apron for an alert aircraft parking apron.

The B-52G/H could be armed with the Boeing AGM-69A Short-Range Attack Missile (SRAM). The SRAM was designed as a follow-on standoff missile to Hound Dog. It had terrain-following, semi-ballistic, pop up and dive, and a combination of inertial and terrain-following flight paths. The B-52G/Hs could be armed with eight SRAMs on an internal bomb bay rotary launcher and six on each under wing pylon. SAC accepted delivery of 1,500 SRAMS. On display at the National Museum of the United States Air Force. *Lt. Col. George A. Larson, USAF (Ret.)*

218

SAC had to accommodate alert bombers and tankers. Tankers were on alert with the bombers. SAC often used existing flight line barracks renovated for updated use and supported by temporary alert trailers. SAC developed a single, permanent readiness crew alert facility located at the head or entrance of the alert aircraft parking apron. These were nicknamed *moleholes*.

Boeing AGM-86B Air-Launched Cruise Missile (ALCM) on display at the National Museum of the United States Air Force. *Lt. Col. George A. Larson, USAF (Ret.)*

AGM-69A SRAM SPECIFICATION AND PERFORMANCE

Wingspan	14 feet
Diameter	17 1/2-inches
Weight	2,240 pounds
Powerplant	(One) SR75-LP-1 two-stage solid-rocket motor
Max Speed	Mach 3.5
Range:	
Low-altitude	35 to 50 miles
High-altitude	100 to 135 miles
Warhead	(1) W-69 Thermonuclear weapon
Yield	200 KT

B-52G painted in Persian Gulf (desert) camouflage colors. The aircraft was removed from the active duty Stratofortress fleet because of reductions under START. The aircraft had its engines removed, along with the nose mounted Electro-Optical (EO) Viewing System. At this time, the aircraft has not been sliced into sections at the 309th Aerospace Maintenance and Regeneration Group managed by Air Force Material Command (AFMC). This is the Department of Defense's single location for regeneration, maintenance, parts reclamation, preservation, storage, and disposal of excess DoD and government aerospace vehicles. B-52G parts support the current B-52H fleet. *Lt. Col. George A. Larson, USAF (Ret.)*

It was developed as a long-range, air-to-ground cruise missile. It was designed to fly at low altitude using a Terrain Contour Matching (TERCOM) guidance system. Its small radar signature allowed the missile to blend into surrounding ground clutter to hide from enemy radars. The missile could be carried on B-52G/H bombers. SAC accepted the delivery of 1,715 ALCMs. Initially production was the ALCM-A, which was then superseded by the larger and more capable ALCM-B.

AGM-86B ALCM Specifications and Performance

Wingspan	12 feet
Carrying Capacity with the AGM-86B ALCM	
Length	20 feet 9-inches
Internal rotary launcher	**External Pylons**
Diameter	24-inches
B-52G	8
	12 (6 on each)
Weight	3,200 pounds
B-52H	8
	12 (6 on each)
Powerplant	(1) 600 pound thrust Williams Research F-107-WR-100 turbofan engine
Max speed	Mach 0.7
Max range	1,500 miles
Warhead	W-80 thermonuclear
Yield	200 KT

Permanent alert facilities were two-story reinforced concrete and concrete block construction, with one story below ground and the top story partially below ground. The alert buildings had a tunnel type entrance/exit covered by a corrugated steel tube. Some variations were authorized:

1. 18,000 square feet to accommodate 70 alert personnel
2. 22,500 square feet to accommodate 100 alert personnel
3. 31,000 square feet to accommodate 150 alert personnel

220

With the operational introduction of the SRAM more changes were made to SAC's alert facilities, as they had been completed after the introduction of ALCMs. SAC had to construct additional auxiliary buildings at the perimeter of the alert apron, including a power facility, surveillance tower, and reserve armed response team and security structure, along with an associated family visitation quarters.[10]

The last B-52H to takeoff from Ellsworth AFB. The Stratofortress makes a low-level pass over Rapid City.
Photo: Ellsworth Air Force Base, 28th Bomb Wing Historian.

Prior to the author's retirement in 1992, his last duty assignment was Headquarters SAC, 544th ARTW, where he participated in providing direct exchange of intelligence and military equipment information to the Soviet Union, and then information on military armaments verification under START, which reduced the number of U.S. and Soviet nuclear delivery vehicles: ICBMs, SLBMs, and SAC bombers.

At this time, the Aerospace Maintenance and Regeneration Center (AMARC) was responsible for the disposition of a large part of the B-52 fleet, receiving additional aircraft to meet START limitations.

B-52 Variant	Number sent to AMARC
B-52C	29
B-52D	87
B-52E	49
B-52F	52
B-52G	61
B-52AG*	87
Total	365

* Modified to carry and launch nuclear warhead long-range ALCMs. These aircraft were accountable as a strategic nuclear weapon delivery weapons system.[11]

B-1B, Ellsworth AFB. *Photo: Ellsworth Air Force Base, 28th Bomb Wing Historian.*

AMARC has four storage methods:

1. Type 1000, or long-term storage. Procedures and preservation methods are directed towards maintaining the functionality and material of systems and components for an extended period of time. Storage is considered valid for up to 48 months.

2. Type 2000, or reclamation. Procedures are directed toward maintaining the functional and material integrity of systems and components pending their removal and return to active service.

3. Type 3000, or flyable hold. Procedures apply to aircraft maintained in active flying status while awaiting either flight schedules or induction to Type 1000 or 2000 storage. This type of storage has a duration of 90 days, which may be extended for an additional 90 days.

4. Type 4000, or disposal. Procedures apply to aircraft which are being turned over directly for disposal. The procedures in this type storage are directed toward making the aircraft maintenance safe while it is awaiting disposal. Aircraft placed in Type 1000, 2000, or 4000 storage are maintained every 180 days. At this time storage integrity is checked, which may include securing tie downs or performing SPRAYLAT touch up.

Before aircraft are placed in storage, AMARC removes guns, ejection seat charges, classified equipment, and items that could be pilfered. Workers drain the aircraft's fuel system and pump it full of lightweight oil, which they again drain, leaving an oil film that protects and preserves the fuel system. After protecting the fuel system, workers cover engine intakes, exhausts, and any gaps or cracks in the upper portion of the airframe with paper and tape. They then spray the covered areas and other easily damaged surfaces, such as fiberglass radomes, fabric control surfaces, and canopies with a vinyl plastic compound called SPRAYLAT. The underside remains unsealed to allow free circulation of air. The black SPRAYLAT is applied with a spray gun to keep out dust and water, preventing occasional dust storms from sandblasting windows and canopies. The white SPRAYLAT, applied over the black, acts as a temperature control. With a white SPRAYLAT covering the internal temperature of the aircraft remains within 10 to 15 degrees of the external temperature.[12]

The method of destruction of 365 B-52s, a program eliminated as a visible nuclear weapon delivery system, went through changes in order to maintain the existing ACC B-52H fleet. The initial destruction process consisted of AMARC employees using a 13,500-pound guillotine to slice through B-52 fuselages and wings. The guillotine blade was lifted to a height of 80-feet by a crane, then released to drop through the prescribed Stratofortress section. Once the B-52s had been rendered non-operational, the open desert storage area allowed START verification by Soviet military satellites. After 90 days, those selected airframes to be completely scrapped were sold to scrap metal processing companies, producing 150,000 pounds of salvageable aluminum and other metals.

In order to save parts for use on B-52Hs, B-52Gs and B-52AGs were sliced through their fuselages with power saws, creating smooth and even cuts to allow the fuselage to be sealed to preserve internal parts and equipment. The SPRAYLAT coating protects the parts while the destruction complies with the provisions of START.[13]

In 1986, Ellsworth and the 28th Bomb Wing started to phase out its B-52Hs and make extensive preparations to accept the arrival of the B-1B Lancer. Construction was required on base at many locations, as contractors completed new unaccompanied enlisted dormitories in March 1986, a new Security Police Group Headquarters in October 1986, and facelift improvements to the 13,497-foot runway. Construction included new aircraft maintenance, testing, and weapons loading facilities, as well as hangars to accommodate the complex B-1Bs. The Wing received the first of its 35 B-1Bs in January 1987.[14]

The first production model B-1 flew in October 1984. The first operational B-1B was delivered to Dyess AFB, Texas, in June 1985, with initial operational capacity attained on October 1, 1986. The final 100th B-1B was delivered on May 2, 1988. The current B-1B fleet is 24 years old, far younger than the B-52H fleet. The B-1B was first used in combat in support of combat operations against Iraq during Operation Desert Fox in December 1988. The next chapter covers the B-1B at Ellsworth AFB.

CHAPTER ELEVEN
B-1B LANCER

USAF News Service.

This chapter will backtrack somewhat to cover the 1972 flood that hit Rapid City, causing great destruction and loss of life. Airmen at Ellsworth AFB assisted in saving lives and in the post-recovery process. Also, it will not be possible to cover all the topics of the B-1B. What follows is a tribute to those airmen and the B-1B Lancer.

Gordon B. Lease was assigned to the 44th SMW, who were at Rapid City on June 9, 1972:

> Following the Rapid City flood of June 9, 1972, the 44th SMW was designated by the USAF to receive the Air Force Outstanding Unit Award. There were other EAFB organizations also designated for service rendered to the Rapid City area in the days following the flood. Many civilians were also recognized and authorized to receive the award.[1]

Rapid Creek flows through Rapid City from west to east, with runoff from higher elevations and Pactola Lake's outflow to the east. A heavy thunderstorm dropped 10 to 15 inches of rain on the Black Hills, with rapid runoff into Rapid Creek. The Canyon Lake dam could not hold back the massive inflow and failed, suddenly releasing 375,000 gallons of water into Rapid Creek and creating an instantaneous flash flood. The damage was tremendous: 236 killed, 2,900 injured, 770 homes and 565 mobile homes destroyed, 2,035 homes and 785 mobile homes damaged, 36 businesses destroyed and 236 damaged, and 5,000 vehicles destroyed at an estimated cost to residential property of $35,900,000 and commercial property damage at $30,900,000. Additional damage was suffered throughout the Black Hills. Overall area damage exceeded $165,000.000.[2] Lease continues:

> My personal actions following the flood was searching for survivors from Canyon Lake to the area of what is now the business location of Abourezk Law Firm on Omaha Street on Saturday June 10th. On June 11th and 12th, I volunteered to work at the Rapid City Water Treatment/Pumping Plant at Mountain View Road. At the time, I still held a State license as a water treatment operator, and I had 12 years of experience in the field, so I was able to contribute.[3]

The B-1A was initially deployed in the 1970s as the replacement for the B-52H, although the B-52H currently remains in the active duty inventory along with the B-1B. Four prototypes of the long-range, high-speed (2.2 Mach) strategic bomber were produced and test flown in the mid-1970s, but the program was cancelled by President Jimmy Carter in 1977 before going into mass production. Flight testing did continue through 1981. The B-1B is an improvement of the B-1A, which was initiated or restarted by President Ronald Reagan in 1981. Major changes included structural modifications to increase payload by 74,000 pounds, an improved radar, and reduction of the aircraft's radar cross section (RCS) by a significant amount. The engine air inlet was extensively modified as part of this RCS reduction, necessitating a reduction in maximum speed to Mach 1.2. The first production B-1B flew in October 1984, with the first aircraft delivered to the Air Force and SAC at Dyess AFB, Texas, in June 1985. Initial operational capability was achieved on October 1, 1986.

North American Aviation.

28th BW B-1B airborne over western South Dakota during a training flight. *Ellsworth Air Force Base, 28th Bomb Wing Public Affairs*

28th BW B-1B on a pass over Mount Rushmore, west of Keystone, SD. *Ellsworth Air Force Base, 28th Bomb Wing Public Affairs*

224

B-1B Lancer

The B-1B's blended wing/body configuration, variable geometry wings, and turbofan afterburning engines combined to provide long range, high speed, and maneuverability with enhanced combat survivability. Forward wing settings are used for takeoff, landing, air refueling, and in some high-altitude weapons employment scenarios. Aft wing sweep settings – the main combat configuration typically used during high-subsonic and supersonic flight – enhances the B-1B's maneuverability in low- and high-altitude environments. The B-1B's speed and superior handling characteristics allow it to seamlessly integrate into mixed force packages. These capabilities, when combined with its substantial payload, excellent radar targeting system, long loiter time, and survivability make the B-1B a key element of any joint/composite strike force. The B-1B weapon system is capable of creating a multitude of far-reaching effects across the modern battlefield.

The B-1B is a highly versatile, multi-mission weapon system. The B-1B's offensive avionics system includes high-resolution synthetic radar capable of tracking, targeting, and engaging moving vehicles, as well as self-targeting and terrain-following modes. In addition, an extremely accurate GPS-aided INS enables crews to autonomously navigate globally without the help of ground-based navigation aids, as well as engage targets with a high level of precision. The recent addition of Combat Track II (CTII) radios permits an interim secure Beyond Line-Of-Sight (BLOS) connectivity until Link-16 is integrated on the aircraft. In a time sensitive targeting environment, the aircrew can receive targeting data from the Combined Air Operations Center (CCAC) over CTII, then update mission data in the offensive avionics system to strike emerging targets rapidly and efficiently. This capability was effectively demonstrated during Operations Enduring Freedom and Iraqi Freedom.

B-1B, nose on view. Maintenance airmen cleaning copilot's window. The air refueling receptacle is located (positioned) on the nose, allowing pilots an excellent view of the approaching refueling flying boom from the tanker above. The tanker's boom operator is guided to the B-1B's refueling receiver by the white guideline markings on the nose. *Ellsworth Air Force Base, 28th Bomb Wing Public Affairs*

The B-1B's self protection electronic jamming equipment, radar warning receiver (ALQ-161), and expendable countermeasures (chaff and flares) system complement its low-radar cross section to form an integrated, robust onboard defense system that supports penetration of hostile air space. The ALQ-161 ECM system detects and identifies the full spectrum of adversary threat emitters, which applies the appropriate jamming technique either automatically or through manual inputs. Chaff and flares are employed against radar and infrared threat systems.

B-1B capabilities are being enhanced through the completion of the Conventional Mission Upgrade Program (CMUP). This program has already improved the aircraft's lethality by adding the capability to carry up to 30 cluster munitions (CBU-87, CBU-89, and CBU-97), a GPS receiver, an improved weapons interface that allows carrying of Joint Direct Attacks Munitions (JDAMs) guided weapons, and advanced secure radios (ARC-210). Aircraft survivability is enhanced through the addition of the ALE-50 Towed Decoy System, which decoys advanced radar guided SAM and AAM systems.

The CMUP adds improved avionics computers which allow the employment of additional advanced guided precision and non-precision weapons: 30 Wind-Corrected Munitions Dispensers, or WCMDs (CBU-103, CBU-104, and CBU-195); AGM-154 Joint Standoff Weapon (JSOW); or 24 AGM-158 Joint Air to Surface Standoff Missiles (JASSMs). The B-1B can carry and employ a mix of these weapons because of its three weapons bay configuration. The B-1B also carries the extended range version of the JASSM. These modifications significantly increase the B-1B's offensive capabilities.

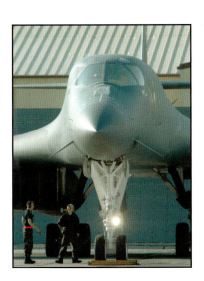

The B-1B nose landing gear can be steered 75 degrees left or right for directional control on the ground. The system is hydraulically operated and electronically controlled with commands sent through the rudder pedals. Nose wheel steering is automatically disengaged when the landing gear is retracted. The landing gear track is 14 feet 6 inches wide and has a wheel base of 57 feet 6 inches. The B-1B has been preflighted for a training flight. The ground crew is still talking with the pilots through their hard-wired headsets. This will be disconnected prior to aircraft rollout onto the taxiway. *Ellsworth Air Force Base, 28th Bomb Wing Public Affairs*

225

Above the EL on top of the tail is a fairing containing an ECM antenna and rendezvous radar beacon. A white tail light and anti-collision strobe light are mounted below the fairing. At the top one-third of the vertical rudder are static discharge wicks to dissipate static electric buildup from the airborne aircraft. *Lt. Col. George A. Larson, USAF (Ret.)*

B-1B on display at the South Dakota Air & Space Museum. *Lt. Col. George A. Larson, USAF (Ret.)*

Future modifications build on this foundation provided by the updated avionics computers. Radar sustainability and capability upgrades continue on the aircraft, in addition to an ultra high resolution capability leading to automatic target recognition features. The addition of Link-16 will give the B-1B capability to operate in the integrated battlefield. Cockpit modifications will relieve reliability problems and increase aircrew situational awareness while providing an integrated flow of information.

The B-1B's nose wheel with the drag link extending from the forward gear bay to the main strut. The nose gear retracts forward and upward into the fuselage. The crew access hatch and retracting ladder is deployed aft of the nose gear strut. Two four-wheel main landing gear retract inboard and aft into the fuselage, Multiple disk brakes are on each wheel, equipped with an anti-skid braking system. The aircraft has fixed geometry engine inlets. The inlet has radar cross section (RCS) vanes placed before the engine to enhance stealth, and prevents radar returns from hitting and reflecting back from the engine compressor. *Lt. Col. George A. Larson, USAF (Ret.)*

B-1B Lancer

B-1B Specifications and Performance	B-1B Crew
Wingspan (full forward)	137 feet
Pilot	Offensive(Weapon) Systems Officer
Wingspan (fully swept)	79 feet
Copilot	Defensive Systems Officer
Length	146 feet
Tail height	34 feet
Powerplant	(4) 30,000 pound General Electric F-101-102 turbofan engines with afterburner
Empty weight	190,000 pounds
Maximum weight	477,000 pounds
Maximum speed	1.2 Mach
Range (unrefueled)	6,500 miles
Low-altitude attack	1,730 miles
with	(8) SRAMs
	(8) B-61s
Service ceiling	30,000 feet
Armament packages	(24) JDAMs
	(24) MK-84s
	(8) MK-85 mines
	(84) MK-82s
	(84) MK-62 mines
	(30) CBU-87s or CBU-89s or CBU-97s
	(30) CBU-103 or CBU-104 or CBU-105 WCMDs
	(24) AGM-158 JASSMs
	(12) AGM-154 JSOWs[4]

General Electric F-101-102 turbofan engine. *Lt. Col. George A. Larson, USAF (Ret.)*

B-1B airborne, view from left rear side. The wings are fully swept back. Operational flight angle is between 65 and 67 1/2 degrees in transonic flight range from 0.8 to 1.2 Mach. The wings minimize the aerodynamic effects of increased transonic drag at 0.85 to 0.90 Mach. *Ellsworth Air Force Base, 28th Bomb Wing Public Affairs*

227

John Williams, assigned to the 28th Organization Maintenance Squadron (OMS), first arrived at Ellsworth AFB in 1982. He provides a different look at aircraft support activities which does not make it into a typical base history:

We had many recalls and practice Operational Readiness Inspections (ORIs). In December 1983 the month was very cold, ending in January 1984. In January 1984, the temperatures were below zero. During this period we were flying at a stepped up pace, and by mid-December, the 28th BW had flown all of its allotted hours for the quarter. However, the commander requested more hours so we could fly through the end of December. SAC approved the request, and we flew all our aircraft night and day. During the summer we hosted the 57th Squadron, RAF. They were at Ellsworth for 90 days, flying with the Wing's B-52Hs in their Victor tankers and Tornado fighters.

In 1985, the RAF returned to Ellsworth. Ellsworth started conversion to the KC-135R variant. During this second RAF presence at the base I was working in Job Control as Senior Weapons System Controller. We received a call that the British Queen Elisabeth was flying to a ranch near Sheridan, Wyoming. However, the aircraft was experiencing high winds and was requesting permission to possibly land at Ellsworth. We had a small dilemma, as what was the correct diplomatic protocol for the Queen of England? I called RAF Flight Leader Lieutenant Envs of the 57th RAF Squadron, which was operating on the base. He called me back, stating proudly that the 57th Squadron would handle the needs of the queen and her aircraft. However, the Queen was able to fly on to Wyoming. By the end of 1985, the B-52Hs were leaving Ellsworth. By 1986 the conversion to the KC-135R was complete, and by early March we had deployed to Malmstrom AFB, Great Falls, Montana, where we operated the KC-135R and the EC-135 until October. The runway at Ellsworth was dug up and rebuilt to handle the new B-1Bs that would soon arrive.

In January 1987, the first B-1B arrived at Ellsworth. Every week a new B-1B was delivered. During a SAC ORI with the IG team, the aircraft generation is underway racing the clock. Our Command Post was called with a real world request of then Vice President George H.W. Bush, who had been flying over the base on Air Force One and wanted to look at a B-1B. For the first time ever, a SAC ORI was put on pause and an impromptu party was organized. People were chosen and pre-positioned at different locations that the Vice President would visit. After the visit and the Vice President had departed the ORI resumed. During October President Reagan visited Ellsworth and Rapid City.

In 1988, the Wing's B-1Bs were combat ready and on nuclear alert. The Chairman of the JCS had invited the Russian military hierarchy to visit Ellsworth AFB. The second ranking Russian general, Marshall Akrameov, led a delegation of several generals and admirals to the base. It turned out somewhat like a summit meeting on the base.

General Colin Powell, Chairman Joint Chiefs of Staff, and to his right, Soviet Marshall Akvameov visited Ellsworth AFB. *Ellsworth Air Force Base, 28th Bomb Wing Historian.*

All of us were briefed that during the tour, if any of us were visited and a Russian general or admiral asked any of us a question, we were to answer his question truthfully but not to disclose any classified information. There was a picture in the base newspaper with Marshall Akvameov seated in the pilot's seat of a B-1B looking out the window with a big grin on his face. The delegation also toured the 44th MW training facility. During a formal reception an F-15 performed overhead aerial aerobatics. There was also an impressive flying exhibition by an Ellsworth B-1B, which at one point startled the Russians by coming in low from behind the way they were facing and then hitting the afterburners and pulling straight up. To say the least, the Russians were a little shook up. The final exhibition was the sound of a klaxon, at which time all the alert crews raced to their assigned aircraft, started engines, taxied down the runway and back to the alert pad, then re-cocked on alert. The general public wasn't invited to see this demonstration of SAC's alert capabilities. I was on duty, waiting to be possibly questioned by the Russians. I looked at the Russians, and they were looking at their watches for the time it took from the time of the klaxon until the aircraft took to the runways. This usually took 4 to 5 seconds.

During November 1988, a B-1B crashed at the south end of the runway. An Air Force investigation team determined it to be pilot error. My people recently completed a 200 hour phase I inspection. We were concerned that it might have crashed because of something that we did, or didn't do. We didn't have much technical data, as the tech orders or manuals were being written and we were guided by Rockwell International engineers on maintenance procedures. The data tapes were recovered and the aircraft wreckage was reassembled in Dock 32 during the investigation. I was at the out briefing and saw the complete presentation of the investigation. There was an animated video of the view from the copilot's seat perspective, broken into fraction of a second intervals of the final minute of this flight.

In 1989, I retired after completing an outstanding SAC MSET inspection. I was hired to work part time at the Ellsworth Transit Alert. During the summer of 1989, a Lockheed SR-71 Blackbird spy plane made an emergency landing at Ellsworth AFB. Being on duty at Transit Alert, we were responsible for handling this aircraft. We parked the aircraft in Dock 82, then helped the two crew members put their space suits on and with air conditioning packs. We had to place trash cans under the aircraft to catch the JP-7 fuel. When the aircraft cools the fuel tanks contract, allowing small leaks; at high speed the heat and air friction expand and seal the tanks. A Lockheed team was flown in on a KC-135Q. It carried JP-7 in its body tanks and JP-4 in its wing tanks for the aircraft's flight operation. A C-141 brought in a replacement engine and the Lockheed team changed out the engine. Once repaired, the SR-71 was towed out onto the taxiway and the KC-135Q towed to a position near the SR-71. A special hose for the fuel transfer was hooked to both aircraft, and the KC-135Q started to pump JP-7 into the SR-71's fuel tank. After it had been at Ellsworth for over a week the SR-71 made a spectacular departure. The afterburners had purple columns of flame with a row of purple rings around them.

Early in 1990 the new Presidential jet, a 747, made a fuel stop at Ellsworth. This was its first landing after being painted at Boeing Aircraft's Everett, Washington, plant. The crew spent the night at Ellsworth, and the next day a T-38 Talon arrived on base with a cameraman in the back seat. Both aircraft took off and together made a photographic pass over Mount Rushmore. During the summer, President George H.W. Bush was on his way to visit Mount Rushmore to take part in the 50th anniversary of the dedication of the four Presidents on the mountain top and landed at Ellsworth.

In 1991 SAC disbanded, with Ellsworth's components going to ACC (B-1Bs) and Air Mobility Command (KC-135Rs). The 44th SMW was under ACC and then Space Command before it was disbanded. In 1992, Air Force One landed with President Ronald Reagan onboard. In July 2007, Air Force One returned with President Bill Clinton onboard. In February 2002, a Russian Air Force Antonov AN-124 landed at Ellsworth to load two fuel trucks for transport to a base in Southwest Asia.

John Williams in front a of Blue Angels F-18 Hornet on the Ellsworth flight line. *John Williams*

229

On Monday, August 14, 2006, the Blue Angels diverted to Ellsworth. Five F-18 Hornets were prepared for takeoff at 5:30 p.m. All pilots arrived at their aircraft for an immediate takeoff. A large storm was near the base. As the pilots were climbing into their jets I heard a warning of lightning within 10 nautical miles of Ellsworth. We were out of our trucks to launch these aircraft, with four actually taking off at 5:41 p.m. The number 7 Blue Angels pilot had to shut off engine number one because of a low oil quantity light. At that time I heard a warning for large hail and high winds. I advised the pilot of this and we decided we needed to tow his aircraft immediately. My partner went to get a tug and I went to Dock 41, opened the doors, and brought a tow bar to the F-18. As we were hooking up the tow bar the pilot said "look there!" I turned to see two tornado funnels that appeared to be off the north end of the runway. Just then the base siren went off as a warning to seek shelter. I decided we were committed in our tow and we would be going to Dock 41, bringing the aircraft with us. Just as we pulled the aircraft inside Dock 41 large hailstones started failing.

The pilot climbed out of his jet and thanked us for saving his jet, and again thanked us on behalf of the U.S. Navy and Blue Angels. The cement block office inside the dock would be our shelter. We were in Dock 41 for over an hour when we heard the all clear for the tornado. I drove the pilot to Base Operations. While there, we were advised that the Blue Angels' support C-130 was inbound and would soon be landing. The Navy wanted to get the F-18 fixed as soon as possible and depart. We parked the C-130 at 7:37 p.m. on slot 23, just outside Dock 41, and at that time the only visible lightning was in the foothills west of Rapid City. We requested fuel for the C-130 and was advised that lighting was still within five nautical miles of the base. We told the crew about the lightning and they acknowledged that.

Meanwhile, approximately 30 Navy personnel had walked over to Dock 41, where they were told about the lightning by my partner. The crew decided they would be out by 6:00 a.m. Tuesday to refuel. Base security personnel came out and roped off the C-130. The C-130 crew stayed at the aircraft, inside the fuselage cargo section and under the aircraft's wing. These were mostly Navy and Marine Corps officers. They were waiting for transportation. I took one to Base Operations and my partner took two more there. I stayed at Base Operations in our office to coordinate our day shift coming in early to refuel the C-130 and find out information about our scheduled inbound of the Vice President's aircraft. At approximately 9:00 p.m. I received a call from the pilot on Air Force Two, stating they would be dead-heading on to Andrews AFB, and he thanked us for waiting for them.

I went downstairs, and one of the C-130 officers asked for a ride back to his aircraft. As we got into our truck Colonel Emig drove up and got out of his car and came over to us. He told the naval officer that the flight line was closed and to stop all activity. My partner had stayed down at Dock 41 in Transit Alert 156 during the time I was in our office. He said he had seen some of the Blue Angel mechanics go out to the C-130 to get tools and parts after they had been advised of the lightning. At approximately 9:20 p.m. the lightning warning was cancelled. The F-18 was towed out of the dock and the engine was run to check the repairs. The pilot had run out of his allotted over duty hours for that day and could not depart as they were trying to do. At 9:10 p.m. the work was wrapped up and both aircraft closed down. We had equipment to put away and paperwork to finish, so we departed at 10:30 p.m.

I finally had a chance to check my emails and the four weather warnings that had been passed throughout the base. What a shock to read these weather warnings to Ellsworth:

5:38 p.m., large hail and high winds
5:42 p.m., lightning within 5 nm
5:47 p.m., tornado sighted on radar
7:08 p.m., cx tornado warning
9:20 p.m., lightning with 5 nm

Interestingly, the only warning I had heard was the lightning within 10 nm and hail and high wind. We were out launching the aircraft and preparing to tow when the lightning was passed out, as I never heard a lightning warning within 5 nm.

Indian Air Force IL-78. *John Williams*

In July 2008, the Indian Air Force flew in two IL-78s and one IL-76 to Ellsworth AFB on a cross-country flight to Nellis AFB. This was a scheduled refueling stop, because the IL-78 tankers had completed their aerial refueling of an accompanying flight of Indian Air Force Su-39s headed to Red Flag, landing at Mountain Home AFB in preparation for the exercise. The two IL-78s and one IL-76 refueled and took off for a direct flight to Nellis AFB and Red Flag. This was the first time that the Indian Air Force participated in Red Flag.[5]

The 12th Air Division (AD) came to Ellsworth on July 15, 1988. This organization was responsible for training B-1B, B-52H, and KC-135R aircrew at Ellsworth and other SAC bases in the region. Headquarters SAC activated a third wing, the 99th Strategic Weapons Wing (SWW), at Ellsworth on August 10, 1989. The Wing assumed primary responsibility for B-1B advanced aircrew training.

B-1B landing with a second B-1B on the flight line. Its landing is assisted by hydraulically operated slats and flaps deployed to improve lift. The seven section leading edge slats extended downward at a 20 degree angle, which retracts after the flaps are retracted. The six section Fowler trailing edge flaps have a maximum deployment of 40 degrees. The leading gear is down and locked for the landing with a slight nose up position. *Ellsworth Air Force Base, 28th Bomb Wing Public Affairs*

B-1B Lancer

The 99th Wing was the Air Force's heavy bomber tactical employment center. It developed tactics, tactical employment training, operational unit support, and employment data clearing house. It developed and disseminated offensive and defensive tactics for weapons system employment across the entire spectrum of possible conflict. Tactics, techniques, and procedures were derived from analysis of current intelligence data, weapon system performance, and aircrew capability. The Wing establishes requirements for intelligence collection and to support weapons system mission development, tactics development, and analysis of adversary defense measures. The Wing prepared unit aircrew for effective weapons delivery by intensive training in dynamic tactical environments.

The Wing was the single point of contact for expert bomber employment instructors. For crisis and conflict conditions, the Wing's response options included focused and heightened tactics deployment and training, tailored intelligence production and dissemination, battle staff deployment, and battle aircrew and aircraft maintenance augmentation. The Wing operated bombing and electronic countermeasures range assets, as well as a dedicated maintenance organization for munitions support and flying operations.[6]

The author was assigned as Deputy Commander for Resource Management, 544th ARTW, Headquarters SAC, Offutt AFB, Nebraska. In 1989, he was responsible for equipment disposition from the Wing's large warehouse stock and made arrangements to transfer equipment to the 99th Wing at Ellsworth: desks, chairs, filing cabinets, lights, artificial plants, office desk top accessories, carpet protectors, and miscellaneous pieces of equipment.

On September 1, 1991, the 99th SWW was redesignated as the 99th Tactics and Training Wing (TTW). On June 1, 1992, the 99th TTW became part of ACC after SAC was inactivated. It became part of the USAF Fighter Weapons Center, headquartered at Nellis AFB.

B-1B forward view. The nose mounted radome opens to the left, revealing the Westinghouse AN/APQ-164 offensive radar for low-altitude, terrain-following precision navigation. The side nose-mounted 30 degree DOWNWARD angle Structural Mode Control System (SMCS) vanes reduce oscillation caused by flight turbulence at low altitude. *Lt. Col. George A. Larson, USAF (Ret.)*

Weapons Officers on the B-1B

The B-1B Weapon System Officer (WSO) is a specialized position in the four man aircraft, besides the Offensive System Officer (OSO) and Defensive System Officer (DSO). USAF OSOs are initially trained in Joint Specialized Undergraduate Navigator Training (JSVNT) at Naval Air Station Pensacola, Florida. Training follows a specific path for all services until being assigned to their primary aircraft weapon system platform:

B-1B WSO's position. Captain Matt Steele during his assignment with the 379th Expeditionary Squadron in Southwest Asia. *Department of Defense*

The B-1B is powered by four 34,000 pound thrust General Electric F101-GE-102 afterburning turbofan engines. The tabs extending from each side of the aft fuselage below the vertical tail are Vortex generators, which smooth air flow around the lower rudder, reducing fatigue and enhancing control surface effectiveness. RFS/ECMS defense antennas are on the tail cone sides and lower surfaces. The blunt tail cone houses the RFS/ECM antennas. The large fairing on either side of the tail cone houses the Towed Decoy System (TDS). This includes the Raytheon AN/ALE-50 Airborne Expandable Decoy System. It is a programmable, computer controlled, expendable countermeasures system controlled by the DSO. The TDS deploys an active Radio Frequency (RF) transmitter to redirect incoming missiles to the decoy and away from the aircraft. *Ellsworth Air Force Base, 28th Bomb Wing Public Affairs*

USN Indoctrination Course
Aviation Preflight Indoctrination (API)
Six weeks of ground school
Basic water survival training

Primary Training
1. Two weeks of ground school
Superfo USN program
SWSO flies in front seat of T-6 Texan II for six control flights
Academics on instrument flying rules
Final instrument check ride
Intermediate Training
1. Visual navigation (VNAV)
Low-level flights at 2,000-foot AGL
2. T-6 Texan II formal flying
Four flights to learn the basics of close-in formation flying
3. Ground school on USAF T-1 Jayhawk
Eight flights
Two-weeks of ground school
Advanced Training
1. T-39, USN twin-engine turbojet trainer
Advanced instrument flying
Radar navigation (A/G radar used in early F-16 variants)
2. Training consists of
VR and IR routes across the SW United States
3. Selection of weapon platform systems
Training for B-1B and B-52H Officers
Parachute Water Survival
Three day course held at NAS Pensacola
Basics of surviving an aircraft ejection over water
How to obtain food and drinkable water
How to successfully be rescued by helicopter/boat
Survivable, Evasion, Resistance and Escape (SERS)
Training conducted at Fairchild AFB, Washington
Three week instruction period
Basic survivable skills
Combat evasion
Techniques to resist exploitation if captured
How to survive an aircraft ejection overland
Electronic Warfare Officer (EWO) for B-1B and B-52H
563rd Flying Training Squadron at Randolph AFB, Texas
12 week training program
Basic radar theory
Basics of electronic attack
Basics of electronic defense[7]

B-1B DSO's position. *Ellsworth Air Force Base, 28th Bomb Wing Public Affairs*

B-1B pilot Kim Black, 37th BS. *Ellsworth Air Force Base, 28th Bomb Wing Public Affairs*

233

B-1B Lancer

B-1B Training

Rockwell Collins was selected by the USAF's ACC to perform the B-1B and B-52 CAT/CWD. The program requires training of all B-1B and B-52H aircrew undergoing training at the Formal Training Unites (FTUs) in the 7th BW at Dyess AFB, Texas, and the 2nd BW at Barksdale AFB, Bossier City, Louisiana. There are twenty-one Rockwell Collins instructors and courseware developers on each of the teams at each base for a total of 42 personnel assigned to the program.

Rockwell Collins instructors provide academic classroom instruction and simulator training for all B-1B and B-52H pilots, navigators, electronic warfare officers, and weapon system officers assigned to classes in the FTUs at each base. There are a variety of training courses, with each course designed to train a specific group of students based on their previous training, experience, and qualifications. The initial qualification course is for those aircrew that have never flown a bomber aircraft. The B-1B Transition Course is for aircrew who have been previously qualified in another bomber aircraft (B-52H). The re-qualification courses are for aircrew that have been qualified in the B-1B and B-52H, but have not flown them for a period of time. The B-52H syllabus also contains an aircraft commander and radar navigator upgrade class.

Academic Training

1. **Aircraft general knowledge**
2. **Specific aircraft systems**

Engine	**Communications/navigation**
Electrical	**Oil**
Hydraulic	**Radar**
Pneumatic	**Egress**
Avionics	**Weapons delivery**
Fuel	**Electronic Countermeasure Systems**

Manning in the simulator and other training devices covers requirements ranging from basic switch logic and emergency procedures to instrument procedures and full profile operation and weapons delivery. Rockwell Collins instructors teach the Annual Instrument Refresher Course to all student aircrew assigned to Barksdale AFB and Dyes AFB. To support the aircrew training effort, Rockwell Collins has a professional staff of personnel at each base that are trained and educated in Instructional System Development (ISD) and/or educational technology. Their responsibility is to create the courseware that is used by the students and instructors in the conduct of training. The courseware consists of firmware. Paper copies on study guides, lesson plans, and various student handouts are used. Most classes are supported by Computer Based Instruction (CBI) or Computer Based Training (CBT). All of this courseware is reviewed and revised on a prescribed schedule.

B-1B Copilot position. Captain David Grasso, 37th BS, assigned to 379th Expeditionary Squadron in Southwest Asia. *Department of Defense*

234

Rockwell Collins instructors and CWD personnel also support the Briefing Room Interactive (BRI) programs at Dyess and Barksdale. The scripts and story boards for each mission briefing are provided by Rockwell Collins via the USAF instructors. The Rockwell Collins personnel then develop and incorporate the graphics, illustrations, animations, photos, and videos, and create links to files and libraries that the aircrew may need to use for reference during the briefing. Links can be provided to weather, maps, range photos, training manuals, and technical documents. On the B-1B program, Rockwell Collins inherited mature CBT/CBI and BRI programs. Rockwell Collins maintains, reviews, and revises the material in those programs.

B-1B on Ellsworth's flight line. Above the engines are canvas-covered bladders, which seal the opening when the B-1B's wings are in full swept back position. The engine exhaust nozzle actuator fairings were removed in the early 1990s to reduce weight and reoccurring maintenance. *Lt. Col. George A. Larson, USAF (Ret.)*

The first B-1B (aircraft number 85-0073) arrived at Ellsworth AFB on January 21, 1987. The first loss of an Ellsworth B-1B occurred on November 18, 1988, when aircraft number 85-0076 crashed while landing. All four engines flamed out while the aircraft was on final approach and it struck a telephone pole. The four crew members escaped, with only the OSO suffering a fractured back during ejection from the aircraft. This occurred because he had leaned forward to see what had caused the bang. The bang was the collision with the telephone pole. At the same time the aircraft struck the telephone pole, the ejection seat(s) sequence(s) was initiated. The WSO's seat was in auto mode, so he was first to eject; unfortunately, he was in an improper body position at the moment of ejection.[9]

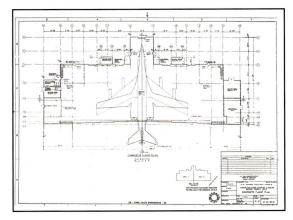

Engineering composite floor drawing of one of the new B-1B docks built at Ellsworth AFB after the exchange of B-52Hs to the B-1B. *Drawing: Ellsworth Air Force Base, 28th Bomb Wing Historian*

B-1B aircraft docks under construction on Ellsworth AFB. *Ellsworth Air Force Base, 28th Bomb Wing Historian*

B-1B with landing gear down, flying parallel to the flight line dock area. B-1Bs are in the background. *Ellsworth Air Force Base, 28th Bomb Wing Public Affairs*

Section of the Berlin Wall removed from the East and West border of the city and shipped to Rapid City, then set up on a concrete pad at Memorial Park, SW of the Civic Center. Internationally, the destruction of the Berlin Wall in October 1989 foreshadowed the imminent demise of the Soviet Union's empire over the next few years. During this transition, the Air Force reevaluated and reorganized its structure and resources to meet the new, rapidly evolving threat presented by a shifting world order. The Russian government slowly rebuilt its military, which by 2011 is modernizing and developing new generations of aircraft, missiles, and naval vessels comparable to that of the United States. *Lt. Col. Larson, USAF (Ret.)*

On January 3, 1990, SAC redesignated the 812th Combat Support Group as the 812th Strategic Support Wing (SSW), which for a short period of time became Ellsworth AFB's fourth wing. The 812th SSW consolidated all combat support activities into one organization. On July 31, 1990, SAC replaced the 12th AO with the Strategic Warfare Center (SWC), which provided operational command and administrative control over Ellsworth's subordinate units. As part of SAC's intermediate headquarters and base-level reorganization plan, SAC renamed the 28th BMW the 28th BW. On September 10, 1991, SAC inactivated the SWC and 812th SSW, with the wing absorbing the functions of the 812th SSW.[10]

Bomber and tanker alert pad off the SW end of the runway as of April 2011, with no active aircraft. In august 1991, in response to a decreasing Soviet military threat in Eastern Europe and the imminent dissolution of the Warsaw Pact military alliance and reduction of military forces along Western Europe's border, President George W. Bush ordered all B-1B squadrons removed from active nuclear alert operations. The 77th BS immediately began the transition, which later was a mandate for ACC, to transition its B-1Bs from a nuclear force to a full-time conventional weapons delivery system. *Lt. Col. George A. Larson, USAF (Ret.)*

B-1B Lancer

The first B-1B operational deployment to the United Kingdom was conducted on June 16, 1993, with two 28th BW Lancers landing at RAF Fairford. The Wing continued to adapt to meet new defense demands in light of the world's changing military and political threats. The Wing's versatile B-1Bs were the first in ACC to transition to an all-conventional munitions delivery role/mission. The 28th's squadron could deploy to anywhere in the world to meet national defense needs and military support requirements.

The Wing tested this concept in 1993 and again in early 1994, during such operational exercises as *Team Spirit*. During this exercise, the first B-1Bs landed in the Republic of South Korea. The author was assigned to SAC's Fast Reaction Team as the intelligence and weaponeering officer, which deployed to OSAN AFB, Republic of South Korea, to support B-52D airborne operations (B-52Ds never landed in the Republic of South Korea) for *Team Spirit* exercises. From 1978 to 1981, and again in 1988-1989, while assigned to the 6th Target Intelligence Groups (TIG), he coordinated U.S. and Republic of South Korea Air Force operations in Team Spirit 1989. Team Spirit exercises always drew negative comments from the Democratic Republic of North Korea, as well as concern from the People's Republic of China for the U.S. show of force in the defense of South Korea. This continues today, made even more serious by the isolation of North Korea and its development of atomic weapons and the missiles to deliver them, currently short range, with longer range intermediate and intercontinental range missiles being tested. Even as North Korea faces food and fuel shortages impacting its population, North Korea continues to take its belligerent path. The North-South Korea contentiousness continues today with periodic clashes of violence. The most recent was the sinking of a South Korean naval vessel and shelling of an island near the DMZ by North Korea. The Wing's B-1Bs took part in *Global Enterprise*, with the mission's intention to demonstrate the B-1B's global capabilities. Averaging just under 48 hours overall flight time, Ellsworth's Lancers set the record for the first B-1Bs to circumnavigate the globe.

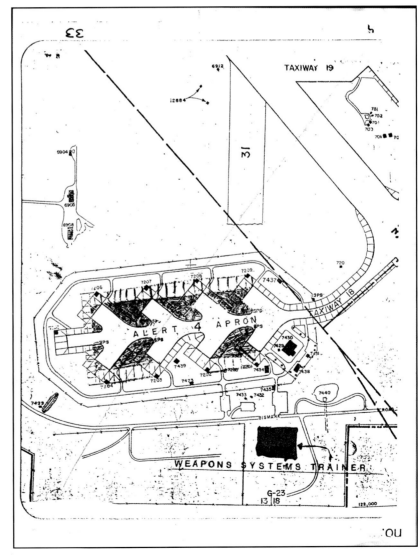

Engineering drawing of Ellsworth AFB's former alert apron at the west end of taxiway number 18. *Drawing: Ellsworth Air Force Base, 28th Bomb Wing Historian*

This operation also included various long-duration round-trip sorties from Ellsworth AFB to bombing training ranges outside the U.S. These were simulated combat mission strikes in support of the U.S. military and its allies around the world. *Bright Star* participation was the Wing's second, but the B-1B's first contribution to a JCS exercise in SW Asia.[11]

In March 1994 Ellsworth welcomed the 34th BS, a geographically separated unit (GSW) that was awaiting airfield upgrades before it could return to its parent organization, the 366th BW at Mountain Home AFB, Idaho. The 34th's B-1Bs are part one of the Air Force's composite wings, which also includes F-15 Eagles, F-16 Fighting Falcons, and KC-135Rs/KC-10s.

Also in 1994, the 77th BS shared in a Wing mission that was by far the most comprehensive tasking ever levied on the B-1B to date. The U.S. Congress mandated that the bomber's mission capability be thoroughly scrutinized over a set period of time. This large scale operational capability assessment came to be known locally as *Dakota Challenge*. The test began on June 1, 1994, and ran through early December. Test ground rules required that the Air Force provide the 77th and 37th BS with all the manpower, funding, and supply support originally intended for the system. The goal was to see if the B-1Bs, given this full backing, could sustain a mission capability of 75 percent over the long haul. Over the next several months mission capability rates increased, first to the established base line and then beyond. At the end of the test the 77th BS deployed to Roswell Industrial Center, New Mexico, for a short period to operate in a bare-base environment and conduct missions as would normally be done in war. By the end of November 1994 and the conclusion of *Dakota Challenge*, the mission capable rate for the entire six months was a sustained rate almost 10 percent higher than estimated![12]

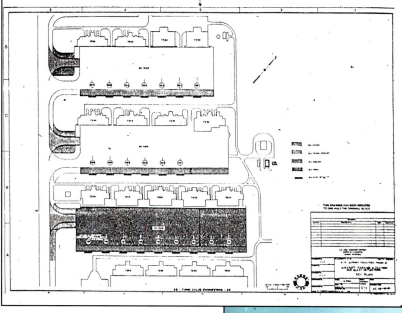

Engineering line drawing of the B-1B dock area. *Drawing: Ellsworth Air Force Base, 28th Bomb Wing Historian*

B-1Bs on the flight line in front of aircraft docks. The second B-1B in the photograph is being readied for takeoff. The hose provides air conditioning to the aircraft. *Lt. Col. George A. Larson, USAF (Ret.)*

238

B-1B Lancer

In early December 1994, the Air Force announced that the 77th BS would soon be inactivated. The public release marked the beginning of a new phase of the unit's service to the Air Force, as plans had already been underway to upgrade the B-1B's precision conventional munitions capability. In these lean fiscal times, however, funds needed for research, development, and production of improved weapons could not be made available without a significant sacrifice from other areas. The 77th BS' inactivation, taking place on March 31, 1995, met just such a need. Even within weeks of closure the squadron continued to carry its share of the Wing's workload in *Dakota Thunder 95*, a composite force training exercise with Air National Guard and Canadian fighter units. On March 11, 1995, the 77th BS flew its last *Global Power* mission, over 17 hours in duration, across the Pacific Ocean.

B-1B prior to touch down on Ellsworth's runway during a cycle of touch and go landings and takeoffs. *Lt. Col. George A. Larson, USAF (Ret.)*

B-1B on the flight line during Dakota Thunder 2000 with a display of conventional munitions and loaders. Close up views will follow. *Lt. Col. George A. Larson, USAF (Ret.)*

The 77th BS launched its farewell *Bone* (similar in appearance to the aircraft in the photo) sortie on March 29, 1995, when Lt. Col John Chilstrom (aircraft commander/pilot), Captain Jeff Kubiak (pilot), Captain Robert Loy (Weapon Systems Officer), and Captain John Wallace (Defensive Systems Officer) took the squadron's flagship aircraft (number 85-0077, called the *Pride of South Dakota*) to the air. As the mighty thunder of its afterburners faded away and the dark jet with the 77th BS' blue tail flash disappeared into the western sky, an era temporarily ended at Ellsworth AFB.[13]

Between September 1966 and September 1977, B-1Bs were upgraded under the Conventional Mission Upgrade Program to carry cluster bomb units. In a joint exercise with the USN, which began on April 21, 1977, and lasted for several days, Ellsworth B-1B crews practiced dropping MK 62 Quick Strike (QS) mines, a first for the B-1B, since MK 62s were previously only carried by B-52s. The mine is a derivative or variant of the conventional 500 pound GP bomb fitted with a fin kit that deploys a drag parachute deployed after the weapon's release, slowing its entry into the water. The B-1B can load a maximum of 84 bottom fused mines. The B-1B is fitted with three weapons bays, each holding 28 MK 62 QS. The mines could be dropped into the water with a maximum depth of 300 feet.[14]

KC-135 modified into special reconnaissance and verification aircraft under *Open Skies* (marked on the tail) for START compliance/verification. On display at Dakota Thunder 2000 at Ellsworth AFB. *Lt. Col. George A. Larson, USAF (Ret.)*

On June 6, 1995, members of the former Soviet Union's military arrived and began checking Ellsworth's B-1B Lancers for compliance with START. They were to complete an ICBM base line inspection. They were briefed at the Dakota Club by the on base Treaty Compliance Office. This was conducted by SMSgt Joseph Kuhn, who recalled:

The team of inspectors first went to the current "Pride Hangar," previously used during the 44th Missile Wing days. They arrived at the Pride Hangar at about 6:30 p.m. They searched the hangar to make sure there were no Minuteman II missiles. After inspecting the Pride Hangar the team headed for building 410, otherwise known as the Mobility Building. Building 410 had been used by the missile wing to store transporter erectors, vehicles used to pull missiles from their silos and transport them. The inspectors were looking to see if there were any transporter erectors in the building and, again, they made sure there were no Minuteman II missiles.

The next day the team left to inspect the railhead, an area near the base firing range. The railhead, like the transporter erectors, was involved in the transport of missiles via rail. After completing their inspection of the railhead, the team traveled to the other side of the flight line to the roll transfer building. The roll transfer building was used to transfer missiles from a transporter erector to an air shipping container. The roll transfer building was the last area inspected. At 10:00 a.m. there was a signing ceremony. Both team chiefs signed documents saying Ellsworth was in compliance with the treaty. Russian Chief Inspector Zoslov and Air Force Major Kocher, assigned to the On-site Inspection Agency in San Francisco, California, signed the compliance documents. Major Kocher and his team were the treaty experts for the United States and were responsible for escorting the inspectors. Everyone on base did what they were supposed to do. There were no snags or hang ups. It went perfectly. The Treaty Compliance Office had to prepare for a follow on B-1B inspection to complete the one B-1B on the base line inspection that had been transferred to Ellsworth from Dyess AFB, Texas.[15]

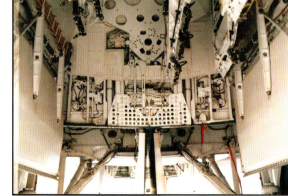

An empty CWM installed inside a B-1B's weapons bay. A perforated spoiler mounted on the bulkhead extends down when the bomb bay doors open in flight. It reduces turbulence inside the weapons bay to allow a clean weapons release. *Lt. Col. George A. Larson, USAF (Ret.)*

240

On May 28, 1997, a B-1B crew from Ellsworth dropped a single CBU-97 Sensor Fuzed Weapon against an array of tanks and vehicles at Eglin AFB, Florida. This was the first time a B-1B had dropped a CBU-97. A record number of B-1Bs from the 7th and 28th BW were deployed to RAF Fairford between May 26 and July 3, 1997, to take part in three NATO exercises.

On September 19, 1997, at approximately 22:20 GMT, B-1B aircraft number 85-0078, 37th BS, crashed 25 miles north of Alzada, Montana, while flying over the Powder River Military Operating Area. All four onboard were killed: Lt. Col. Antony Beat (aircraft commander/pilot and vice commander 28th BW), Major Clay Culver (Assistant Operations Officer, 28th BW), Major Kirk Cakerice (copilot), and Captain Garry Everett (Weapon Systems Officer). An Air Force Accident Investigation Board determined that the aircraft impacted the ground while conducting a defensive maneuver which involved slowing down and turning sharply to avoid a threat.

B-1Bs supported two Air Expeditionary Force (AEF) deployments to Bahrain during 1977. Two B-1Bs from the 28th BW joined the 347th AEF on November 18th as tensions with Iraq rose once again. In February 1988 a third B-1B was sent to Bahrain, as military action against Iraq looked likely. B-1Bs continued to take up previous missions performed solely by the B-52. In early 1998, Wing B-1B crews simulated Soviet Air Force TU-22M Backfires attempting to penetrate southernm U.S. East Coast defenses. This was part of the 20th Fighter Wing Operational Readiness Exercise. The B-1Bs flew four exercise sorties simulating airborne attacks by nuclear armed Russian Air Force long-range bombers against North Carolina's coast and Shaw AFB.

Two B-1Bs assigned to the Kansas Air National Guard (aircraft numbers 85-0064 and 86-0136), supported by one from the 28th BW (aircraft number 86-0128), arrived at RAF Fairford on July 23, 1988, to take part in the 1998 Royal International Air Tattoo, departing for the U.S. on July 27th. The first Block D B-1B, aircraft number 85-0091, arrived at Ellsworth on October 29, 1998. The 77th BS received all the *Fast 7* Block D aircraft. The other six were 85-0075, 85-0073, 85-0074, 85-0075, 85-0083, 86-0097, and 86-0104. The upgrade also included the Towed Decoy System. B-1Bs were alerted for possible military action against Iraq in mid-November 1988. Two Ellsworth and two Dyess AFB B-1Bs flew to Oman. The planned air strikes were cancelled.

The 77th BS' Block D B-1B flew its first test sortie on November 25, 1998. The Lancer crew consisted of Majors Troutman and Rodriguez and Captains Brunner and Humphries from the 53rd Evaluation Group, Det. 2. They flew to the Utah Test and Training Range from Ellsworth, dropping four inert BDU-56 2000-pound bombs with live JDAM tail kits.

The B-1B made its combat presence for the first time on December 18, 1988, with aircraft number 86-0096 (nicknamed *Wolf Pack*) from the 37th BS and aircraft number 86-0135 from the 9th BS, both from the 7th BW. These two B-1Bs attacked targets in Iraq with MK 82 500-pound conventional bombs during Operation Desert Fox. The next night the 37th BS' aircraft, along with a 28th BW B-1B (aircraft number 86-0102, named *Black Hills Sentinel*), completed another bombing mission. The commander in Oman reported that Iraqi anti-aircraft fire during these two nights was classified as heavy. One 28th BW B-1B (aircraft number 85-0084) and one 7th BW B-1B (aircraft number 85-0109) were ordered from the U.S. on December 20th to support the previously deployed B-1Bs. They were never used in combat. The six B-1Bs returned to their home bases at the end of 1998.

A conventional Weapons Module (CWM). It can be fitted with a variety of weapons. On display at Dakota Thunder 2000, Ellsworth AFB. *Lt. Col. George A. Larson, USAF (Ret.)*

CVM loaded with (28) 500-pound bombs. The CVN can also carry ten cluster bomb units (CBUs) or a 1,000-pound CBU-89 Gator. It can be loaded with 72 BLY-91/B anti-tank mines and 22 BLU-92B anti-personnel mines. The FZU-89/B proximity sensor detects enemy armor and explodes the mine at the optimum height above the tanks. The anti-personnel mines are designed to eliminate mine fields so they do not have to be cleared by personnel. *Lt. Col. George A. Larson, USAF (Ret.)*

CVM loaded with CBU-87 Combined Effects Munitions (CEM), each containing 202 bomblets. *Lt. Col. George A. Larson, USAF (Ret.)*

On March 5, 1999, changes were made to the support for the Expeditionary Air Force. This led to a reorganization of functions:

13th BS, Dyess AFB, TX (six B-1Bs in a training unit role)
77th BS, Ellsworth AFB, SD (six B-1Bs added to the Wing as a training aircraft contingent)
184th BS, McConnell AFB, KS (lost two B-1Bs)

MK 82 AIR bombs fitted with BSU-49/B AIR fins. These bombs can be dropped in either high-drag under ballute deployed or low-drag. Bomb release is controlled by the WSO. The bombs have a wide release range from 200 to 700 knots. *Lt. Col. George A. Larson, USAF (Ret.)*

The changes for the EAF created a designated *lead wing* that provided contingency leadership at the tactical level. The lead wings provide pre-designated commanders should the AEF have to provide a Group or Wing level leadership to the combat area. Two AEFs share a 90-day on-call cycle, and are responsible for providing rapid response within a maximum response time of 48 hours. The cycle process was initiated on January 1, 2000. The 28th was designated as lead wing in the EAF.

500-pound conventional bombs on a flat bed trailer, secured to a wood platform by chains to prevent movement during transport from the bomb dump to the aircraft. *Lt. Col. George A. Larson, USAF (Ret.)*

Five Ellsworth B-1Bs landed at RAF Fairford on April 1, 1999, to support Operation Allied Force. The 77th BS committed aircraft numbers 85-0073, 85-0075, 85-0083, and 85-0091, with the 73 BS adding aircraft number 86-0102. The aircraft would be operating in a high-threat environment, so the 77th BS' aircraft had completed the Block D upgrade. This gave the DSO the capability for the onboard electronics suite to correctly and with great accuracy identify ground based threats and select the appropriate electronic countermeasures to defeat these radars.[16]

Weapons bay with a load of (28) 500-pound MK 82 AIR training bombs on a CVM. An operational bomb can be released at higher air speeds than the previously carried Snakeye folding fin 500-pound bombs. *Lt. Col. George A. Larson, USAF (Ret.)*

The Kosovo Crisis was not like the major coalition effort of the Gulf War in 1991. Back then clear military plans had been built over a period of months, greatly aided by a firm consensus that Iraq was the aggressor and all measures necessary had to be taken to evict Iraqi forces from Kuwait. In Kosovo, the NATO partners brought contending opinions to the table. Commanders feared that losing aircraft could crumble NATO's will to continue the campaign.[17]

Inert training 2,000-pound Joint Direct Attack Munitions (JDAMs) on a munitions loader. The JDAM adds center body strakes and a new tail unit to the standard MK 84 conventional bomb. A GPS receiver in the tail guides the weapon via four pivoting tail fins. The guidance computer uses GPS and INS to determine its location and guide the bomb to pre-programmed geographical coordinates. The target can be entered any time prior to weapon release or automatically entered through the target designation with onboard sensors. Once released from the B-1B, JDAM operates autonomously without further impact from the WSO. It is an all-weather weapon. *Lt. Col. George A. Larson, USAF (Ret.)*

While the air campaign was gearing up in intensity, talk of a ground invasion began. However, it was clear from the beginning that NATO had to keep discussion of ground force options off the table. President Clinton stated, "I do not intend to put our troops in Kosovo to fight a war." U.S. Army General Henry H. Shelton, Chairman JCS, said "the military reality is that it would take anywhere from 20,000 to a couple hundred thousand ground troops to carry out a NATO military action in Kosovo, numbers well beyond what NATO was willing to contemplate."[18]

Most likely, the experiences of Bosnia and the ambivalence about political elements of the Kosovo crisis meant that NATO would never agree as an alliance to fight Slobodan Milosevic's army and special police with ground forces. The Russian Politburo and leadership would not permit the introduction of U.S. or NATO ground forces into Kosovo. Russian President Boris Yeltsin, on State Television, warned NATO not to commit ground troops into Kosovo. After Yeltsin's television broadcast, White House spokesman Joe Lockhart stated in the White House Press Corps Room "We've been officially reassured at a high level that Russia will not be drawn into the conflict in the Balkans."[19]

To carry out a sustained air campaign, NATO primarily tapped the resources of the U.S. Air Force. For the Air Force, the commitment to the Kosovo campaign quickly went from a contingency operation to a major theater war. The Air Force had undergone a 40 percent downsizing after 1999. That meant that Kosovo strained the Air Force's leaner force and tested in concept, for the first time, *Expeditionary Force* operations. By percentage, the Air Force deployed a higher share of its active and reserve force than at any time in the last 30 years (as of 1999). The Air Force's commitment during the Vietnam War used 15 percent of the Air Force's aircraft. Desert Storm required 30 percent of the Air Force's aircraft. During air operations in Kosovo, the Air Force allocated 50 percent of its aircraft in country and other worldwide commitments.

243

B-1B Lancer

High-demand command, control, and communications, along with computers/intelligence surveillance and reconnaissance (C4ISR) assets were represented at a 45 percent rate of active aircraft. For the Air Force, it committed 22 percent of its bombers (B-1Bs and B-2s) and 44 percent of its fighters. ISR and jamming/electronic warfare assets were nearly all committed. Forty percent of the Air Force's tankers were dedicated to Kosovo air support operations. This was only made possible by doubling aircrew: one flying and one on crew rest! President Clinton, after advice from the Chairman JCS, called up the reserves to maintain the pace of Kosovo air operations.

Air operations started with continuous combat air patrols over Kosovo and Bosnia. Suppression of enemy air defenses would be assigned to specific missions. Operation Allied Force included combinations of NATO and U.S. aircraft, and some U.S. only packages. NATO seized and maintained air dominance from the start of air operations. Crews of B-1B bombers reported thirty SAM firings during their first 50 missions they flew from their in-theater base at RAF Fairford, England. They were supported by ALE-50 towed decoy pods of electronic countermeasures reeled out behind the aircraft. Ten SAMs locked onto the B-1Bs but were diverted by the decoy pods. Despite remarkable caution during air strikes, there was unintentional loss of life. NATO aircraft released 23,000 bombs and missiles, and of those there were 20 incidents where bombs went astray from their targets, causing collateral damage and casualties. By far, the most serious international incident came from the accidental bombing of the People's Republic of China's embassy building on May 7, 1999. Several JDAMs hit the building, crashing through several floors and killing three Chinese nationals. The U.S. apologized, indicating intelligence sources used an outdated map of Belgrade which pin-pointed the wrong location. Every sortie flown and target struck during Operation Allied Force was designed to force Milosevic into acceptance of conditions laid down by the international community.[20]

MJ-1 lift truck. The driver, with a bomb on the forward lift, approaches the B-1B from the rear to raise the bomb into one of the B-1B's three weapons bays. *Lt. Col. George A. Larson, USAF (Ret.)*

B-1Bs launched just before midnight on April 1, 1999, shortly after their arrival at RAF Fairford, were assigned to bomb Serbian Army staging/depot areas in Kosovo. An Ellsworth B-1B (aircraft number 84-0074) landed on April 8th to replace aircraft number 85-0075, which required maintenance, landing back at Ellsworth on April 11th. Another swap-out occurred on April 11th with aircraft number 85-0073, which landed back at Ellsworth on April 26th. A third change in aircraft was made on May 15th, with aircraft number 86-0129 relieving aircraft number 86-0102, which returned to Ellsworth on May 18th. The fourth aircraft exchange

B-1B Lancer

occurred on May 27th, with aircraft number 85-0075 switching out aircraft number 85-0083, returning to Ellsworth on May 29th. The final aircraft swap-out occurred on June 3rd, with aircraft number 85-0075 taking over for aircraft number 85-0083, returning to Ellsworth on June 6th.

By the end of Operation Allied Force in Kosovo, five B-1Bs at RAF Fairford had flown 18 combat sorties with 74 actual bomb releases, dropping 1,260 tons of ordnance (MK-82s and CBU-87s). The destruction and damage caused to Serbian military installations was significant:

(7) helicopters	**(6) runways damaged**	**(4) POL storage facilities destroyed**
(4) MiG-21 Fishbeds	**(12) SAM launchers destroyed**	**(1) tank company attacked[21]**
(3) Galeb aircraft	**(8) ammunition storage buildings destroyed**	
(2) aircraft staging areas hit	**(3) barracks destroyed**	

Map from Andy Louder, Army Air Force Exchange Service (AAFES). He was assigned to AAFES Europe.

In 1995, after the first Allied air strikes penetrated Bosnian air space on August 30th, NATO had been involved in supporting the Serbian function in the Bosnian Civil War for three years. In the fall of 1992, the U.N. Protective Service asked NATO to enforce its ban on unauthorized Serbian military aircraft operations over Bosnia and Herzegovina. After Serbian military aircraft continued to fly in a defiant response to the U.N., Operation DENY Flight was initiated. This included the aerial engagement of Serbian aircraft openly disregarding the U.N. no fly zone, along with no air operations (flights) allowed over six Bosnian cities.

In April 1995, NATO aircraft attacked Serb military installations near the declared *Gorazde Safe Area*. In August they struck Serb radar sites, anti-aircraft missile launchers, radio-relay communications facilities, and anti-aircraft gun positions. By September, the majority of 357 identified Serb military installations had been destroyed by PGMs. These surgical air strikes limited collateral damage, providing a glimpse of U.S. attacks during the First Gulf War. It took less aircraft to reach some level or amount of destruction than the mass air bombings during the Vietnam War by SAC Arc Light sorties.

U.S. Air Force General Ronald R. Fogleman, Chief of Staff, USAF, commented in March 1966:

> Throughout this bloody conflict, Sarajevo remained a symbolic reminder of U.N. resolve to achieve peace in the region thanks to the efforts of professional flight and ground crews of the U.S. Air Force.

In Bosnia and Herzegovina, with a mixed society of Bosniacs (Muslims), Croats, and Serbians, the issue was highly complex. Bosniac voters overwhelming voted for independence from Yugoslavia in early 1992. The Bosnian Serbs boycotted the vote, forming their own separate government. The Bosnian Croats followed suit. It did not take long for vicious fighting to begin between these three factions, referred to as the Bosnian Civil War.

The MJ-1's width allowed it to drive between the B-1B's main landing gear, approaching from the rear of the aircraft, positioning itself under one of the three weapon bays, and then uploading the carried bomb. *Lt. Col. George A. Larson, USAF (Ret.)*

245

On March 3, 1992, armed citizens classified as *Irregular troops*, Bosniacs in Sarajevo, and Serbs surrounding the city faced off against each other. The Serbs blockaded the city. The U.N. arms embargo prevented the Bosniacs from acquiring weapons, while the Bosnian Serbs acquired weapons form the Serb-dominated former Yugoslavian Army depots and warehouses. This allowed the Serbs to gain and force control over 60 percent of Bosnia and Herzegovina. The U.N. imposed economic sanctions against the Serbs and authorized peacekeepers to enter Bosnia to assist in delivery of humanitarian relief by force if necessary.

However, under the reality of the conditions of the roads and control of the same into Sarajevo by Serb military forces, airlift resources became the primary method to deliver humanitarian supplies to 380,000 people inside the encircled city. The multi-national U.N. humanitarian airlift was code named *Provide Promise*, and started with an initial airlift delivery of supplies into the Sarajevo airport by French Air Force Lockheed C-130 Hercules on July 2, 1992. The airlift continued for three and one-half years, three times longer than the Berlin Airlift. To support relief of Sarajevo the USAF added more airlift assets, with the first C-130 landing on July 3rd. Regular Air Force operations were augmented by Air National Guard and Air Force Reserve aircraft delivering food, medical, and relief supplies into the city, as well as to rural and isolated areas of Bosnia and Herzegovina. If the aircraft could not land, supplies were dropped during low-level passes over isolated communities.

MK 82 500-pound bomb being loaded in a B-1B's weapons bay. The bomb has been hoisted up into the CVM by the MJ-1 lift truck, Two lugs on the bomb's upper surface are captured by an ejection rack, which is installed on one of the CVM's four swing arms. Ejector racks must be cleaned before weapons can be reloaded on the CVM. *Ellsworth Air Force Base, 28th Bomb Wing Public Affairs*

Continued Serb resistance to the delivery of humanitarian efforts forced the U.N. to impose no fly zones, with precision air strikes to force the issue. Additional diplomatic pressure on the Bosnian Serbs resulted after the United States mediated an agreement between the Bosniacs, Bosnian Croats, and Croatian government to create a Federation of Bosniacs and Croats. This agreement ended the fighting between these factions. The Bosnian Serbs retained their stance of belligerency, attacking unarmed aircraft trying to deliver humanitarian supplies. One Italian Air Force C-130 was shot down, with all crew on board killed. Due to increased small and medium arms fire around Sarajevo airport supplies were delivered by parachute drops or low-level ground skimming cargo extraction on cargo pallets.[21]

MHU-111 munitions trailer. *Ellsworth Air Force Base, 28th Bomb Wing Public Affairs*

246

The Army Air Force Exchange Service (AAFES) moved into the area almost immediately after the signing of the Dayton Peace Agreement. This is a little known part of the U.S. presence in former war-torn Bosnia and Herzegovina, except for the thousands of U.S. military and NATO forces served by AAFES exchanges or that supported those directly by the U.S. Army in the middle of the destruction. Mr. Andy Louder, a friend and neighbor, lives in the Chapel Valley residential section of western Rapid City, South Dakota, and was the manager in charge of AAFES operations in the Balkans. He mobilized available resources in only a few days after receiving orders to make it happen. Louder retired in 2010 after serving 35 years with AAFES. His last assignment was as manager of theEllsworth AFB Main Exchange. His following account of AAFES activities in the Balkans looks at the ground situation after Ellsworth AFB 28th BW strikes throughout the area:[22]

Before leaving the Balkans as General Manager, I spent a lot of time working on pre-planning for deployment. Joining me in this were the initial nine employees who were first to go to the Balkans. On December 7, 1995, I left AAFES Europe Headquarters at Mainz-Kastel and flew to Budapest, Hungary, with orders from Mr. Michael Cunningham, AAFES-Europe Vice President and Colonel Ralph Tucillo, former AAFES-Europe Commander, to rent a car, go and find Taszar, and see what was going on. I spent the next four days in Hungary, taking a look see, trying to locate real estate for our stores. These was none to be had. On December 12th, I headed back home to sort out my personal business. Two days later I was back on the road with Kevin Boucher. Together we drove through a blinding snow storm, heading for Hungary again.

The rest of the original nine employees arrived on December 19th. Together we worked toward one common goal, to take care of the troops and make money to continue AAFES operations. Working virtually non-stop with two to three hours of sleep a night, this group (Ross Armstrong, Kevin Boucher, Eva Bowles, Tom Komons, Mike Mrksic, Stacie Redd, Nick Vigiland, and Sam Winegar) gave everything they had to get the three stores opened by Christmas Eve (1995).

AAFES shipping containers and trucks used to store and deliver merchandise into the Balkans for AAFES stores.
Andy Louder

We had no warehouses; our merchandise was stored in shipping containers on a runway. Facing the worst winter to hit the area in 70 years, our team had to use every ounce of energy it had working in the snow and sub-zero temperatures. But we did it! The first couple of months were really quite tough. Getting little or no sleep, we rapidly expanded the number of facilities to support the troops. I made countless trips from Hungary to Croatia and Bosnia. For me, the most memorable trip was when Chris Weisshaar (AAFES Europe Chief), Mr. Cunningham (Loss Prevention), and I drove to Gradacac, in Northern Bosnia, in early January 1996 to meet with the 1st Brigade Command.

247

Destroyed Sava River Bridge. *Andy Louder*

As a result of the 17-hour drive in the snow, we became the first AAFES employees to cross the floating Sava River Bridge at Zupanja. U.S. Army engineers were tasked by their command in Europe to replace the destroyed Sava River Bridge. Army engineers had to ford a full flowing river in winter, with muddy roads hindering pontoon river bridge construction and in bitter cold weather. Army engineers started building the 2,043-foot long pontoon bridge on December 30, 1995, completing the pontoon bridge in early January 1996.

At the time the situation wasn't really secure; many of the roads hadn't been cleared of mines and the bunkers hadn't been destroyed. When we arrived at the 1st Brigade Headquarters, we found a three story building which had been heavily damaged by mortar shells during the war. Signs were posted instructing us to wear helmets and flak vests inside the building, even in the shower. Needless to say, we passed on a shower that night. We slept in what passed for the 1st Brigade conference room. The next day we found a location to set up our forward storage and support area. This, too, was a building which had been heavily damaged, but the location was perfect. Our mission completed, we headed back to Hungary.

I recall another trip with Mr. Cunningham and Colonel Tuccillo. They both teased me because I slept like a rock on a helicopter ride from Taszar to Camp Harmon, Croatia, and all the way back. The ride was in below zero temperatures because of the open windows that had .50 caliber machine guns sticking out of them. But when you're tired, you'll use every chance you can to get some sleep.

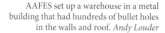

AAFES set up a warehouse in a metal building that had hundreds of bullet holes in the walls and roof. *Andy Louder*

There are many memories I have of the group I worked with. I remember Stacie Redd running a cash register outside in the snow for hours on end, with her head wrapped in a scarf and only her glasses showing. I remember Eva Bowles shivering in the "Scud Warehouse," so named because of all the bullet holes in the roof and walls, and Tom Komons, Jack Winegar, and Ross Armstrong freezing all day as they fought to unload shipping containers in the open. I remember others running a register and counting money, day in and day out. Many of these people came in the first few groups. The bottom line is that our AAFES team was made up of hundreds of dedicated people from AAFES Europe and Headquarters staff back in Dallas, Texas, to the people in the States who had to pick up the slack as we left our day jobs to support the deployment in Bosnia and Herzegovina. We at AAFES did what we do best – take care of our armed forces wherever they may be.[23]

Andy Louder in Bosnia. A U.S. Army 155mm self-propelled howitzer is in the background. *Andy Louder*

The American Forces Press Service news release of February 7, 1966, read:

Operation Provide Promise's mission complete. The longest running humanitarian airlift history officially ended on January 9th during a ceremony at Sarajevo airport according to the Department of Defense officials. Throughout nearly four years of war in Bosnia, U.S. forces helped keep a multi-national life line of food and medical supplies flowing into Bosnia. Over the past three years, Joint Task Force Provide Promise airlifted more than 160,000 metric tons of food and supplies into Sarajevo despite severe weather, snipers and political turmoil. U.N. convoys were unable to reach isolated areas or regions surrounded by Serb forces, task force aircrews dropped nearly 18,000 metric tons of bulk food, medical bundles and individual meals. Deliveries began in February 1993 and continued for 19 months.

The Joint Task Force began deactivating on February 1, 1995, turning over its remaining responsibilities to US Army Europe, according to Department of Defense officials. By March 15th, all personnel and facilities will have been returned to their parent agencies and units. Air operations were only part of the Provide Promise mission. U.S. forces also ran a field hospital in Zagreb, Croatia: provided reconnaissance information gathered by unmanned aerial reconnaissance vehicles; and patrolled the borders of Serbia and Macedonia. A 60-bed hospital treatment center provided emergency care for U.N. Protection Forces. U.S. Army, Navy and Air Force units on six month rotations staffed the field hospital. The hospital served more than 47,300 U.N. military personnel, admitted more than 2,000 patients and performed more than 1,200 surgeries. A Czech facility replaced the U.S. medical facility.

From July to early November 1995, US unmanned aerial vehicles collected high resolution video and still images. NATO and U.N. officials used the images to monitor ongoing operations. One part of Provide Promise mission continues. Task Force Able Sentry continued to monitor and report troop movements along the Serbia and Macedonia border. The U.S. Army 1st Battalion, 15th Infantry Regiment from Schweinfurt, Germany manned observation posts and conducted patrols. U.S. Army Europe took up operational control in mid February 1996.[24]

Bare bones AAFES exchange set up in the Balkans with metal shelving units and a new metal roof. U.S. Army soldier with full combat gear and M-16 rifles are shopping. Locals were hired to stock shelves and unload cargo containers. *Andy Louder*

Throughout AAFES operations in the Balkans, U.S. and U.N. forces were exceptionally supported in their hard military and political tasks. General Dennis Reiner, Chief of Staff, U.S. Army, wrote in a letter to AAFES personnel:

Once again, the Army and Air Force Exchange Service has answered the call of duty by providing a professional, all-volunteer team to serve our soldiers in the Balkans as part of Operation Joint/Endeavor/Guard. I have seen firsthand and the outstanding job they are doing to bring a little bit of home to our soldiers, with popular merchandise and fast food. AAFES associates have become a valuable part of the American team and have shared the good and bad of the experience with those they serve, eating the same food, sleeping in the same tents, and trading their business suits for BDUs and Kevlar. All the while they have enthusiastically and unfailingly provided value, service and support to America's army during a tough mission. On behalf of all our soldiers who have benefited from AAFES dedication serving them, I want to say "Thank you AAFES!" for being there for us. Your contributions toward the success of Joint Endeavor/Guard will never be forgotten.[25]

Multi-purpose launcher (MPL) on its traveling/mounting dolly. The MPL can be mounted in any of the B-1B's three weapons bays. On the right of the MPL, the rectangular box is the power control panel, which is always mounted toward the front of the aircraft's weapons bay. *Lt. Col. George A. Larson, USAF (Ret.)*

250

The 28th BW returned to RAF Fairford very quickly after fighting over Kosovo on July 22, 1996, with two B-1Bs from the 77th BS (aircraft numbers 85-0083 and 85-0091). They stood static display and flew during Royal International Tattoo as part of military aircraft aerial demonstrations. The B-1Bs returned to Ellsworth on July 26th. During the month of July, Ellsworth gained the USAF Weapons School B-1B Division's training on JDAMs. Instructors provided aircrew and maintenance personnel on the JDAMs, preparing (at that time) for the weapon's integration and deployment on/with the B-1Bs. In October, an Ellsworth B-1B dropped seven live GBU-35 JDAMs on the Nellis AFB Bombing Range against target munitions bunkers. After the B-1Bs returned to Ellsworth, the Weapons School students reloaded simulated munitions into the aircraft weapons bays to practice weapons reloading on a combat turn around. The students trained and practiced on the new electronics suite, an upgraded self-targeting mode on the radar for obtaining ground targeting information to reprogram (re-target) the JDAMs to destroy previously undamaged targets.

AGM-154 Joint Standoff Weapon (JSOW). It uses a GPS/INS navigation system to lock on to its target, and is able to be released from the aircraft at a distance from the target of 17 miles at low level and 46 miles at high level launch. The GPS/INS allows the weapon to follow and complete a pre-programmed flight path to the designated target. The INS used gyroscopes and accelerometers to determine changes in relative direction and position. The drift inherent in an INS is too high to allow its use as the sole guidance, so the GPS makes correction to the INS system. During flight, the GPS receiver updates the weapon's navigation system and keeps the imagery infrared seeker pointed at the target. When the imagery infrared seeker goes active it sends back a video image to the B-1B. The WSO uses a monitor to select and aim point on the target, allowing a precision strike. *Lt. Col. George A. Larson, USAF (Ret.)*

The 37th BS sent four B-1Bs to participate in Bright Star 99 between the USAF and Egyptian Air Force aircraft. With permission from the Egyptian government, over an Egyptian Air Force bombing range a B-1B dropped what an Air Force press release stated was "a JDAM equivalent weapon." This was the first time this new weapon had been carried outside the U.S. and was released to strike a designated ground target.

The new century marked further integration of the B-1B into a major conventional weapons platform, demonstrating its flexibility and importance to U.S. forces in Southwest Asia.

Ellsworth B-1B moves out of its parking spot in preparation to begin a training sortie under the direction of the aircraft's crew chief. The photograph shows the aircraft's clear aerodynamic shape. The aircraft carries all weapons inside the weapons bay, with no drag from externally mounted weapons. *Lt. Col. George A. Larson, USAF (Ret.)*

251

On February 14-18, 2000, the 37th BS completed a series of ten non-stop, 21 1/2 hour simulated long-range combat sorties during Global Power missions called "Cornet Spider 28." The 37th BS was tasked with flying two B-1Bs each day in the following pattern:

Took off from Ellsworth AFB
Flew north to penetrate the Alaskan Yukon Range
Flew southwest to Hawaii (no landing)
Flew east from Hawaii to the U.S. West Coast
Penetrated the coast over southern California
Flew over the southwestern U.S.
Turned north, landing back at Ellsworth AFB

During this five day exercise ten B-1Bs flew a total of 87,000 miles, requiring aerial refueling by KC-135Rs, which offloaded approximately 321,700 gallons of aviation fuel.

From February 29 through March 9, 2000, the 77th BS was tasked to fly two B-1Bs each day in the All Services Combat Identification Evaluation Team (ASCIET). Once again, the squadron's B-1Bs were to simulate Soviet Air Force Tu-22Ms attacking U.S. Navy and British Royal Navy warships conducting joint sea defense exercises off the southwest coast of Florida. The two B-1Bs then headed toward the southern Georgia coast, flying low over the Gulf of Mexico and trying to penetrate the U.S. Army installation at Fort Stewart, Georgia, and its Patriot SAM batteries and associated radars. On September 8th and 9th, two 77th BS B-1Bs (aircraft numbers 86-0095 and 86-0111) flew to South Africa, the first appearance by B-1Bs at the South African International Airshow held at Waterkoos South African Air Force Base.

The next combat deployment for Ellsworth's B-1B was during "Operation Enduring Freedom." Prior to the implementation of U.S. air power, one 34th and five 77th BS B-1Bs flew to Diego Garcia on September 19, 2001. President George W. Bush ordered a U.S. military response to the multiple terrorist attacks on September 1, 2001. Terrorist's highjacked four U.S. passenger jets: one each crashing into the two New York Trade Towers, one into the Pentagon, and one crashing into an empty field in Pennsylvania after its passengers tried to force their way into the locked cockpit to take back control of the aircraft from the highjackers. The terrorists nosed over the jet, sending it into the ground and killing all on board, thus preventing the mass slaughter of another target in Washington, D.C.[26]

During Operation Enduring Freedom, B-1Bs from the 37th and 34th BS joined to become part of the 34th Expeditionary Bomb Squadron. The B-1Bs were stationed at Diego Garcia, a small strategically located island in the Indian Ocean and part of British Indian Ocean territory. Eventually, the 7th BW from Dyess AFB, Texas, added five more B-1Bs. Also on the runway at Diego Garcia were B-52Hs and KC-135Rs,

B-1B taking off from Diego Garcia to perform ground support for coalition troops on the ground in Afghanistan on October 4, 2001, during Operation Enduring Freedom. Five B-1Bs are on the flight line, with three KC-10 tankers across the flight line. *Ellsworth Air Force Base, 28th Bomb Wing Public Affairs*

augmented by B-2s flying non-stop and home to Whiteman AFB, Missouri. B-1B missions averaged 12 to 15 hours, a round trip distance of 5,500 miles. B-1Bs struck Al-Qaeda and Taliban target areas, including early warning radars, ground forces, command and control facilities, and airfields and aircraft.

Aerial view out the window of an Ellsworth B-1B over Afghanistan during Operation Enduring Freedom. *Ellsworth Air Force Base, 28th Bomb Wing Public Affairs*

The results of the B-1B combat sorties were impressive, again with no losses:

B-1Bs had a combat mission effectiveness of 95 percent
B-1Bs flew 5 percent of the attack sorties
B-1Bs dropped 39 percent of total bomb tonnage
Weapons released:
2,974 JDAMs
1,471 MK-82s
135 MK-84s
70 CBU-87s[27]

As a cost saving measure the Air Force, after Operation Enduring Freedom, was directed by the Pentagon to begin a reduction in the number of operational B-1Bs. Ellsworth's 77th BS was inactivated, with the 34th BS replacing the 77th BS. Fortunately this was only temporary, but the base nearly ended its historic presence in western South Dakota, since it was removed from the active Air Force base list. Once again, fortune smiled on Ellsworth AFB.

USAF

B-1B Lancer

DEPARTMENT OF THE AIR FORCE
HEADQUARTERS 28TH BOMB WING (ACC)
ELLSWORTH AIR FORCE BASE, SOUTH DAKOTA

Colonel James M. Kowalski
Commander, 28th Bomb Wing
1958 Scott Dr., Suite 1
Ellsworth AFB SD 57706-4710

Dear Friends

On behalf of the men and women of Ellsworth Air Force Base, I welcome you to Dakota Thunder 2003. For more than 51 years the professionals at Ellsworth and the 28th Bomb Wing have been privileged to be your neighbor here in the beautiful Black Hills. Throughout our history, you have been our friends and unfailing supporters in the defense of our great nation.

Your unfailing support and friendship have been paramount in helping us accomplish our mission of providing rapid, decisive, and sustainable combat airpower – Anytime, Anywhere!

Today we say thank you and offer you a chance to view some of our country's military equipment and meet the professionals behind the scenes who accomplish our mission. I encourage you to spend some time talking with our people, they're anxious to answer your questions and hear your comments.

We hope you enjoy the events and displays we've arranged. We have worked hard to present a variety of aviation and combat support activities including a U.S. Air Force A-10 flight demonstration. This day is a small token or our appreciation for your friendship and continued support.

I encourage you to take this opportunity to learn about our mission and meet the fine people stationed among you here in the Black Hills. We hope you enjoy the day.

Sincerely

JAMES M. KOWALKSI, Colonel, USAF

Global Power For America

Press release: Ellsworth Air Force Base, 28th Bomb Wing Public Affairs.

THOMAS DASCHLE
SOUTH DAKOTA

COMMITTEE
AGRICULTURE
(202) 224-2321
TOLL FREE 1-800-424-9094
Email: Tom_Daschle@daschle.senate.gov
Internet: http://www.senate.gov/~daschle

320 SOUTH FIRST STREET, SUITE 101
P.O. BOX 1536
ABERDEEN, SD 57402-1536
(605) 225-8823

1313 WEST MAIN STREET
P.O. BOX 8168
RAPID CITY, SD 57709
(605) 348-7551

320 NORTH MAIN AVENUE, SUITE B
P.O. BOX 1274
SIOUX FALLS, SD 57101-1274
(605) 334-9596
TDD (605) 334-4632

United States Senate
WASHINGTON, DC 20510-4103

September 13, 2003

Dear Friends:

Welcome to Ellsworth Air Force Base and Dakota Thunder 2003! I am confident that you will enjoy the amazing skills of the pilots and the impressive capabilities of the planes that will be on display today.

This year, we are especially proud of our hosts – the men and women of Ellsworth. Many of them have been deployed around the world in support of Operation Iraqi Freedom and Operation Enduring Freedom. In fact, four Ellsworth crew members received the Distinguished Flying Cross for outstanding service in striking a target in Baghdad that was believed to house Saddam Hussein.

You will have an opportunity to see the B-1 bomber in action – the plane that has played a key role in both Afghanistan and Iraq. The military had considered retiring a number of B-1s from the force, but its stellar performance in these operations has inspired the Senate to pass an amendment that I introduced that will effectively roll back the B-1's retirement plan and help ensure that our nation has an adequate number of these increasingly vital planes.

A sophisticated plane like the B-1 requires a highly skilled and motivated support team. That is where the people of Ellsworth come into the picture. Today, you will have the opportunity to interact with some of the people whose professionalism and commitment was critical to the United States' effort in Iraq and Afghanistan. I encourage you to join me in thanking them for a job well done.

I also want to salute the pilots who will be performing today, including the A-10 East Coast Demo Team and the Air Force's Heritage Flight. Seeing their aerial skills further reinforces my belief that America has the world's most-talented and best-trained airmen. They are a testament to our nation's determination to defend our freedoms and our way of life.

My thanks again to those who worked hard to put together this year's Dakota Thunder. Please enjoy the show.

With best wishes, I am

Sincerely,

Tom Daschle
United States Senate

STATE OF SOUTH DAKOTA
M. MICHAEL ROUNDS, GOVERNOR

August 2003

The *Dakota Thunder* Air Show is truly an astounding exhibition. On display for you here today are the world's best machines and personnel, a strong and steadfast branch of the most formidable armed forces in the world. The United States Air Force is truly extraordinary, both for the tremendous technological strength at its disposal and the excellent personnel who wield it.

In Operation Iraqi Freedom, Ellsworth B-1 B's flew 220 combat sorties and dropped 4.56 million pounds of munitions with an 80 percent mission capable rate. Ellsworth B-1 B's dropped 22 percent of all the guided munitions in the conflict, and their participation was vital to the stunning success of our Armed Forces in Iraq.

We are here to acknowledge these dedicated individuals and their exploits. We will witness performances by a number of pilots, including a demonstration by the A-10 demo team. As a pilot, I am inspired by the skill and precision these men are going to put on display for us.

Now more than ever we should be grateful for the perseverance of these individuals. In trying times, we rely most on the devotion and strength of those who have committed themselves to the defense of our country, of our way of life. These brave and noble soldiers have stood, unwavering, in our defense, and their self-sacrifice should be lauded here today far above any feat of piloting or aeronautics, no matter how impressive.

I hope you all enjoy this tremendous exhibition and the valor and courage of those who have made it possible.

Sincerely,

M. Michael Rounds

MMR:ls

Press release: Ellsworth Air Force Base, 28'h Bomb Wing Public Affairs

Press release: Ellsworth Air Force Base, 28'h Bomb Wing Public Affairs

B-1B Lancer

Left to right: B-2 Spirit, B-52H Stratofortress, and a B-1B Lancer during a low-level pass over Ellsworth AFB during the Dakota Thunder air show and open house. *Lt. Col. George A. Larson, USAF (Ret.)*

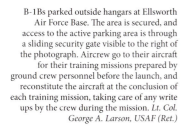

B-1Bs parked outside hangars at Ellsworth Air Force Base. The area is secured, and access to the active parking area is through a sliding security gate visible to the right of the photograph. Aircrew go to their aircraft for their training missions prepared by ground crew personnel before the launch, and reconstitute the aircraft at the conclusion of each training mission, taking care of any write ups by the crew during the mission. *Lt. Col. George A. Larson, USAF (Ret.)*

Dakota Thunder 2003 was Ellsworth AFB's annual air show and open house, which reflected its continuing appreciation for communities around the base, in addition to celebrating the Air Force's 100 years of flight. This was a significant event, even under the threat of news from the Base Closure Commission (which, as indicated previously, eventually was side stepped and remained on the Air Force's active base list), and demonstrated the base airmen's dedication to maintaining the defense of the United States. The local population responded with a huge crowd traveling to the base for the September 13th event. The author walked through the crowd, asking for comments on the importance of the base. Overwhelmingly the comments were "Ellsworth Air Force Base is important to the security of the United States and leads the attack on those around the world who want to destroy the country," followed by "The base has a significant economic impact to the Rapid City and surrounding cities which we cannot afford to lose." The big draw was the United States Air Force Aerial Demonstration Team, better known as the Air Force Thunderbirds. For two days, at the conclusion of each day's air show, the Thunderbirds flew their famous tight formations to the delight of the crowds watching along the edge of the flight line.

During the 2004 U.S. Senate campaign in South Dakota, the Republican Party challenger John Thune was pitted against the Democratic U.S. Senator Tom Daschle, who on April 16th made an election campaign speech at Ellsworth AFB. He stated "if elected over Senator Daschle, it puts Ellsworth in a lot stronger position than having someone who's going to be in the minority and someone who doesn't have a relation with the President of the United States." The President at that time was Republican George W. Bush. In a live debate between the two men broadcast on KSFY-TV and KOTA-TV on October 17th Thune stated "I think we have got to have somebody that has a relationship with the President of the United States, can work constructively across party lines in the Congress to get this done if we're going to save Ellsworth." On October 17th the author was in the audience at a Rapid City political rally when Thune stated "an all-Democratic Congressional delegation would

B-1B Lancer

have little political influence if President Bush is elected to a second term" (which did happen in the 2004 November general election). In a press release from Thune's Rapid City campaign headquarters, Republican U.S. Senator Bill Frist, Majority Leader, stated:

> Who is the President going to listen to more? The majority leader of the Senate, who he works with almost on a daily basis, or a Senator from another party who every day is saying things on the floor that demonstrate a lack of support? This time around, the President is appointing on the BRAC Commission, all of them.[28]

John Thune defeated seated U.S. Senator Tom Daschle. On May 13, 2005, the Department of Defense BRAC Commission recommended Ellsworth AFB be closed and removed from the active Air Force installation list. This set off all sorts of statements of economic doom and gloom in Rapid City and the surrounding communities. Newly elected U.S. Senator Thune did not agree, indicating the importance of the base to the people of South Dakota and the defense of the United States.[29]

With the decision to close Ellsworth AFB and draw down the B-1B fleet in numbers and base all aircraft at Dyess AFB, Texas, AMARC started receiving B-1Bs for storage.

AMARC B-1B LANCER INVENTORY

Serial Number	Arrival Date at AMARC	Departure Date from AMARC
84-000056	August 20, 20902	
86-000131	September 6, 2002	
85-000062	September 21, 2002	
85-000071	August 23, 2002	
84-000058	August 26, 2002	
85-000070	August 26, 2002	
84-000055	September 3, 2002	
85-000086	September 11, 2002	
86-000096	September 17, 2002	
85-000092	September 17, 2002	
84-000054	September 18, 2002	
85-000082	September 19, 2002	
84-000050	November 1, 2002	
84-000053	January 6, 2003	
85-000067	March 25, 2003	
86-000128	March 31, 2003	
86-000097	April 30, 2003	September 2, 2004[30]

In a stunning turn about, on August 26, 2005, the nine member BRAC Commission voted 8 to 1 to remove Ellsworth AFB from its recommended base closure list. BRAC Commissioner Harold Gehman noted "We have no savings, we're essentially moving airplanes (B-1Bs) from one very, very good base to another very, very good base, which are essentially equal."[31]

Interestingly, while the uncertainty of the existence of Ellsworth AFB blew in the wind of Washington, D.C., politics, a 34th BS B-1B flew into the heart of the Russian Federation to attend the Moscow International Aviation and Space Salon held at Ramenskoye Airfield outside Moscow from August 16 to 18, 2005. Captain Steve Jones (Pilot) noted:

> It's an honor to fly the first B-1B into Russia. We're all humbled that the Russians invited us to the air show and allowed us to participate. I think it shows how much progress out two nations have made since the Cold War. The fact that the United States would bring one of its strategic bombers

256

into this country and that the Russians would allow us to not only display the aircraft, but that we're able to fly it here…it shows how diffused the whole Cold War has really become. We're parked about 300 feet from Russian weapons systems, some of which were designed primarily to shut this aircraft down, and here we are parked right next to them at their air show. It's pretty cool.

Captain Ryan Sweeney (Weapons System Officer) recalled:

Both the aircraft and the U.S. service members were well received. The Russians have been very accommodating and welcoming. After flying our profile here and landing, they thanked us for our demonstration. The Mayor of Moscow also formally thanked us and presented us with a token of appreciation for our participation. It's obvious that they want us to feel welcome and we definitely do.

Captain David Black (34th BS Maintenance Unit Officer in charge of overseeing the B-1B's maintenance) stated:

With its capability to go more than 900 mph, the Lancer is considered to be one of the premier fly-over jets; it represents American horsepower and makes the most noise. People seem to be very curious about it. They want us to tell them the difference between our aircraft and their bomber. Although the B-1B and its crew normally average 14 to 15 air shows a year, it didn't make their participation in this year's Moscow show any less exciting. We couldn't have been more warmly accepted. It's just a great honor, and it's amazing to be held in Russia. I didn't think we'd ever be doing this in our lifetime.[32]

On May 2, 2008, a milestone was reached for the B-1B: twenty years in the Air Force as a strategic and conventional bomber. During its expanded conventional role, the B-1B has supported ground forces in Afghanistan and Iraq, where U.S. Army ground commanders indicated it is their most valuable aircraft. Beginning in 2003, the former nuclear designed strategic bomber has maintained a continuous air presence in Southwest Asia after modifications to deliver a variety of conventional munitions. As a result, due to expanding world-wide conventional support requirements and missions, Ellsworth AFB commanders held a series of information meetings on the expansion of its Powder River Bombing Range. The author attended some of these meetings while collecting information for this book. During these meetings the Air Force commanders and affected landowners exchanged comments on these training flights that would have B-1Bs flying at high speed and low altitude, creating an on-the-ground sound level of 130 to 135 decibels, but only at periodic times. B-1B crews needed a larger area to train, expanding the existing flight area from 7,000 square miles and adding another two areas to create a unified training area of 28,000 square miles. Not all this is in South Dakota; the area extends into the neighboring states of Montana, Wyoming, and North Dakota. Farmers and residents expressed their concerns over expected noise levels impacting their sleep and animal production, especially during the early spring calving season and stock production, which is vital to their economic bottom line during tough economic times. Air Force officials stated that B-1B pilots have been instructed about their flying near 17 specified noise sensitive areas, which B-1B crews would fly around and not through. Over and over again, Air Force officials commented that the B-1B over-flight noise, with an intensity of 113 decibels in indoor measurements, would probably not occur more than once or twice per training mission. The expanded Powder River Range has a relatively low population density area according to the Air Force. This makes the occurrence of an overflight profile effecting residents below a rare event. When the author was interviewing B-1B aircrew on the B-1Bs' "show of force" at low-level over enemy units in Afghanistan, this type of training in the United States was saving U.S. military soldiers' lives!

The B-1B's effectiveness in Iraq for urban ground force combat is peculiar, considering it was designed for nuclear and not conventional warfare, according to USAF Lt. Col. James Johnson, ACC B-1B Weapons System Chief:

The bomber was designed to make flights around the globe from four garrison bases to deliver nuclear ordnance. Now it's an inter-theater platform, trading long-range capability for long loiter time in Iraqi air space.

Air Force Senior Master Sgt, Chuck Klien, ACC B-1B Aircraft Manager, noted:

Upgrades enabled it to use smaller bombs and more of them. It's no longer this big guy we faced in the Cold War; now the B-1B uses precision smart bombs to support troops in contact. Operating at approximately 20,000 feet, the B-1B waits patiently with up to 35 tons of precision guided bombs. When ground troops encounter the enemy it can engage in minutes because of its readiness and speed.

B-1B on the taxiway, heading to the south end of the runway for takeoff on a training sortie.
Lt. Col. George A. Larson, USAF (Ret.)

In the first six months of Iraqi Freedom, B-1Bs delivered 40 percent of the bombs dropped, but only completed five percent of the combat operational sorties. By March 2007, B-1B sorties had increased by 25 percent. Other Air Force bombers have their own advantages. The B-2 uses stealth, approaching its target undetected until bombs strike their targets, while the B-52H's operating costs are less while operating over target areas. Klien:

The B-1B's speed, range, and payload makes the bomber a perfect fit for the war in Iraq. The airplane can do the job continuously and has the right mix of weapons, and its loiter time and massive payload makes it a fast life saver for ground troops.[33]

Air Force Senior Airmen James Pyeatt (28th Munitions Squadron) recalled:

Our days in Southwest Asia to support the B-1Bs is very important. We first load the munitions trailer, check the munitions to be certain they are secured for transport, and slowly pull out to the flight line. The munitions are delivered to specific aircraft depending on mission load requirements. The bomb load is determined by that day's combat sorties. We do not know that information until we get to the flight line. This changes from one sortie to the next, This is usually a mix of conventional precision munitions. We also load countermeasures for the specific mission to be flown. Weapons are now at the aircraft. The loader takes over to correctly configure the munitions in the weapons bays. The loader signs for the munitions. The expediter at the aircraft handles all the required paperwork to direct which munitions go to what aircraft. I also collect any unused munitions from aircraft which complete a mission if not retained on the aircraft. Once the loaders place the munitions, the expediter makes sure the bombs are properly positioned and ready for operational release. The cycle is repeated throughout the day as required. During a 24 hour cycle, our munitions personnel normally handle 4 to 6 sorties per day. I worked the 12 to 12 shift (midnight to noon) from January to July 2007. During this time period our squadron never had nor encountered any problems with the sophisticated precision guided munitions. But it was the aircraft Crew Chief who made sure the bomber was ready for that day's launch.[34]

258

Air Force Staff Sgt. Ryan Warkler (34th Aircraft Maintenance Squadron, Crew Chief) noted:

From wake up to aircraft recovery my day is full. I like to get up early, go to the gym, then head for the bus stop to go to work. Others I worked with just barely get out of bed and head straight for the bus stop. Once to the work area I check out my normal tool set, wait for that day's aircraft work schedule. On my aircraft I may have to take with me additional tools, which I sign for at the tool desk. A "Bread" truck drives us out to my aircraft on the flight line, drops us off at the aircraft. I do a pre-flight check, either to complete a generic flight check or possibly heavy maintenance as required. This varies on the aircraft from day to day. Every day is different, so then you've gotten all that accomplished and your jet is ready for the crew (ERED) so you wait there, get your jet pinned, and wait for the crew to show up. They get there and you brief them on any current maintenance that has occurred on that jet, so if it happens to reoccur they will be ready for it. You do your walk around with the crew and then they tell you you are good to go. You hand over the jet and you operate the headsets to the crew inside the aircraft until they clear you off and they taxi out to the main runway.

There are a whole slew of things that can go wrong on launch with no "Red Balls" (maintenance write-ups) or many, just depends, and the crew will try and help you work through them and get the jet fixed so it is ready to be launched. An average launch could take an hour or four hours. So you launch the jet and they complete their sorties, usually 12 to 14 hours for the airborne crew, and if you didn't catch the jet you launched, the other shift would do it in that time. You would do other maintenance that needed to be done as a crew chief. You seldom had any dead time while at work. You always strive for a 100 percent FMC (Full Mission Completion) rate. That way, if anything were to pop-off, we were always ready, and that is usually how I spent a day during deployment to Southwest Asia.

There are always some difficulties required to be faced each day. Everything that breaks on the aircraft over there you don't expect to break, so we don't have it in the kit, or all the jets have gone through that same item and depleted the kit. It always seemed that what we needed for some reason we just didn't have. This work is conducted in unforgiving heat. Sometimes the workload can get to the people, or the fact you are away from your family or the constant 12 hour shifts. Those things can wear on a person, making their willingness to work less. And that makes it hard on that person and everyone else, because you have someone not pulling their weight, and this is a comment from a person who has been to the island of Diego Garcia. Living a life that has many restrictions during a six month combat tour. I've been to Iraq and I've seen how everyone has to dress during off work hours. But I think that if you can run around in swim shorts and whatever T-shirt you want within taste, it kind of helps you feel that you're away from work and gives you peace of mind. Just the small things over there make life better.

I did not see or feel there were too many personal hardships when deployed to Southwest Asia. Personally, if I am being pushed then I can push right back. I will make sure my work is good. Some people, though, may work faster and mess things up to keep schedules with the pace of combat activities. The potential is there for that to happen. There is a constant pressure on everyone over there to get the work done fast because the ground pounders rely on us, but just as badly as our jets need us. My family takes it differently every time I go overseas. I go to Iraq in the heat of battle no matter how many times I tell them I'm not. I always think I am so obvious and they worry. My wife knows the situation and takes it well, but still hates it. When we talk we seldom have much to say. She will talk about her day and life that is going on just as it was when I left, minus me. I will say, well I launched a jet, worked on a jet, went to the gym, went to sleep just like I did last week when she called, and the week before. For me that's the worst, because I never have much to say, so she takes that the hardest when I don't chat with her.[35]

Major Brian Witkowsky (nicknamed Sea Bass) (34th BS, 28th BW) described a mission:

A typical mission during a state-side training week helps us prepare for overseas combat deployments. We will have a formal briefing for the entire squadron, where we will get a chance to get caught up on current events or pressing matters that can be accomplished at a Group level. At that point we will go to our vault (classified work area) area to execute our mission planning. We usually mission plan one day and fly the next. During mission planning we do initial coordination, where we determine the scenario, take inputs from intelligence officers (real world, or one which has been created for our training scenario), and objects for the day or for the mission. The rest of the day is pretty well set for the mission. In case of a close air support (CAS) mission, you are trying to contact the Joint Tactical Air Controller (JTAC) or get intelligence on the area of operations, or imagery, maps, or charts, and all those types of things at some point complete an attack meeting on which you are trying to do. Weaponeers look at specific targets and what kind of fuses to mate to what type of weapons. The end of the day concludes with a formal mission brief. The mission lead will conduct the briefing, review all the objects and scenarios, and how the mission is to be tactically executed.

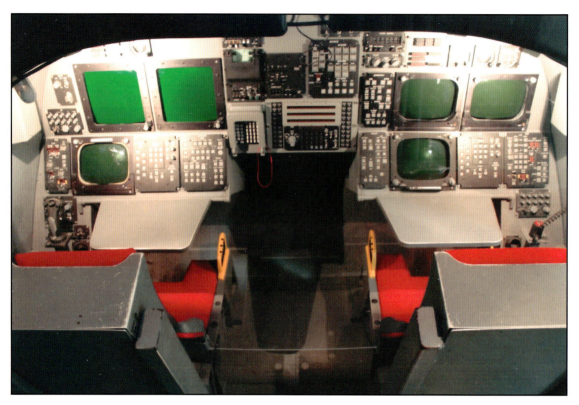

Interior of a B-1B simulator trainer on display at the South Dakota Air and Space Museum. The trainer is for WSO (left) and DSO (right). *Lt. Col. George A. Larson, USAF (Ret.)*

The next day you show up for the mission at the Step Desk. That's where the Squadron Supervisor provides weather briefing guidance for the mission on what aircraft you will be flying. You get out to the aircraft about one hour prior to take off, and we depart and execute a mission. For missions around South Dakota, it is about an hour in a military operating area, 50 percent at low altitude and 50 percent at high altitude simulated bombing. For a close air support sortie we do things such as a show of force on a low-altitude, high-speed pass over the enemy, not kinetic engagement against a potential enemy or hostile force. We also try to get an air refueling in and then come home to do some practice approaches (touch and go landings) on Ellsworth's runway. Once landed, you begin a mission debrief.[36]

Sometimes the B-1B is so effective it doesn't even need to drop a bomb. U.S. Army officials from the 82nd Airborne Battalion, 505th Parachute Infantry Regiment who returned from Iraq in November 2007 indicated a B-1B show of force was more than enough to put the enemy's heads down during gun battles. A show of force occurs when an aircraft flies by low and fast, releasing flares to intimidate enemy combatants. The B-1B can perform an intense low-level fly by.[37]

Witkowsky described mission planning:

In Southwest Asia, take off was basically similar to mission planning back at Ellsworth. A mission planning cell does most of that planning for us. We step to the jet where we have pre-flight crews. En route routes depend on what theater we are going to and at that point we prepare the aircraft, get the radio set up, make sure our computer's area is all set up the way we like it to be, and make sure the aircraft systems are top notch. Once we are in country, we have to check in with the command and control element there. They will either provide us with a new tasking or continue on with the current mission. In theater, we are there to provide over watch for those ground forces, and again, typical tasking may not use or release weapons. The JTAC will be contacted and given an update on our status. Then we get an update on the situation on the ground. Once we do that, if there are no troops in contact and no hostile situation, they will usually give us information on a specific area in the country to familiarize us with that area. If there is a troop contact situation, the command and control element can move to any area of operations there and we go support from there. We check in with JTAC. This will be a little more expeditious. JTAC will give us a close air support brief, tell us what the target is and location, elevation, and nearest friendly troop elements. At that point our Weapons System Officer uses his radar to get a picture of that area based on the coordinates of that area and with the JTAC talking to the Weapons System Officer, and then we will reconfirm that target area. The Defense System Officer works his laptop computer and imagery. We try to minimize kill time over target. Once the target is located and confirmed by the crew and the ground, at that point the JTAC will give us clearance and the aircraft will turn into the target and we will release the weapons. If at that point the desired result is not met there is an option for a re-attack. Also non-kinetic to include show of force. Those would be used in situations maybe where the enemy location is not precise, but we can let them know there is a B-1B on station. When we go low and fast we get their attention.[38]

Air Force Major Joseph Kleeman (nicknamed Harpoon, 34th BS, 28th BW) and Witowsky comment on missions:

When we get a call from the JTAC, we try to get the mission done. The Weapons Systems Officer double checks any possible hostile threats, with a whole suite of electronics just dedicated for that. Even though it does not frequently happen, it is always in the back of one's mind.[39]

Witkowsky:

Any mission where you can provide that support to those guys who need it. We are like an airborne on call artillery. Especially in a lot of those crews, you cannot get in artillery due to the elevation. We can help those guys, and you can hear the relief in their voices even if you don't release weapons. This always means a lot. JTAC personnel will often come to us when we are at our bases and tell us how much they appreciate our mission. There have been specific missions when you could hear gun fire in the background over the radio, and when you provide that support and you achieve the effects that guy is looking for. When you hear the gun fire stop, you can hear the relief in their voices.[40]

Kleeman:

I would have to agree. It is when you know no doubt you helped them out. You can hear the sigh of relief over the radio in their voice. It makes it all worthwhile.[41]

Witkowsky:

A lot of guys, including me, did a tour with the troops on the ground as an air liaison officer (as of 2008). There have been eight to ten officers from the squadron, and that experience working with the ground commander helps us understand what is going on and share with other members of the squadron, opens their eyes to what is going on, especially on the ground. Up in the sky, you sort of get a displacement as to what is going on. There was one time when we were flying around and we were the only asset available to the ground commander. It was a support for the troops. Working as a team, this is important.[42]

Kleeman:

The best thing about our job is supporting troops on the ground, ours or Coalition. Just to know that they are safe, maybe just because of our presence, that is a very satisfying feeling. I do not see how it can get any better than that.[43]

Witkowsky:

Even if the guy on the ground is not actually engaging hostile combatants, just to hear the voice and relationship is unique. Everyone has everyone else's back. Away from combat, our life has gotten a whole lot better with access to the internet, phone calls home, sending photographs, and with squadron support my family is well cared for, even from the members of the 37th BS. At home we take care of deployed members' families.[44]

Air Force Major Bob Liebman, B-1B Weapons System Officer (WSO), 28th BW, flew numerous combat missions in Southwest Asia (three OIF and OEF flying deployments between 2003 to 2004). Major Liebman provides a close look at the war in Southwest Asia:

In Southwest Asia, B-1Bs were stationed at undisclosed locations so the host countries would have non-plausibility the bombers were stationed on their home soil. B-1B crews flew 10- to 16-hour combat missions. Even during the Iraq War B-1Bs flew missions over Afghanistan. This was a coordinated effort between aircrew, maintenance, and munitions personnel to keep the bombers mission ready and airborne. There could be up to eight of the ten assigned B-1Bs airborne at the same time. For instance, for operations into Iraq and Afghanistan (one grouping for each country of operations):

One prepared to take off for its in country mission
One entering the country to transmit to the assigned patrol area
One aircraft orbiting in its assigned patrol area
One aircraft exiting the country for return flight back to home base

There has been a great political debate over the existence of Weapons of Mass Destruction (WMD), primarily chemical weapons. Troops on the ground found FROG-7 (Free Rocket Over Ground) rockets viewed by troops in Aviation Week & Space technology (AW&ST) photos which appear to be equipped with chemical warheads. Before and during the Coalition attack, large convoys of trucks moved north out of Iraq into Syria and possibly Iran. They were possibly carrying out WMD from Iraq. This was of great concern to President George W. Bush and one of the reasons for the invasion of Iraq, as well as removing terrorist training facilities in that country. Do not believe the *New York Times* or *Washington Post* versions for one minute. B-1B attacks destroyed the Salman Pak terrorist training camp outside Baghdad. Overhead photographs showed a derelict Russian Tupolev passenger jet in the camp with the wings off. I suspect the wings were cut off to enable easier transport of the airframe hulk to the middle of their training facility. It's in the middle of a field. The camp was also a probable source of equipment and funds for terrorists. In my opinion, Saddam was probably more of a regional terrorism enabler and facilitator than an operational leader.

B-1Bs in Southwest Asia had multiple roles. This included Non Traditional Intelligence Surveillance (NTISR). Non-traditional for bombers I suppose, but bombers did do this same stuff in the Pacific in WWII and Korea. B-1Bs hunted Scuds; interdiction and strategic strikes; time sensitive targets (TST) and dynamic targets; and close air support (CAS). B-1Bs and other Coalition aircraft were striking Baghdad even as the ground offensive started. It took the Coalition ground forces about two weeks to get to Baghdad. B-1Bs traveled at high speed, working on TST as directed. A B-1B from Ellsworth took out Saddam Hussein's staff early on in the war. On April 7, 2003, as described in a _Time Magazine_ article, "The Hit." We believed we might have taken out Saddam, until he reappeared later. The Saddam strike only took 12 minutes. We were referred to by ground commanders as the "Roving Linebacker." I believe the CENTCOM (Central Command) Air Commander/CENTAF (Central Air Force), Lt. General Michael Moseley, said this, "His HQ was the CAOC."

When hunting for SCUDs, B-1Bs patrolled specific areas of Iraq. The aircraft's Doppler radar assisted the crew in locating ground targets. Before a crew released a weapon, it required positive identification so as not to hit friendly forces. Even when receiving targeting information based on ground controllers, the crew carefully plotted and determined where the bombs would fall, sometimes preventing accidental friendly fire casualties. B-1Bs conducted interdiction and strategic strikes to block or reduce the flow of supplies from rear depot areas to front line troops. Strikes were conducted against Iraqi command and control centers, and attacked concentrations of ground forces and support infrastructure. After the first day of air strikes, the B-1Bs only hit one pre-programmed target. The normal airborne operating procedures were to strike targets on the ground by information passed in flight. The real time, in-flight strike re-planning was mostly accomplished by the aircrew in the jet; an important point in a world where someone would entrust everything to RPVs (remotely piloted vehicles). We had airborne AWACS for deconfliction as well as direction by the Combined Air Operations Center (CAOC). CAS information also came from ground controllers to the Command and Control Center/AWACs. If nothing happened (i.e., if TST or CAS were not tasked), then there was a specific (default) target set for the B-1B to strike. Again, real time targets took priority.

B-1B crews also completed Collateral Damage Assessment (CDE, as defined by the ROEs), which was also part of Battle Damage Assessment (BDA). During CAS operations, the challenge was to minimize collateral damage because of geopolitical impact/significance. This is the fundamental reason we have the ROEs, in addition to protecting the good guys. During the air war, everyone ("everyone" may be too general, but the spirit of this comment is probably close to the mark, as an aircrew perception, anyway) at one time or another wanted to impose authority over selection of targets and release of weapons. The impact was reduced/eliminated somewhat by a series of Weapons and Tactics Conferences, with all parties involved in striking targets to prevent delays in effective employment of advanced weapons. Interestingly, this process again slipped back to the restriction on targets, referred to as ROEs. It has heavily tilted toward geopolitical implications/significance, especially as developed in Afghanistan as of the summer of 2010. The term "ROE" has been in use to define these constraints and procedures from the beginning of the OIF & OEF campaigns. CENTCOM drafts, promulgates, evaluates, and updates the Theater ROEs continuously. I suspect that the level of external interference in their execution goes through seasons defined by the current environment, directives, and sensitivities.

Major Liebman, from June to December 2007, served a tour of duty on the ground with the U.S. Army in Afghanistan:

Before that, I had three OIF and OEF flying deployments between 2003-2004, to include 40 combat missions in the B-1B, plus a brief stint supporting mission planning and ATO production at the CAOC as an Electronics Warfare Officer (EWO) assigned to the U.S. Army in country with Improvised Explosive Devices (IEDs). These Air Force deployments in support of the Army were called "ILO" by the Pentagon, deployed "In Lieu Of" Army resources. IEDs are particularly nasty weapons, designed to kill U.S. and Coalition troops, and independent Afghanistan truck drivers who supply goods and services to villages throughout the country. After the end of the Cold War, the U.S. Army had to make budget considerations, cutting its electronics warfare capabilities. The Army was not equipped or manned to do the job on the new explosive devices/dangers found in Afghanistan.

We assisted, and eventually the Army will rebuild its EW capabilities. I was at a Forward Operations Base (FOB) at Sharana, Afghanistan. Sharana was a base originally built and occupied by the Russian Army during its occupation of that country. It was upgraded and enlarged by U.S. forces. I was to assist in countering the IED threat as part of Joint Task Force PALADIN.

I worked with the U.S. Army with the 36th Engineer Brigade on the base. One of their many missions in Afghanistan included the RCP (Route Clearance Package), where the team hunted for, located, and removed (destroyed) IEDs. Air Force Explosive Ordnance Disposal (EOD) personnel assisted Army personnel in the dangerous IED removal. There was a whole fleet of specially designed and constructed vehicles to complete the dangerous mission, such as the Joint EOD Rapid Response Vehicle. The terrain is very rugged in the country, with high mountains down to low valleys. Many tribal villages in some of the isolated and remote areas date back to the time of Alexander the Great.

The Taliban's reign of terror is less of a local issue, but more one of refugees who previously had lived in Afghanistan and escaped to Pakistan during the Russian occupation and brutal attacks by the Soviet Army. They operate somewhere out of the rugged mountains of Pakistan/Afghanistan border regions. This is a fertile area for Taliban recruitment, assisted by the "Madrarras" religious schools, which spread the Jihadist philosophy during transit into and fighting in Afghanistan. The Pakistan government is the key U.S. ally in the area and faces turmoil from the Taliban in their country, which also affects nearby India's political and military leadership. The situation is even more intense due to the presence of nuclear weapons and delivery systems (missiles) in Pakistan and India. Much of the equipment (weapons) used by the Taliban appear to come from external sources, including the Peoples Republic of China Army (referred to as the Peoples Liberation Army, or PLA), with Al-Qaida part of this thing too, but I'm not sure how.[45]

B-1B pilot and copilot simulator on display at the South Dakota Air and Space museum. *Lt. Col. George A. Larson, USAF (Ret.)*

With all the B-1B aircrew from Ellsworth AFB on various deployments to Southwest Asia, it is impossible to comment on or tell all their stories. The following is a snapshot of some of the Airmen provided by Major Liebman from his personal logbook and recollections about these missions in Southwest Asia:

Mission number 1, January 21, 2003: Classified as OEF/Afghanistan. Aircraft number 85-079. Pilot. Lt. Col. Rich Clark; copilot, Captain Sloan Hollis; WSO, Captain Bob Liebman and DSO, Captain Ty Neumann. The mission lasted 9.9 hours, a XCAS reaction to Khoust airfield rocket attack.

Mission number 2, January 30, 2003: Classified as OEF/Afghanistan. Aircraft number 85-079. Pilot, Major Dave Ruffin; copilot, 1st Lt. Matt McDaniel, WSO, Captain Danny Coddington and DSO, Major Bob Liebman. The mission lasted 8.5 hours.

Mission number 3, February 2, 2003: Classified as OEF/Afghanistan. Aircraft number 86-121. Pilot, Lt, Col, Joe Winterseen; copilot, Lt. Col. Rich Clark; WSO, Major Bob Liebman and DSO, Captain Danny Coddington. The mission lasted 9.4 hours. CAS practice.

Mission number 4, February 5, 2003: Classified as OEF/Afghanistan. Aircraft number 86-138. Pilot, Major Doug Bodine; copilot, Captain Marc Chiasson; WSO, Danny Coddingron and DSO Major Bob Liebman. The mission lasted 8.4 hours. Low-altitude pass over SOF and targets in Hindu Kush.

Mission number 5, February 8, 2003: Classified as OEF/Afghanistan. Aircraft number 85-094. Pilot, Lt. Col. Joe Winterseen; copilot, Captain Jerry Brown; WSO, Major Bob Liebmann and DSO, Captain Danny Coddington. The mission lasted 8.8 hours. Low-altitude sonic passed over old fort north of Bamian.

Mission number 6, February 11, 2003: Classified as OEF/Afghanistan. Aircraft number 85-087. Pilot, Lt. Col. Joe Winterseen; copilot, Captain Jerry Brown; WSO, Captain Danny Coddington and DSO, Major Bob Liebman. The mission lasted 6.5 hours, The mission was SCUD hunting practice.

Mission number 7, March 1, 2003: Classified as OEF/Afghanistan. Aircraft number 85-087. Pilot, Major Al Nixon, copilot, 1st Lt. Abizer Tyabji; WSO, Major Bob Liebman and DSO Captain Dirk Benedict. The mission lasted 9.2 hours.

Mission number 8, March 4, 2003: Classified as OEF/Afghanistan. Aircraft number 86-138. Pilot, Lt. Col. Rich Clark; copilot, Major Doug Bodine; WSO, Captain Dirk Benedict at DSO Major Bob Liebman. The mission lasted 8.2 hours.

Mission number 9, March 9, 2003: Aircraft number 85-079. Pilot, Lt. Col. Jim Kowalski; copilot, Major Al. Nixon; WSO, Captain Bill Dobbs and DSO, Major Bob Liebman. The mission lasted 8.4 hours. It was GFAC practice with JTAC and Northern Alliance GFAC students.

Mission number 10, March 12, 2003: Classified as OEF/Afghanistan. Aircraft number 85-083. Pilot, Major Al Nixon; copilot, Lt. Col. Pat Harrison; WSO, Major Bob Liebman and DSO, Captain Bill Dobbs. The mission lasted 6.8 hours.

Mission number 11, March 15, 2003: Classified as OSW/Iraq. Aircraft number 85-087. Pilot, Major Al Nixon; copilot; Lt. Col. Pat Harrison; WSO, Captain Bill Dobbs and DSO, Major Bob Liebman. The mission lasted 7.4 hours. The aircrew attacked AAA and radars in western Iraq. IAW Operation Southern WATCH. Lead aircraft flown by Lt. Col. Jim Kowalski; Captain Jeff Haynes, Captain Steve Landsberger and Captain Mike Coppola.

Mission number 12, March 17, 2003: Classified as OEF/Afghanistan. Aircraft number 86-129. Pilot, Major John Stachnik' copilot; captain Marc Chiasson; WSO, Major Bob Liebman and DSO, Major George Chappel. The mission lasted 8.5 hours.

Mission number 13, March 21, 2003: Aircraft number 85-087. Pilot, Major John Stachnik; copilot, Captain Marc Chiasson; WSO, Captain Bill Dobbs and DSO, Major Bob Liebman. The mission lasted 9.7 hours. NTISR in western Iraq, flex to XCAS, unable to contact two JTACs. Dropped 6 JDAMs on Secret Police Headquarters. Saw the initial big OIF strike package inbound to the country during egress from the country while returning to base.

Mission number 14, March 23, 2003: Classified as OIF/Iraq. Aircraft number 86-115. Pilot, John Stachnik; copilot, Captain Marc Chiasson; WSO, Major Bob Liebman and DSO, Captain Bill Dobbs. The mission lasted 10 hours. NTISR in western Iraq, flex to XINTO, struck 3 bridges in northwestern Iraq. Flex to XCAS to kill 10 tanks in revetments near An Najaf w/JTAC.

Mission number 15, March 25, 2003: Classified as OIF/Iraq. Aircraft number 86-095. Pilot, Captain Chris Clark; copilot, Captain Aaron Juhl; WSO, Major George Raihala and DSO, Major Bob Liebman. The mission lasted 16 hours. Preflight plan was for multiple strikes on West Bagdad, NTISR in western Iraq' flex to XINT- attacking ammunition bunkers located there. Dropped six JDAMs on the ammunition bunkers and 4 JDAMs on ammunition storage depots. We evaded AAA and rocket/missile air defenses. Flex to XCAS for JRAC at An Najaf. Delivered several JDAMs on Iraqi Republican Guard T-72 tanks attempting to cross a river and overrun U.S. forces under cover of a big dust storm. Broke the attack with confirmed armor kills. BDA from JTAC.

Mission number 16, March 28, 2003: Classified as OIF/Iraq. Aircraft number 85-079. Pilot, Captain Chris Clark; copilot, Aaron Juhl; WSO, Major Bob Liebman and DSO, Major George Chappel. The mission lasted 12.4 hours. NTISR in western Iraq, dropped 3 JDAMs on Baath Party Headquarters facility, evaded AAA. Flex to XCAS, dropped 19 JDAMs on Iraqi armor and artillery located north of Bagdad-evading AAA and rocket/ missile air defenses, SEAD support in conjunction with F-16CJ and EA-6B.

Mission number 17, March 31, 2003: Classified as OIF/Iraq. Aircraft number 86-125. Pilot, Captain Chris Clark; copilot, Marc Chiasson; WSO, Major George Chappel and DSO, Major Bob Liebman. The mission lasted 10.7 hours. NTISR in western Iraq, flex to XINT with numerous bomb runs in northwest Bagdad-including one dropping 11 JDAMs dropped on targets. Flex to XINT bomb run ingress Bagdad from the southwest, SEAD with EA-6B.

Mission number 18, April 3, 2003: Classified as OIF/Iraq. Aircraft number 85-085. Pilot, Chris Clark; copilot, Captain Marc Chiasson; WSO, Major Bob Liebman and DSO, Major George Chappel. The mission lasted 12.5 hours. NTSIR in western Iraq, flex to XCAAS and four AAA defending airfield and Iraqi troops in the field. Flex to XCAS with JTAC, dropped 12 JDAMs on a north to south heading on ridge top artillery emplacements-noticed secondary explosions along the ridge line-northeast of Bagdad.

Mission number 19, April 6, 2003: Classified as OIF/Iraq. Aircraft number 86-095. Pilot, Major Ben Williams; copilot, Captain Dru Dickerson; WSO, Captain Chad Rubino and DSO Major Bob Liebman. The mission lasted 7.6 hours. NTISR in western Iraq, flex to drop 21 JDAMs in Bagdad-which was terminated by CAOC, flex to default targets in Bagdad-also terminated by CAOC. Penetrated Bagdad/Meztwice without weapons release consent from CAOC-which continually contacted aircraft that permission was coming for weapons release-but it did not. Returned to base with all weapons.

Mission number 20, April 11, 2003: Classified as OIF/Iraq. Aircraft number 86-113. Pilot, Lt. Col. Jeff Paterson; copilot, Matt Brown; WSO, Major Bob Liebman and DSO, Captain Ken Biollot. The mission lasted 7.7 hours. NTISR in western Iraq, returned to base with all weapons onboard the aircraft.

Mission number 21, April 25, 2003: Classified as OIF/Iraq. Aircraft number 86-095. Pilot, Lt. Col. Bob Namks; copilot, 1st Lt. Jim Smith, WSO, Major Bob Liebman and DSO, Captain Jeanette Ho. The mission lasted 7.8 hours. NTISR in western Iraq, flex to CAS to attack northwest of Bagdad, returned to base with all weapons.

Mission number 22, April 29, 2003: Classified as OEF/Afghanistan. Aircraft number 86-138. Pilot, Lt. Col. Rob Banks; copilot, 1st Jim Smith; WSO, 1st Lt. Ben Allen and DSO, Major Bob Liebman. The mission lasted 7.9 hours.

Mission number 23, May 1, 2003: Classified as OIF/Iraq. Aircraft number 86-113. Pilot, Captain Jason Worley; 1st Lt. Jim Smith; WSO, Captain Jeanette Ho and DSO, Major Bob Liebman. The mission lasted 9.5 hours. CAS stack northwest of Bagdad-returned with all weapons.

Mission number 24, May 7, 2003: Aircraft number 86-111. Pilot, Captain Jason Worley; copilot, 1st Lt. Jim Smith; WSO, Major Bob Liebman and DSO, Captain Patrick Snyder. The mission lasted 11.1 hours. Border town of Afghanistan. Last B-1B sortie over Afghanistan for the deployment. Flex to XCAS, top cover for JTAC at Qalat, low-altitude for a high value pass at Kandahar airfield enroute to returning to base.
(AEF Rotation)

Mission number 25, December 13, 2003: Classified as OEF/Afghanistan. Aircraft number 86-113. Pilot, Lt. Col. Joe Winterseen; copilot, Captain Richard Millard; WSO, Captain Jeremy Stroh and DSO, Major Bob Liebman. The mission lasted 14.8 hours. Road show mission and flex top cover for JTAC at Shkin after rocket attack.

Mission number 26, December 15, 2003: Classified as OEF/Afghanistan. Aircraft number 86-125. Pilot, Lt. Col. Joe Winterseen; copilot, Captain Richard Millard; WSO, Captain Mike Cornelius and DSO, Major Bob Liebman. The mission lasted 16.9 hours, Road show, flex to XCAS top cover for Afghanistan President Hamid Karzai at the formal re-opening of Afghanistan ring-road system.

Mission number 27, December 20, 2003: Classified as OEF/Afghanistan. Aircraft number 85-085. Pilot, Lt. Col. Joe Wintersteen; copilot, Richard Millard; WSO, Major Bob Liebman and DSO, Mike Cornelius. The mission lasted 14 hours. Road show, flex to XCAS, top cover for JTAC and convoy movement.
(Emergency leave back to Ellsworth through January 2004)

Mission number 28, February 17, 2004: Classified as OEF/Afghanistan. Aircraft number 86-130. Pilot, Captain Sloan Hollis; copilot, Captain Tom Bowman; WSO, 1st Lt. Mark Radio and DSO, Major Bob Liebman. The mission lasted for 13.3 hours. Top cover for JTAC at Khiwst.

Mission number 29, February 19, 2004: Classified as OEF/Afghanistan. Aircraft number 86-094. Pilot, Lt. Col. Joe Winterseen; copilot, 1st Lt. Jim Winning; WSO, Captain Mike Cornelius and DSO, Major Bob Liebman. The mission lasted 14.3 hours.

Mission number 30, February 27, 2004: Classified as OIF/Afghanistan. Aircraft number 85-083. Pilot, Major Doug Bodine; copilot, Captain Lou Pine; WSO, Major Bob Liebman and DSO, Captain Bob Patchen. The mission lasted 15.5 hours. Top cover and ISR for JTAC at Khowst. Tanker weather cancelled at t/o. emergency aerial refueling with the alternate tanker; picked through bad weather system during return to base.

Mission number 31, March 5, 2004: Classified as OEF/Afghanistan. Aircraft number 86-138. Pilot, Captain Derek Leivestad; copilot, Captain Sloan Hellis; WSO, 1st Lt. Jim Couch and DSO, Major Bob Liebman. The mission lasted 15.9 hours.
(AEF rotation)

266

Mission number 32, September 10, 2004: Aircraft number 86-121. Pilot, Lt. Col. Tom Curran; copilot, 1st Lt. Brian Peterson; WSO, Major Marc Shipman and DSO, Major Bob Liebman. The mission lasted 15.4 hours. Top cover for JTAC in western Afghanistan.

Mission number 33, September 12, 2004: Classified as OEF/Afghanistan. Aircraft number 86-121. Pilot, Lt. Col. Jeff Mikesell; copilot, Dan Dorson; WSO, Major Bob Liebman and DSO, Captain George Stone. The mission lasted 14.6 hours. Flex to XCAS with JTAC; southeastern Afghanistan; pursued two groups in high terrain in accompany with AC-130 gunship. Major armed combatant's ground battle occurred southeast of Kandahar after we returned to base.

Mission number 34, September 25, 2004: Classified as OEF/Afghanistan. Aircraft number 86-116. Pilot, Major Doug Bodine; copilot, Captain Jeff Dill; WSO, Captain Ryan Sweeney and DSO, Major Bob Liebman. The mission lasted 15.4 hours. Top cover for JTAC in western Afghanistan; number one engine malfunctions and returned to base.

Mission number 35, September 28, 2004: Classified as OEF/Afghanistan. Aircraft number 86-138. Pilot, Captain Jim Winning; copilot, Captain Steve Wright; WSO, Major Bob Liebman and DSO, 1st Lt. John Verbanick. The mission lasted 8.6 hours. In flight mission abort for number 2 VIB HI short of initial checkpoint, after aerial refueling with tanker. Returned to base on three engines.

Mission number 36, October 4, 2004: Classified as OEF/Afghanistan. Aircraft number 85-083. Pilot, Captain Bridget McNamara; copilot, 1st Lt. Chuck Richards; WSO, Captain Ben Kessler and DSO, Major Bob Liebman. The mission lasted 15.6 hours. Top cover for JTAC in western Afghanistan. Number one engine low on oil in country.

Mission number 37, October 25, 2004: Classified as OEF/Afghanistan. Aircraft number 86-130. Pilot, Jim Winning; copilot, 1st Lt Jess Hamilton; WSO, Captain Luke Baker and DSO, Major Bob Liebman. The mission lasted 14.9 hours. XCAS in central and southwestern Afghanistan, NSTR.

Mission number 38, October 31, 2004: Classified as OEF/Afghanistan. Aircraft number 86-130. Pilot, Lt. Col. Jeff Mikesell; copilot, Captain Ryan Venhuizen; WSO, Major Bob Liebman and DSO, Captain Mike Chavannes. The mission lasted 14.9 hours.

Mission number 39, November 11, 2004: Classified as OEF/Afghanistan. Aircraft number 86-138. Pilot, Captain Kevin Kohl; copilot, Captain Steve Wright; WSO, Major Bob Liebman and DSO, Captain Dan Haggerty. The mission lasted 14.7 hours. CAS practice in central Afghanistan.

(Tanker ride on KC-135R, aircraft number 59-1480) Tanker pilot, Melendex-Cruz; copilot, Scot Stewart; boom operator, TSgt. Alex Bloom; Squadron Commander, Lt. Col. Hartford. I do not remember how long the mission lasted. The tanker refueled B-1B, aircraft number 86-138. The crew of the B-1B consisted of pilot, Major Brian Witkowsky; copilot, 'Bambi' Peterson; WSO, 'Gunny' Been and DSO, 'Charlie' Nguyen.

Mission number 40, November 18, 2004: Aircraft number 86-130. Pilot, Lt. Col. Tom Curran; copilot, Captain Dan Dorson; WSO, 1st Lt. Mike Brazda and DSO, Major Bob Liebman. This mission lasted 15.3 hours. XCAS/NSTR.

Mission number 41, November 24, 2004: Aircraft number 86-130. Pilot, Captain Jay Worley; copilot, Captain Jeff Dill; WSO, 1st Lt. John Verbanick and DSO, Major Bob Liebman. The mission lasted 14.8 hours. XCAS in central and southeast Afghanistan, contact with JTAC and unknown number of AH-64s.[46]

The B-1B is constantly being upgraded. A highly accurate sensor-based technology allows the B-1B to strike targets with even more precision (first tested on June 10, 2008, at Ellsworth AFB, and now on all B-1Bs), called the Sniper Pod. The pod identifies coordinates for almost any target, even though not picked up by the aircraft's radars. A 28th BW crew tried the pod with technical support from the 337th Test and Evaluation Squadron, 7th BW, Dyess AFB, Texas. The aircrew and support personnel tested the Sniper Pod

MQ-9 Reaper (right). *USAF.*

on the aircraft for the first time. The Sniper Pod was developed for more precision delivery of PGMs in Southwest Asia to support ground troops in urban warfare and close terrain operations.[47] Air Force Lt. Col. Lucien Case, Chief of Plans, 28th Bomb Wing Operations Group, described the pod:

Basically, it's another sensor we can use to improve our accuracy and the timeliness of a strike we are performing in support of the ground force. This bringing technology to the B-1Bs has gone more quickly than expected, because B-1B program officers were able to configure it with laptop computers onboard planes, rather than having to reprogram it into the bomber's avionics system. They basically took a three year plus integration program and reduced it to 15 months. One difference between the Sniper Pod and (at that time) current B-1B technology is that it can pick up targets that don't show up on the regular B-1B radar. Using the Sniper Pod and its infrared sensors, aircrew can more easily track targets identified by teams on the ground. That will reduce the chance of friendly collateral damage and increase our chances of finding the target he's actually wanting us to strike, and doing it in an accurate manner.[48]

Air Force Major Eric Upton, 37th Expeditionary Controller Squadron, 28th BW, stated:

I am about (in 2008) to culminate my one-year tour in Southwest Asia. It was an outstanding year. I want to share what I believe to be one of the most significant aspects of our service as airmen. The significance of my tour is not marked by a piece of cloth on my chest, or even the work I did while I was here, but by the relationships I forged with fellow airmen in a time of war and the indelible mark that their relationships left in my heart. You see, it's relationships that enrich an individual achievement, bring a team victory, and give life meaning. A hallmark of our relationships as airmen is that we take care of one another. From fellow commanders leaning forward their support to each other and airmen cooperating in their daily efforts, to friends and family back home sending care packages or words of encouragement and love. I've seen what generates air power, wingman! Let me give you an example of this relationship making. We bring more than 3,500 active duty, guard, and receive airmen together and swap out an entire wing over a 30 day period. We do this all while executing the air tasking order, keeping food on our plates, running chapel services, constructing and occupying new facilities, moving cargo north, guarding the perimeter, trans-shipping blood units, delivering communications theater-wide, and contracting and paying for every single requirement. We do a remarkable job taking care of each other and never give those seats a second thought. How is this possible? All of the above accomplishments and more are made through team work, enthusiasm, selflessness, and trust. And when you test an airmen you get an "A" for air power every time. We can bring sword or shield at a moment's notice. It's truly impressive, and it's marked by our individual nature coupled with organized and disciplined manner with which we carry out our daily missions.

"We can" is the attitude of the day around here. Airmen are self reliant, expected to figure it out. Here, airmen participate in moveable teams and community service, and have a sense of urgency in getting materials and services delivered to form the nucleus of sortie generation. They do it knowing other airmen are counting on them. Wingmen at each echelon, airmen, non-commissioned officers, senior NCOs, officers, and civilians mentor through action, encouragement, and good, old-fashioned hard work. We've got a big job to do, and we know no one else can do it. Our friends and family back home are counting on us. The support we receive through their gifts, prayers, and words remind us that we've got to keep them flying, get the bombs on target, and endure a grueling sustained effort to achieve victory.

My experience here over the past year taught me a lot, but I will reflect most often on the mark. Airmen here and those back home make everyday special. See airmen lending helping hands to make work manageable, exuding positive attitudes, putting others first, and relying on each other to get the job done right makes my heart happy. We are friends, family, and fellow fighters. Wingmen make me so proud, and fellow fighters. I sing our Air Force song as loudly and proudly as I can each chance I get! I am absolutely convinced as long as there are wingmen, there will be airmen, and as long as there are airmen, there will be the United States of America.[49]

The Air Force Song
Off we go into the wild blue yonder,
Climbing high into the sun;
Here they come to meet our thunder,
At' em boys, give'er the gun!
Down we dive, spouting our flame from under,
Off with one helluva roar!
We live in fame or go down in flame. Hey!

268

Nothing'll stop the U.S. Air Force!
Minds of men fashioned a crate of thunder.
Sent it high into the blue;
Hands of men blasted the world asunder,
How they lived God only knew!
Souls of men dreaming of skies to conquer.
Gave us wings, ever to soar!
With scouts before and bombers galore. Hey!
Nothing'll stop the U.S. Air Force!
Here's a toast to the host of those who love the vastness of the sky,
To a friend we send a message of his brother men who fly,
We drink to those who gave their all of old,
Then down we roar to score the rainbow's pot of gold.
A toast to the host of men we boast, the U.S. Air Force!
Off we go into the wild blue yonder, keep the wings level and true;
If you'd live to be a grey-haired wonder keep the nose out of the blue!
Flying men, guarding the nation's border,
We'll be there, followed by more!
In echelon we carry on. Hey!
Nothing'll stop the U.S. Air Force.[50]

Master Sgt. Russell Johnson, 28th Aircraft Maintenance Squadron, is Assistant Non-Commissioned Officer in Charge and Aircraft Systems Reliability Superior Team. As noted by the ACC Inspector General, he specializes in electronic warfare systems and assists flight chiefs with their duties. He supervises 70 airmen and NCOs, helping them with their daily tasks. During the 2006 inspection, he had the responsibilities associated with an NCO ranked directly beneath the rank he held in the recent inspection, so he said he knew more or less what to expect. During the recent inspection, he was responsible for ensuring cohesion and coordination among five different specialties simultaneously being performed by the 28th AMX airmen. Johnson:

We spent countless days of numerous months testing and practicing aircraft to ensure we worked well together and were able to achieve the flow of the show correctly and with ease. Over the last few months, we saw where we were heading and what we needed to do to get there. I saw a 180-degree turn-around, with everyone working hard and coming together as a team. Without teamwork, we wouldn't have been successful. We had people performing five different specialties all at once, so we really had to pull together for it to work.

Aerial view of Ellsworth Air Force Base from east to west. Base housing is visible at the center bottom and middle right. The runway and hangars are visible at the top of the photograph. *Ellsworth Air Force Base, 28th Bomb Wing Historian*

B-1B Lancer

I deployed twice to Southwest Asia, and the scenarios of the inspection and long days were not only challenging, but also mentally, physically, and emotionally draining. The ORI is much like a real deployment, because you work with the same people 12 hours a day for months. Spending that much time together encourages a feeling of brotherhood and sisterhood, which by default improves teamwork. The intent of the level of stress the inspection imposes on you is to expose individual weaknesses so that we can adapt, get better, and be more prepared for real life situations.

Staff Sgt. Eli Barnett, 28th AMXS weapons loader, who went through a permanent change of station to Ellsworth on July 19, 2001, experienced two of the base's preparatory exercises and two ORIs overseas conducted at Aviano Air Base, Italy, as well as Osan Air Base, Republic of South Korea. He also endured a six month deployment. Barnett:

Inspections provide airmen with a base line for deployments by allowing them to develop teamwork skills, explore different experiences first hand, practice responding to simulated casualties, and learn how to endure long hours. The exercises leading up to the inspection provided a good simulation of a deployed situation and helped with getting into the mind set of what was to come. Being new does make forming cohesion among the team difficult! It all came down to trusting my team. I really had to trust everyone I work with, because without that trust, nothing would have gotten done and the mission would never have gotten accomplished. The inspection gave my teammates and me a good understanding of our limitations and abilities, and what we need to do in order to overcome obstacles in real life. I learned what my team and I could do, and it's a lot. We were able to keep our heads clear and overcome all the tasks set before us.[51]

After the near closing of Ellsworth AFB by the BRAC Committee, the base received good news of its gaining an additional mission for the base. On June 21, 2010, the Air Force announced that Whiteman AFB, Missouri, and Ellsworth AFB, South Dakota, would host ground control stations for the MQ-1 Predator and MQ-9 Reaper remotely piloted aircraft (RPAs). Each base was to gain approximately 280 personnel (civilian and military) for the new mission. Initial operational capability with the new Predator squadron at Whiteman was February 2011, and Ellsworth by May 2012.[52] The addition of these two bases is part of the Air Force's drive to acquire and operate 65 round-the-clock airborne RPAs by the end of 2013. Ellsworth's squadron will begin being manned by January 2012, to be operational by May. The new personnel assigned to Ellsworth will operate the RPA reconnaissance and attack aircraft at overseas locations.[53]

In a significant step in the historic process of bringing the remotely piloted aircraft mission to Ellsworth, Detachment 1 of the 28th Operations Group was activated on April 1, 2011. Major Chris Clark assumed command of the detachment during the ceremony on base. Colonel William Eldridge, 28th OG Commander, was officiator of the ceremony, and said "The Air Force could not have selected a better leader than Major Clark. The activation of Detachment 1 marks the start of the RPA mission for the 28th OG." Eldridge also said the mission of Detachment 1 is vitally important both to the Air Force and to the security of the United States, stating "RPAs fly missions in combat operations and patrol U.S. borders to assist law enforcement. The incoming RPA squadron will complement the mission of the 28th Bomb Wing."

Clark said the ceremony marked a significant milestone in the history of Ellsworth. Detachment 1 will serve as a precursor to the reactivation of the 432nd Bombardment Squadron, which was one of the four original Doolittle Raid squadrons. Clark noted, "We are proud to welcome home another Doolittle squadron. Just like our predecessors, we are making a brighter future in uncertain times." The 432nd Attack Squadron is scheduled for reactivation next year (2012), with MQ-9 Reapers assigned to it. Until then, Detachment 1 will be working for the incoming squadron from the ground up. The detachment will be responsible for coordinating meetings with the various organizations on base who will assist with the squadron's activation. Once activated, the 432nd ATKS will operate 24 hours a day, seven days a week, 365 days a year. Clark noted, "My hope is that the MQ-9 mission will complement the B-1 mission. To find and kill the enemy 24/7."[54]

View from the east side of the Pride Hangar. It was large enough to hold two B-36s.
Lt. Col. George A. Larson, USAF (Ret.)

MQ-9 REAPER SPECIFICATIONS AND PERFORMANCE

Wingspan	48.7 feet
Wing area	167.1 square feet
Length	66 feet
Tail height	12.5 feet
Empty weight	4,900 pounds
Combat weight	10,500 pounds
Takeoff weight	10,500 pounds
Powerplant	(1) 900 shp Honeywell TPE 331-10 turbo-prop engine
Max speed	300 mph
Cruising speed	175 to 200 mph
Range	1,150 miles
Service ceiling	25,000 feet
Max ceiling	50,000 feet
Armament	(7) hard points under the aircraft's wing (maximum of 3,750 pounds) Two inboard weapons stations Two middle weapons stations Two outboard weapons stations Center station not used
Type	Up to maximum of (14) AGM-114 Hellfire air-to-ground missiles Or (4) AGM-114s and (2) 500 pound GBU-112 Pave Paw II laser guided bombs Or the 500 pound GBU can be carried Planned ability to carry to AIM-92 Stinger air-to-air missiles
Avionics	AN/APY-8 LYN II radar AN/DAS-1 MTS-B multi-spectral targeting system
Endurance	
Maximum	14 to 28 hours (reconnaissance missions)
Fully loaded	14 hours[55]

The citizens of Box Elder, to insure the continued existence of Ellsworth AFB, moved ahead and created the South Dakota Ellsworth Development Authority in 2009. It was approved by the State of South Dakota Legislature to "Protect and promote the economic impact of the base and to promote the health and safety of those living and working near the base. It has the power to raise money to buy out residents and businesses that are in incompatible uses." Interestingly, this covers about one third of the town's residents and 25 percent of its area. The Authority purchased 230 acres of land to build a new mobile home residential area, complete with updated lot dimensions, roads, water, and utilities. The area will also include locations for multi-family housing, such as apartments. At this point, there is no planned effort to force people out of existing mobile homes, but the plan is to assist people to relocate. It is part of a long term plan to maintain the viability of Ellsworth AFB by clearing land areas in high-noise runway approaches/takeoffs and possible aircraft impact areas. The author attended some of the land use meetings on planning and development options.

The sacrifice of Ellsworth AFB's men and women continues today (as of 2010) with service in Southwest Asia. The following is just a brief snapshot in time of the commitment and their professionalism. On July 28, 2010, 190 airmen assigned to the 28th BW came home after a 189-day deployment to Southwest Asia. They returned from duty with an Air Expeditionary Force personnel rotation in Southwest Asia in support of Operation Iraqi Freedom and Operation Enduring Freedom. During their deployment the 28th Maintenance Squadron, one of several squadrons within the 28th BW to deploy, met mission requirements and established records for putting bombs on target faster and with less ground time.

28th BW B-1B being refueled from a KC-135R over western South Dakota during a training flight. The B-1B is approaching from behind and below the tanker. The tanker's boom operator has lowered the flying boom, and when the B-1B closes to the proper position and distance, the boom operator will fly the boom into the nose mounted fuel receiver, connect, and begin the transfer of fuel. *Ellsworth Air Force Base, 28th Bomb Wing Public Affairs*

Senior Airmen Candice Thomas (28th MXS Aircraft Electrical and Environmental Systems journeyman) described their work:

We improved mission capabilities and set a new standard. I am looking forward to celebrating my homecoming with friends next weekend in Deadwood, South Dakota. The airmen who have been deployed there before take you under their wing and tell you what to bring and not to bring, as well as what to do and what not to do. I felt prepared and everyone was willing to help each other out.

As well as personnel assigned to the 28th MXS, the 28th BW also deployed from the 34th BS, 28th Operations Support Squadron, 28th Aircraft Maintenance Squadron, 28th Munitions Squadron, 28th Logistics Readiness Squadron, and 28th Medical Operations Squadron. Colonel Steven Hiss, 28th BW Vice Commander, said "It's important to come and show support for our returning airmen. I'm proud of what they have accomplished, and it's good to be here to let them know all the hard work they've put in is appreciated."[56]

The following looks at the base's preparation for an ACC ORI. It is not the same Operational Readiness Inspection (ORI) as the author had to prepare for as an Intelligence, SIOP, and nuclear weapons officer while assigned to a SAC BW from 1970 to 1992:

Headquarters SAC IG did not provide a pre-warning, and conducted a no-notice ORI within a specified period. It placed fear and terror in those who were on a SAC base, with one's career literally hanging in the balance if you did not pass. For most of us in the intelligence career field a Satisfactory was not acceptable, with the goal of excellent or outstanding to maintain on promotion track with other SAC officers. As a non-rated officer the author faced many tough and comprehensive SAC ORIs.

The current IG exercises under ACC are comprehensive for the purpose of testing a bomb wing's preparation for its wide-ranging missions. Ellsworth AFB personnel had to get ready for a command inspection. On August 24, 2010, the men and women of Ellsworth went through an Operational Readiness Exercise (ORE). This was an opportunity for Ellsworth to practice various scenarios in simulated deployed environments.

Staff Sgt. Keith Stetser, 28ᵗʰ Medical Group Ambulance Services, described it this way:

The exercise was practical for both new and veteran. This is effective for some of our newer airmen who have never been deployed. It's also good practice for those of us who have experienced a deployment, by helping us to keep our skills sharp and adapt to changes in procedures.

Second Lt. Jake Sabin, 28ᵗʰ Contracting Squadron Team Chief and Exercise Evaluator, said:

I observed the squadron's airmen perform protective procedures against a simulated chemical attack. Everyone improves with practice. The purpose of this exercise is to utilize proper procedures in response to each possible scenario, so our airmen will be prepared for all operations in a deployed environment and be confident in knowing they are ready for anything.

Technical Sgt. Matthew Paxton, 28ᵗʰ BW Contracting Squadron Superintendent, noted:

One aim of the OER was to ensure the safety of deployed airmen and enable them to effectively accomplish the mission and return home. We want to make sure our airmen's emergency management skills are raised to the highest standard. The ORE gives us the opportunity not only to prepare for the upcoming ORI, but to ensure our airmen are properly trained and prepared to get the job done and come back.

Staff Sergeant Calvin Courtney, 28ᵗʰ BW Contracting Squadron, Simplified Construction Contracting Officer, said this:

I assisted airmen in honing their skills by playing a casualty during the ORE. Acting as uninjured airmen, your wingmen will simulate Self-Aid and Buddy-Care by treating for shock and going through the different steps of healing for the specific injury. Acting as an airmen killed in action, however, you will be removed from the office in order to allow the second in command to step in and take over. This role playing allows everyone in the office to be ready to assume command responsibility. I was deployed alongside Army soldiers in Afghanistan, and it is crucial to practice donning protective gear. With this training and preparation, airmen can do what needs to be done without thinking. They learn to feel what is right, rather than just throwing on their equipment when a chemical attack is imminent. In a deployed environment, you're never very far from your bag of protective gear against chemical warfare. Practicing to don it quickly and properly can save your life. Diligent practice of procedures in routine exercises, such as the ORE, can yield improved success in both the upcoming ORI and in the deployed environment.[57]

B-1Bs on Ellsworth's flight line with supporting ground equipment.
Ellsworth Air Force Base, 28ᵗʰ Bomb Wing Public Affairs

On September 16, 2010, late at night, 28ᵗʰ BW airmen began processing through the Pride Hangar during the base's final ORE before the scheduled upcoming ORI to be held October 12ᵗʰ through 18ᵗʰ. The ORI would be packed with more than 300 graded items, all designed to gauge the skills airmen have been practicing during the ORE. Master Sgt. Kathy Evans, 28ᵗʰ BW Law Office Superintendent, said:

The OREs have really helped me to brush up on my Self-Aid and Buddy Care training, team building, and deployment scenarios specific to my office. They have also motivated me to be a good wingman by maintaining a positive attitude. A positive attitude can mean the difference between a successful ORI or a failed one. If an airmen comes into the inspection with a poor attitude it can affect the performance of their entire shop. However, a positive attitude can actually have the opposite effect and make all try harder. In addition to maintaining a positive attitude, airmen are given the opportunity to learn from the mistakes in previous OREs through continued practice and repetition of skills. Whether practicing SABC (simulated atomic, biological, chemical), conducting post-attack reconnaissance, or reviewing the Law of Armed Conflict, airmen are expected to make mistakes throughout the exercises. However, as time goes by the number of mistakes decrease as airmen move closer and closer to the perfection of their deployment skills.

Captain Heather Alwin, 28th BW Assistant Staff Judge Advocate, noted:

Its good to make mistakes early on. This allows evaluators to identify problem areas that might occur during the ORE and implement procedures to fix them. I approached the exercises with an open mind, willing to learn new skills and experience my career field in a different way. It's not every day I get to practice techniques, which aren't routine. The OREs have been a great way for me to keep my deployment skills fresh. Colonel Jeffrey Taliaferro, 28th BW Commander…The OREs have shown marked improvement on the part of airmen. The ORI is an opportunity to show the Air Force that we are the best at what we do. Every airmen is given the opportunity to take the skills they have learned and out them into practice during the inspection.

A B-1B fly by kicked off the celebration in honor of a successful ORI completion, as crowds of airmen, accompanied by their family members, gathered outside the Ellsworth AFB Dakota Club on October 22, 2010. Airmen and their spouses came together and were treated to a free lunch, live music, and the fly by at the club after successfully accomplishing a series of strenuous war fighting scenarios during the base wide inspection.

The IG team arrived at Ellsworth on October 12th to find out how prepared airmen at Ellsworth are to protect themselves, their wingmen, and fight back against the enemy when necessary. This is determined by guidance from Air Force Instruction 90-201, which evaluates whether airmen are able to meet mission requirements when deployed. The overall Satisfactory grade Ellsworth airmen achieved for their sweat, efforts, emotions, and exhaustion is one that lets the world know Ellsworth's airmen have what it takes to put bombs on target and come back home.

Senior Airmen Kristi Leaman, 28th Security Forces Controller and ORI Excellent Performer, said:

I was responsible for answering calls to domestic disturbances and dispatching patrols. Some of my duties as a dismounted security patrol during the inspection included securing and defending the Installation Control Center and my sector against hostilities. Even without an outstanding rating, I was happy with the overall experience gained and confidence in my team and leadership. Great leadership counts 100 percent toward success. As Colonel Taliaferro mentioned before in briefings, it's important to have a base knowledge of a range of fighting techniques. We know we're not fighting in the same kind of combat as we did in Korea or Vietnam. We're in the desert fighting a different kind of warfare, but it's still important to hone these skills.

The inspection and the preliminary exercises also nurtured closeness among my team, and this bond helped me stay alert and keep my morale up to maintain security and keep an eye on what needed attention while defending my sector. It makes the job easier when you have great people working with you. As cliché as it sounds, they become your family. You train hard with them. You get to know each other really well. The inspection is mentally, physically, and emotionally challenging, involving wearing equipment which adds between 20 to 30 pounds of body weight, sweating in mission oriented protective posture gear, and enduring long hours. All of those small hardships combined make this bond among members of the team critical.

B-1B Lancer parked in front of the Pride Hangar on the flight line. *Lt. Col. George A. Larson, USAF (Ret.)*

The exercises leading up to the inspection not only allowed for such a bond to be created, but also allowed

274

responses to tasks to become as automatic as muscle memory. Practice makes you better. Everyone did a good job. We worked so hard together and so well as a team. I feel more prepared to deploy after the inspection. It increased my awareness and background knowledge of different fighting positions, increased my confidence, and has helped me to be well-rounded to secure and defend. You never know where a war is going to take you. It's important to learn all kinds of skills, tactics, capabilities, and defense technique.[58]

B-1B Mishaps

There have been numerous B-1B accidents for Ellsworth AFB assigned aircraft and aircrew. The following looks at these mishaps. Aircraft number 85-0088, on February 9, 1980, was approaching the runway at the end of a training mission under deteriorating weather conditions. Major Dan Ford, DSO, contacted the base to determine weather (400-foot overcast, one and one half mile visibility, with light rime ice). The onboard ILS system not fully tested, and the aircraft was only authorized to fly the localizer to a full stop. The light rime icing condition was under the legal limits for penetration and landing. The B-1B descended in the clear until 5,800-feet mean sea level (MSL), 12 miles from the runway. The descent down to Minimum Descent Altitude (MDA) of 3,660-feet MSL was started at seven miles from the runway at 5,200-feet MSL. In the descent the Ice Detector Light came on, the wind screen started to ice up, and the aircraft was flying in blowing snow and ice. The instructor pilot, Lt. Col. Ross Younkin, turned on the defog anti-ice while the aircraft commander, Captain Randy Davenport, continued to fly the approach.

Four miles from the runway, Lt. Col. Younkin was able to see the ground and started calling out land marks. The aircraft leveled out at the MDA and continued to drive inbound to the runway threshold. The runway lights came into view and the aircraft set up for a visual descent to landing. Two seconds later the entire B-1B went dark: there was no Vertical Situation Display (VSD), no course guidance, no Master Caution Lights, no oxygen, and no interphone. Captain Davenport continued the landing and pushed the throttles up, not expecting any response because the throttles are electrically connected to the engine fuel control, as he tried to improve the approach with the engines stuck slightly above idle. At the same time, Lt. Col. Younkin was reaching for the overhead panel on which the electrical system controls were located. The aircraft

was slightly fast and slightly high, so making the runway was assured. Seconds later aircraft electrical power returned the aircraft accelerated. The result was that the aircraft had too much thrust, requiring Captain Davenport to retard the throttles to idle. The aircraft was still 20 knots too fast passing the runway threshold. A Solid touchdown was at 3,500-feet from the end of the runway. The aircraft braked to a stop. Lt. Col. Younkin scanned the main caution panel and informed Captain Davenport the Nose Wheeling Steering and Electrical Bus #3 had failed, resulting in loss of electrical power. Captain Davenport applied differential thrust and breaking to clear the runway. However, because of the Bus #3 and Bus Tie # being open the engines could not be shut off. Forty minutes after landing a combination of maintenance and aircrew actions shut down all four operating engines.

Aircraft number 85-0076, on November 17, 1988, at 10:40 p.m., crashed during a TACAN approach to the runway in poor visibility because of ground fog. The B-1B hit a utility pole prior to crashing near the south end of the runway. All four crew members ejected. One crew member suffered a minor concussion in landing. Another crew member injured his back, requiring crew medical evacuation to Brooks AFB Hospital, Texas. Weather conditions were deteriorating at the time of the crash (300-foot ceiling, with fog and snow). The SAC Accident Board investigation indicated the aircraft struck fixed ground objects after the aircraft commander allowed the aircraft to descend below the published minimum descent altitude approach being flown. These objects included three wooden monitor poles, a high voltage power line, and an approach light stanchion. The monitor poles, about 14-inches in diameter, were 2,900-feet from the approach end of the runway. Suspected deviations included a failure of the aircraft commander to establish sufficient visual cues to determine the runway environment before departing the published minimum descent altitude. The copilot failed to advise the aircraft commander that he was descending the aircraft below the minimum descent altitude before establishing the landing environment. Neither aircraft nor engine icing were contributing factors to the accident.

Members of Ellsworth's 28th Munitions Squadron assemble 2,000-pound bombs converted to JDAMs for use in Libya for Operation Odyssey Dawn on March 26, 2011. *Ellsworth Air Force Base, 28th Bomb Wing Public Affairs*

275

Aircraft number 86-0099, on September 19, 1990, struck a flock of ducks at low level over Montana's Powder River Range. The training mission profile consisted of a night air refueling, low-level night terrain following on IR-478, with direct routing to Ellsworth for one approach and landing. The flight was normal for the first half of the low-level route. The aircraft was at 1,000-feet AGL and 560 knots when a loud explosive sound was heard throughout the cockpit. The aircraft had struck a flock of ducks. Sparks erupted at the pilot's station. Major Tom Owskey was onboard as IP, sitting in the pilot's station. He said the sparks came from a utility light knocked loose. The Air Refueling/Nose Wheel Steering Indicator light broke loose. Captain Paul Dow aborted the low-level flight, starting a climb. Major Owskey squawked emergency on the IFF and obtained Ellsworth clearance for a direct approach. Due to darkness the pilots could not evaluate the aircraft damage. However, both pilot windscreens were fractured, forward visibility through the left windscreen was zero, and limited through the right windscreen, with only a four-inch space clear of cracks.

The crew declared an in flight emergency through the Salt Lake Center Air Traffic Control Center and evaluated the aircraft's flight controls. No flight control problems were noted. Lt. Bob Distaolo, DSO, called Ellsworth Command Post, advised them of the situation, and coordinated effort. Major Owskey identified a hairline fracture on the inside of the windscreen into the crew compartment, increasing the possibility of an implosion of the windscreen into the crew compartment. To reduce the risk airspeed was reduced, lengthening flight time to Ellsworth by 60 minutes. The pilots executed a low approach to runway number 13 to determine which pilot had the best visibility for landing. Captain Dow flew on ILS approach down to 200-feet AGL, at which point Major Owskey would take control of the aircraft and visually land while Dow monitored the runway edge to detect any lateral drift from the runway center line. Major Owskey requested the Runway Supervisory Officer call out the distance remaining to the end of the runway as the aircraft approached touchdown. Major Tom Dyer, WSO position and avionics instructor, called out the remaining altitude above the ground as read on the Radar Altimeter. Major Owskey took control of the aircraft from Captain Dow, easing the aircraft down onto the runway at the optimum touchdown zone. As the nose lowered, a larger portion of the windscreen without cracks allowed the pilots to complete their landing rollout. The B-1B successfully stopped on the runway, with the crew climbing out of the aircraft.

On March 24, 1992, aircraft number 86-0093, along with KC-135R aircraft number 62-3540, collided over Nebraska. Both aircraft were at 22,000- to 24,000-feet. They hit at 11:22 a.m. MST. Both aircraft had completed air refueling and were practicing time prior to SAC's bombing competition, *Proud Shield 92*. The B-1B had damage to the top of its vertical stabilizer, one-third of the left horizontal stabilizer was gone, and the aft radome was crushed. The KC-135R suffered a 15-foot tear in the lower right fuselage, forward of the leading edge of the wing, rupturing the main fuel cell. Both were able to safely return to Ellsworth.[59]

On September 19, 2007, four B-1B crew members were killed at 2:25 p.m. near the Powder River Military Training Range along the Montana and Wyoming State lines: Pilot and 28th BW Vice Commander Colonel Anthony Beat; instructor pilot Major Kirk Cakerice; WSO Major Clay Culver; and WSO Captain Gary Everett. At the time of impact, the crew was performing an authorized and often practiced defensive maneuver in which the crew slows down the aircraft and turns sharply to evade a threat. Lt. Col. Clete Knaub, 37th BW Maintenance Officer, noted the following:

I was temporary duty to Nellis AFB, Nevada. My airmen and I were at McCarron Airport. Just as we were getting ready to board the civilian airliner to return to Ellsworth one of my troops came to me, literally pale. "Sir, I just talked to my mother and she said we just lost a B-1." I must have looked at him in a very stern manner because he took a step back, raised his hands, and said, "Sir, I am not kidding." As soon as he said that we boarded the aircraft. I sat there thinking it must be a Dyess B-1. I don't like to lose bombers, or it could be a B-52H from Minot. The press sometimes gets the two mixed up. Once we got to altitude I used the air phone and, for some reason, called my wife first. The phone rang and a friend answered. When I asked to speak to my wife our friend said, "Sharon is not here, there has been an emergency." At that point, I knew the unthinkable had happened.

Mr. James Ramp, 28th BW, 28th Comptroller Squadron remembered:

I was in the 37th BS Specialist Maintenance Flight at the time. I remember getting the phone call that we had lost a jet. I remember feeling sad because I knew the Air Force had just lost some good aviators, and more importantly, that Air Force families' lives had just been changed forever; that husbands, fathers, brothers, and friends were lost and could not be replaced. As a whole, the tone of the base immediately following the accident was shock, disbelief, grief, and a determination to stand next to the families who lost loved ones.

The following day, the base flew two B-1B missions and then stood down for several days and held a memorial service in honor of the crew.[60]

ACC flight lines were silent on Monday, September 22, 2007, as routine training and exercise operations ceased for a safety stand down day. General Richard Hawley, ACC Commander, said:

Stand down days give us an opportunity to step back from the high pace and intensity of our day-to-day operations, where we can get consumed by the pressure of the moment. They give us a chance to think about exactly how we're doing our business, what the risks are out there, and what can we do to get rid of them.

Six military accidents in a week led Secretary of Defense Williams S. Cohen to direct a Department of Defense wide safety stand down day. The Air Force chose September 26th as its stand down day; however, General Hawley moved ACC's date to Monday, following the B-1B bomber accident on September 19th in Montana. Hawley noted:

We lost a B-1 Friday afternoon that had four great airmen onboard. We're a family, so everybody in the Air Force feels the loss of those four people. It's that sense of family, that personal sense of loss I think makes the timing of this day particularly important.

The general said he hoped to accomplish two things during the stand down. First, he wanted to increase awareness in every person in ACC. This awareness involved knowing the role people play in maintaining a safe organization and understanding the risks. Second, he wanted to focus attention on Operational Risk Management, a process that encourages people to be sensitive to risk. He emphasized that people need to use ORM to identify

risks, lessen those risks where possible, and if they must take a risk, make sure the appropriate people make the decision to accept that risk.

Hawley continues:

The thing we can all do to make this a safer Air Force is be aware of risks, be sensitive to risks, and when we see them, take some kind of action. Bring them to our supervisors' attention, do something to get us started along the path of mitigating that risk, so that it won't come up and bite us down the road someplace and cause an accident.

Hawley emphasized that these exercises did not just focus on airplanes, but on all aspects of operations. Throughout the day, ACC people reviewed air and ground accidents from the past year and looked for ways each could have been prevented. Commanders also led small-group training sessions to identify risks in daily operations. "Each wing will forward a list of items requiring headquarters attention to the ACC Office of Safety," said Colonel Gregory Marshall, ACC Chief of Safety. "A team will then compile the reports and present them to Hawley for action by the appropriate agencies. Hawley continued:

This is serious business defending our country is perhaps the most important job anybody can have. It's difficult. It is dangerous, and we'll take every step we can to make sure we have the safest, most professional operation this country can have.[61]

Lt. Col. Clete Knaub, 37th BW Maintenance Officer, commented on the relationship of those killed to the men and women who worked maintenance:

The accident occurred during the time period when the bomb squadron commander commanded flight line maintenance, working so closely with these airmen. I counted two of the men lost as good friends: Major Culver and Major Cakerice. These two gentlemen went out of their way to get to know maintenance personnel and procedures. They reached out to maintainers to make them feel a part of the squadron. They truly were officers and gentlemen. I honor their memory and was proud to call them my friends. I miss my comrades and often replay the accident in my mind, wishing it hadn't happened. I miss my friends and pray that all is well with their families and loved ones. It is important to note that we learned from the crash and changes were made to aircrew training, which has paid dividends, in that no B-1 has crashed since then flying the same mission profile.[62]

Captain Bob Russell, 28th BW, Chief B-1B Weapons System Trainer stated in September 1997:

Chances are you haven't noticed Ellsworth's little known aircraft that is whisper quiet when it takes off, even though four massive engines are throttled up to full afterburner! It might impress you to know that this vehicle can reach destinations halfway around the world in less than 15 minutes. Sound too good to be true? Well, Ellsworth isn't home to any super secret aircraft that I know of, but it does have a B-1B Simulator System that can do all of that and more. With a full motion, exact replica cockpit, complete with a dazzling new visual system, this aerospace vehicle can fly. That's just one of many tasks that quietly takes place in the big brick building recently renamed "Flight Training Center." You may not have noticed this impressive building located at the south end of Ellsworth's gigantic runway, but it's directly across the street from the Airmen Leadership School. It's in this 28th Operations Support Squadron facility that B-1B aircrew from throughout the Air Force and Air National Guard and Reserve train to fight our next conflict. This kind of simulated training is known as "Mission Rehearsal."

B-1B pilots and weapon systems officers are pitted against some ominous air and ground threats from the most hostile parts of the globe, and we have not lost an airplane yet (in an active combat theater of operations). In preparation of this experience B-1B aircrew also spend a full day each year reviewing procedures and information to allow a B-1B to land in some of the worst weather imaginable. A full day of classroom training is complimented by a grueling 90 minute simulator session to test each pilot's and weapons systems officer's reactions in rain or fog so thick you can't see the runway until just seconds from touchdown. Traveling at only a few hundred feet above the ground at 600 knots in the dark of night is exhilarating to say the least, but with NVGs you can see everything.[63]

Colonel Will Fraser, 28th BW Commander, had this to say:

The crash site, located about 25 miles north of Alzada, Montana, and about 12 miles from the capital, is about one mile long and about one-half mile wide. The crew was scheduled for a routine defensive tactics training mission that was to last several hours. They launched about 2 p.m. out of Ellsworth, headed for the Powder River Military Operating Area. A few minutes after 3 p.m., I received a call from the command post stating that they had been called by the Carter County Sheriff, who said that an aircraft had gone down. We knew that we had an aircraft in the Powder River MOA, so I responded to the command post and we began calling the Denver Center and Salt Lake Center, who did an outstanding job of responding and letting us know they were not in contact with our aircraft. We additionally called the Belle Fourche training site, which we train against, and they said they had seen an aircraft go down. There was no previous indications that the aircraft or crew members were experiencing any problems. We had no prior indication until we got that first phone call that an aircraft had crashed. No in flight emergency had been declared, and we had no after indications of any kind of problems with this aircraft.

The weather in the Powder River MOA was clear, so they were operating under visual flight rules. The aircraft had completed a successful mission earlier that day prior to completing the engine running crew change. The first sortie launched and then returned back here to Ellsworth. Having completed a successful flight, they turned the aircraft over to its new crew members and reported no problems. Despite the tragic loss, operations at Ellsworth won't change, including senior leadership flying regular training missions. I personally think that flying is very safe. We are very conscientious, and do everything that we can to minimize our risk by making sure that we maintain and then fly in a safe manner. But flying is inherently dangerous. We do everything we can to mitigate that risk, and therefore, we become safer and safer, as is indicated by our safety records over the years. I am personally ready to fly any time.[64]

The author attended the memorial service for the four Ellsworth airmen on September 22, 1997. He was among over 1,500 in attendance at Dock 33. Four tables were set up, each one for a fallen airmen, with a U.S. flag, helmet, photograph, boots, and flowers. Colonel Fraser said "Four brace men who proudly wore the United States Air Force uniform have sadly been taken from us. They have flown their last earthly mission. Our loss is indeed heaven's gate. From their examples, we will draw strength. Because of them, we will be better husbands, brothers, and fathers." Lieutenant General Phil Ford, Commander Eighth Air Force, Barksdale Air Force Base, LA, said "The four men died doing what they all loved to do, flying a B-1 bomber. It's like losing a member of the family. These men had two roles, that as warriors and family. I've heard it said that it is the warrior, not the reporter, who gives the freedom of the press."

The crowd filled all the set up chairs, with a large overflow standing at the rear and to each side in Dock 33. It was an emotional, 70 minute memorial service with tears flowing on many faces. General Ford stated "That it is the warrior, not the poet, who gives us freedom of speech. That it is the warrior, not the poet, who gives the right to a fair trial. But most importantly, it is the warrior who gives us courage and faith for today and hope for tomorrow. They made life better and they made life safer for all of us." Captain Hodges Vicellio noted "God did not send the pain to those hurt by deaths. Those who have lost these men must move past the questions of why to questions of how. How can we deal with the reality of death? This winter of life will pass as all seasons do, and the families of the fallen airmen will not be left to heal alone. We will be like Christ, holding your hands and hearts until you can walk again." The most moving and emotional part of the memorial service was at the end, when the airmen's flight helmet visors were covered over by a gray cloth. The assembled throng slowly moved past the four tables in a solemn goodbye.

The author talked to members of the B-1B Crash Recovery Crew on the recovery and removal of wreckage in Montana, with the pieces loaded onto flatbed trucks for the trip back to Ellsworth. The wreckage was assembled in order of the aircraft's outline on a hangar floor. The recovery work collected all small pieces of the wreckage. All recovered pieces were tagged and locations on the ground marked with corresponding numbered flags prior to removal to assist accident investigators in determining what happened after the bomber struck the ground.

On December 12, 1977, the base newspaper printed an article on the cause of the B-1B crash. The author also obtained a copy of the official Air Force Accident Investigation Report that indicated pilot error as the contributing cause for the bomber's crash. The report indicated "the September 19th B-1B crash near Alzada, Montana, which killed all four crew members has been thoroughly investigated by a board of officers and the results have been announced." According to Colonel Ned Schoeck, investigating officer, "The crash was the result of the two pilots, who were trying a technique they didn't often use. On this particular maneuver, the crew of Fury 02 practiced a SCRAM with speed brakes. Although that's an accepted technique, neither pilot flew SCRAMS with speed brakes regularly."[65]

The Air Force Accident Investigation Report indicated that according to the crew's flight plan, the crew planned to twice perform the defensive maneuver, essentially a U-turn to carry the aircraft away from threats. After an uneventful flight to the Powder River Training Range, the pilot performed the first SCRAM using the B-1's terrain following computer. The B-1B's computer can perform a SCRAM at a maximum side-to-side angle, or roll, of 45 degrees. The crew wanted to compare its performed SCRAM to one done manually at an angle of 65 degrees. At the steeper angle, the bomber could turn faster for evasion. Colonel Schoeck recalled:

Instructors continually work to expand their ability. It allows them to pass on their experiences. A SCRAM is a defensive turn, where an aircraft sees a threat and then makes a quick turn out of the area. It can be flown without the speed breaks, and also visually or with the aid of the aircraft's terrain following radar. Although SCRAMS are common maneuvers, neither pilot had flown one using speed brakes in the past year. Both pilots were highly qualified instructor pilots. In the last 15 seconds, though, there were numerous things that contributed to their inability to recover from the maneuver. By reconstructing the accident from information gathered both at the scene and through analyzing the aircraft's Crash Data Recorder, the Accident Investigation Board was able to determine the crews' last flight maneuver. During the final maneuver, the crew visually followed

279

the terrain of the Powder River Military Operating Area. Before the pilot made the fatal turn he started to slightly descend.

The report continued to say that fourteen seconds before impact, the pilot appropriately set his engines to idle, beginning the deceleration from a speed of 598 mph, or 520 knots, to the planned completion speed of 492 mph. The altitude was 534-feet. The earlier SCRAM had been performed at 300-feet. The pilot also set the bomber's speed brakes, raised simultaneously to slow the plane or raised independently to aid in turning. The pilot's unfamiliarity with speed brakes would be cited as a major factor in the crash. Colonel Beat or Major Cakerice did not commonly use speed brakes, although both were authorized by the Air Force to use them. Another factor in contributing to the crash was the slight descent at which the pilot began the second SCRAM. Between 14 seconds to impact and 10 seconds to impact, the plane banked from 3 degrees right angle to 65 degrees right angle, slowing to 590 mph. At the 65 degree tilt, the plane had reached the maximum angle at which it could perform level flight. The plane was in an established descent, and with the bomber at maximum tilt, the pilots could not stop the descent. Colonel Schoeck continued, "Because of some of the safety features on the B-1B, they weren't able to completely correct the situation. The aircraft continued to bank until it was almost perpendicular to the ground."

At nine seconds before impact the pilot sensed the descent, moving the control stick left, trying to flatten out the angle of the plane, while also pulling the nose up to increase "G" forces. With speed brakes extended the aircraft did not respond. During the next five seconds the aircraft titled further right, varying 69 to 77 degrees. The bomber side slipped down to 200-feet. At 3 to 5 seconds to impact the bomber's nose was at 11.5 degrees, increasing "G" forces to 3.87, 1.37 "G"s beyond maximum mission profile settings in the computer and setting off the Red Stall Warning Indicator. At the same time an audio warning sounded in the pilots' helmets. The onboard computers automatically became limited to pilot actions that would not cause a stall and immediate loss of flight, limiting the pilots' control. Three seconds before impact, the bomber's engines were set on idle and speed brakes fully extended, with the bomber tilted at 80 degrees. The right wing was the first to hit the ground, followed by the aircraft's fuselage. The cockpit section was thrown to the left, while the rest of the aircraft continued straight ahead while breaking apart from the impact. Debris from the crash covered an area approximately one-half mile wide and one and one-half miles long.

Colonel Schoeck continued:

The crew began the SCRAM correctly. Unfortunately the bank angle, the use of speed breaks, and descent angle all contributed to the outcome of the accident. Although the B-1B has a Zero-Zero ejection system, meaning people can safely eject from the aircraft at any speed or altitude, there was no evidence anyone tried to eject. Upon hitting the ground, the aircraft was traveling at 429 knots and banked right 80 degrees, with a "G" load of 4.01. During peacetime operations, the B-1B is limited to 2.5 "G"s, which requires a 67-degree of bank for level flight. At the start of the maneuver, the aircraft was about 500-feet above the ground. Previously, the crew had completed a SCRAM using the Terrain Following Radar at 300-feet. They were well above the minimum altitude of 300-feet for that area. Each crew member was highly qualified to fly the aircraft and Colonel Anthony Beat and Major Kirk Cakerice were both instructor pilots.[66]

On December 12, 2001, aircraft number 86-0114 crashed into the Indian Ocean near the island of Diego Garcia. All four airmen ejected and were recovered at sea, returning to the air base on Diego Garcia. The B-1B took off from the air base with a destination of Afghanistan for an assigned airborne patrol area to support U.S. and Coalition troops on the ground. On September 24, 2002, the ACC Investigation Board's final report indicated that the cause of the crash as unknown. The board indicated that aircraft malfunctions affecting the reliability of the pilots' altitude information might have made it difficult for the pilots to maintain control of the aircraft.

On January 4, 2005, the Air Force released a Public Affairs announcement that all B-1Bs had been grounded after one of the bomber's nose gear collapsed at a forward deployed base (unidentified) location in Southwest Asia supporting Operation Enduring Freedom and Iraqi Freedom. The ACC released a statement that after landing safely, the pilot taxied the B-1B to its assigned parking area, shut down the engines, and the nose gear collapsed on the aircraft. After reviewing the problem for six days B-1Bs were cleared to resume training flights and combat air operations.

On March 8, 2008, a 28th BW B-1B struck two Andersen AFB emergency response vehicles after landing because of an in-flight reported emergency.

The B-1B crew, after takeoff from Andersen, reported a hydraulic leak. Once back on the ground at Andersen, the crew exited the aircraft because of possible fire concerns, and the B-1B started to roll and struck the two parked fire trucks. The B-1B had been deployed from Ellsworth, en route to appear and perform at the Singapore Air Show. On March 20, 2008, fire crews from the Box Elder Fire Department and Ellsworth Fire Department combined their efforts to extinguish several grass fires caused by a B-1B during an emergency landing. The area of the Black Hills, in 2008, was in the midst of a severe drought, with fire danger extremely high. On April 4, 2008, a 28th BW B-1B crashed at Al Udeid Air Base, Qatar. The crash location was reported by the Associated Press and not through official Air Force channels, which only identified the crash location as somewhere in Southwest Asia. This was followed by an Air Force Public Affairs news release that a U.S. Air Force B-1, while taxiing after landing at Al Udeid Air Base, was involved in a ground accident and caught fire. The four crew members evacuated the aircraft safely. The fire was quickly extinguished. It appears the B-1B hit something during taxi operations, causing the explosion.[67]

One of two 28th BW B-1Bs taking off from Ellsworth AFB on March 27, 2011, to drop bombs on Libyan targets in support of Operation Odyssey Dawn. The base had to do all the work arming bombs and preparing the aircraft after a snowfall and cold weather. *Ellsworth Air Force Base, 28th Bomb Wing Public Affairs*

On January 21, 2011, 24 years after the first B-1B (called the "Wings of Freedom") arrived at Ellsworth AFB, delivered by SAC, Commander John T. Chain, after deplaning, said:

Too often people think of SAC bombers only as carriers of nuclear weapons. We have to change that perception. The B-52 and B-1 have an enormous impact to contribute to this nation's conventional capability. Their firepower is awesome. They could strike a terrorist type target anywhere in the world, flying non-stop from the United States and return with air refueling.

Colonel Tom Vitamvas, former Ellsworth AFB Deputy Commander of Maintenance when the first B-1B landed at Ellsworth, stated:

The object in those days was to train for alerts with nuclear weapons. Since then, the B-1B has gone from the nuclear business to carrying conventional smart weapons. A lot of hard work and training went into preparing the base for the arrival of the aircraft. New hangars and facilities had to be constructed at the same time updates to existing buildings went on. We were all very excited to get a brand new airplane. Everything had to be perfect before the B-1B arrived.

Marian Tolley, 28th BW Commander's Secretary, said "After the B-1B arrived, the Wing Commander would call us every time there was a take off and ask if we had heard it." Colonel Vitamvas noted:

Despite the initial excitement, the B-1B was not without its naysayers. Many people were not convinced a supersonic, nuclear bomber was what the nation needed. There were operational difficulties that caused people to question its value. However, after the crews flew in the aircraft they were all very pleased with it. The B-1B has been a truly incredible aircraft, in multiple theaters of operation. I valued the time I was able to spend with it.[68]

Colonel Jeffrey Taliaferro, 28th BW Commander, said:

The President's budget (2012) cuts six B-1B bombers. We're just now beginning to take a look at what the impacts would be and how it would impact home station training and our comments overseas. These decisions are primarily based on the needs of the nation. It's not a matter of B-1s that are broken and we need to get rid of them. These are strategy decisions on what the nation needs going into the future. If there were changes in the mission, there may be changes in the number of positions that would support that mission. But it's still way too early to know what the size or nature of these impacts would be. We're still just beginning to get into the details of what

281

those impacts would be, and are waiting for the Secretary to make that decision about where those aircraft would come from.

Pat McElgunn, Director, Ellsworth Air Force Base Task Force said;

We've asked the State's Congressional delegation to take a look at it. We've been told there won't be any details forthcoming for about three or four weeks. It's a line item in the budget proposal right now. Ellsworth has about 1,500 airmen assigned to B-1 maintenance and 400 airmen in the Operations Wing. Personnel reductions wouldn't necessarily be abrupt or dramatic. There may be a kind of attrition, with new people not coming to Ellsworth. The Ellsworth Task Force is working with South Dakota's Congressional delegation to head off some of the cuts. The delegation is very keen on getting good answers. We'll assess the situation from that point forward.[69]

Airmen of the 28th Aircraft Maintenance Squadron load a 2,000-pound bomb into the weapons bay of a B-1B on March 27, 2011. With less than two days from receiving the first notice to take off, Ellsworth airmen generated hundreds of weapons for the two B-1Bs to support Operation Odyssey Dawn. *Ellsworth Air Force Base, 28th Bomb Wing Public Affairs*

B-1B Lancer crews returned to Ellsworth on Wednesday, March 30th, following combat strikes in Libya in support of Operation Odyssey Dawn. Airmen from across the base joined together to celebrate the successful mission and safe return during an event on the flight line.

One of the two B-1Bs on the taxiway at Ellsworth after completing a non-stop flight from the base to Libya and back on March 30, 2011. The B-1B is moving under the water spraying from two Ellsworth 28th BW Civil Engineering fire department trucks. Ellsworth personnel lined the flight line to the right as the aircraft went through the water. *Ellsworth Air Force Base, 28th Bomb Wing Public Affairs*

This marked the first time B-1s were launched from the Continental United States, struck targets in support of a military operation overseas, and returned home to a hero's welcome.

B-1B on the flight line at Ellsworth AFB, prepared for a training mission. *Ellsworth Air Force Base, 28th Bomb Wing Public Affairs*

"It was amazing to see the airmen of the 28th Bomb Wing come together as one cohesive unit for this historic mission," said Colonel Jeffrey Taliaferro, 28th BW Commander. "When the nation called, they answered, and I couldn't be more proud of their accomplishments." The B-1s flew two missions over Libya, striking all intended targets before returning home. Colonel Taliaferro continued, "Our objective was to destroy nearly 100 military targets to protect the civilian population of Libya. Of note, this was the first time B-1s were launched from the United States to strike targets overseas." The Air Force committed two B-1Bs to attack Moammar Gadafi's military forces and military support facilities. The attacks were completed prior to the March 31st hand-over to NATO command, with the RAF and French Air Force aircraft picking up the attacks in Libya. The Air Force continued to provide air support under command of their officers through NATO. The weapons dropped by the two B-1Bs were expensive, and part of the Air Force's estimated cost of four million dollars per day.

Colonel Taliaferro continued, "This was a base wide effort. And while our airmen make airpower look easy, it's actually the product of a lot of hard work – blood, sweat, and commitment – that got the mission done." As the aircrew disembarked they were met by their families, friends, and fellow airmen, who welcomed them home as heroes. Captain Matt Tull, 34th BS WSO, stated:

This is a really humbling experience. This was such an amazing effort for everyone to come together to make the mission a success, and offer help to the people in Libya. Looking back, I'm still awestruck by the coordinated effort that made it happen in such a short time. I don't think we were really thinking about it too much. We were busy doing our job back home, flying training missions and preparing. We've all got upcoming deployments and other things to worry about. But when the orders came down, that's what we're trained to execute. We're ready for it at any time.

Following the welcoming ceremony, maintenance crews set to work recovering the two aircraft and returning the base to normal day to day operations. Colonel Taliaferro noted, "Now we've got to get right back to work to ensure we're ready for already scheduled combat deployments. Even today, airmen of the 28th BW are still putting bombs on target in Southwest Asia."[70]

283

CHAPTER TWELVE
THE ELLSWORTH MEMORIAL PARK WALK OF HONOR

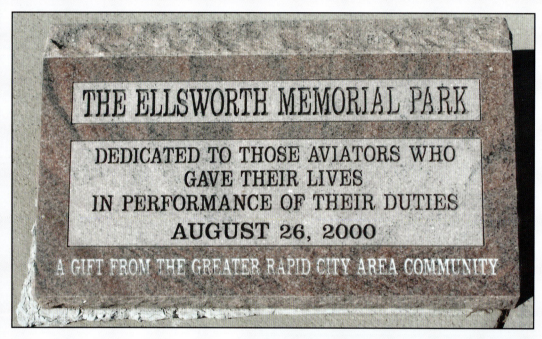

The park is located west of the the Rushmore Center on the base and is lighted. A concrete walk winds through the area of markers dedicated to those who lost their lives while assigned to the base, defending the security of the United States. The memorial park is for those airmen who paid the ultimate price in our country for protecting our freedom. The motto *Freedom Is Not Free* is important as the United States fights in Southwest Asia and in other hot spots around the world. *This and subsequent images Lt. Col. George A. Larson, USAF (Ret.)*

View of the Ellsworth Memorial Park, from the south to the north. The following pictorial walk through the park starts at the upper right hand corner of the top of the photograph and progresses around the walkway, ending at the top left of the photograph. The plaques on each pedestal tell the story and names of those killed during their assignment to the base.

285

The Ellsworth Memorial Park Walk of Honor

CHAPTER THIRTEEN
ELLSWORTH AIR FORCE BASE 2011: A PHOTO PRESENTATION

Newly constructed School or Doolittle gate at the northeast entrance to the base. The road curves into and out of the gate, with a steel pop-up barrier past the gate to prevent a bolting vehicle from speeding through without stopping. Heavy steel fencing keeps vehicles on the road. There are two other gates. In April 2011, the Commercial gate at the south end of the base was also operational, including a large enclosed commercial vehicle inspection building. The third gate, referred to as the Main Gate, was being rebuilt to include a new pass/registration building. The gates are staffed by civilian guards augmented by regular Security Police, often with security dogs. *This and subsequent photos: Lt. Col. George A. Larson, USAF (Ret.)*

Civilian gate guard to the right of the photo, with a military working dog and its handler. Security is very important on the base because of the combat ready B-1Bs, the Weapons Storage area, and the newly assigned Predator squadron.

The 28th Security Forces tested new hand-held equipment for scanning ID cards at Ellsworth's gates. The test period started on February 14, 2011, for a newer version of the Defense Biometric Identification System, or DBIDS. Ellsworth is one of two military installations in the United States to be a testbed for the new identification system. Technical Sgt Jon Proffit, 28th Squadron Forces Squadron Visitor Control Center NCO in charge of pass and registration, described it this way:

DBIDS is a Homeland and Department of Defense initiative used to manage personnel, property, and installation access. Previous versions of DBIDS required everyone to be manually registered into the system prior to entry into the installation. This process was extremely labor intensive and inconvenient for ID cardholders, but necessary to meet DoD requirements. Ellsworth and Grand Forks AFBs will participate in this beta test of the DBIDS 4.0 version, which eliminates the need for all ID cardholders to pre-register. [Registration into the system will happen automatically when an ID card is scanned at the gate entrance.] The testing period will last about six months [until August 2011]. The exact end date of the test will be determined by the performance and functionality of the equipment and data provided.

DBIDS 4.0 allows 28th SFS airmen to register personnel at Ellsworth who are in possession of a Common Access Card (CAC), or any other approved DoD ID card with a barcode, with a DBIDS hand held scanner. The process only takes three to five seconds.[1]

Army Air Force Exchange Service (AAFES) Shoppette at the north side of the Doolittle Gate. Self service gasoline station with a shoppette inside for basic food, drink, and entertainment DVD rentals. It is ideally located to support the nearby barracks and military housing areas.

Family camp area, which allows parking of campers and recreational vehicles. Similar to the facilities offered at private campgrounds such as KOAs located around the Black Hills.

Base housing office from the Family Camp area

Child Development Center

Transient Lodging Facility

Entrance sign to Ellsworth Rushmore Heights housing area

288

Rushmore heights housing area, with wide streets, sidewalks, lights, and play areas throughout for military families assigned to Ellsworth AFB.

Centennial Estates military housing area, north of Rushmore Heights. Currently direct access from Ellsworth AFB has been blocked, with the housing area fenced to close it off from the base due to privatization of Ellsworth's housing units. Access is now through the city of Box Elder.

Wide and modern streets in housing area, torn down as no longer usable. Located at the edge of Centennial Estates.

Transient quarters/barracks

Hospital, Dental Clinic, Pharmacy

Officers/Enlisted Dakotas Club

Main Exchange, food court, flower shop, barbershop, All Seasons, and Credit Sentinel Credit Union

Base library

Commissary

Theater

Air Force Finance Center

Freedom Chapel

Headquarters Air Force Financial Center

Bowling lanes (Bandit lanes)

Base car wash bays

Ellsworth AFB 2011: A Photo Presentation

Base water tower.

Base recreation center, crafts center, and auto hobby shop.

Black Hills Chapel

Chapel Activities Center

Black Hills Bandit Inn, an on base dining facility.

Teen center

Black Hills Center. Family Support, Red Cross, and Air Force Aid.

Base swimming pool complex.

Bellamy Fitness Center

Pine Tree Inn

28th Bomb Wing Headquarters building.

28th Bomb Wing Operations Group building, located east of and attached to the 28th Bomb Wing Headquarters building by a covered hallway.

Maintenance Training Flight Detachment 8.

28th Civil Engineering Squadron.

Air Force Office Special Investigation (AFOSI) Detachment 226.

290

Military Working Dog Area

28th Bomb Wing Maintenance Squadron Area.

Base football field and parade area.

34th Bomb Squadron

Entrance to flight line through security gate at the south end of the 34th Bomb Squadron.

28th Maintenance Group Headquarters

37th Bomb Squadron/37th Aircraft Maintenance Unit.

Aviation fuel tanks located northeast of the bomb squadron area, surrounded by security fence.

RV storage area north of the aviation fuel tanks. Access to rental spaces by padlock gate.

B-1Bs on the flight line between their hangars to the west of the squadron buildings. The entire area is secured by access controlled security access gates, which are lighted at night.

View to the south of the above B-1B parking area, with two additional B-1Bs on the ramp. One of the flight line security access gates is visible in the right center of the photograph.

View of area behind the B-1B hangars, with one of the hangars to the right and the Pride Hangar (former B-36 hangar-concrete arch roof) at the center of the photograph.

28th MX Squadron building

Base traffic Management Office, better known and referred to as TMO.

28th Logistics Readiness Squadron

291

Pride Hangar, currently used as a base athletic facility, but former Cold War B-36 maintenance hangar.

Last row of B-1B maintenance hangars at the south end of the aircraft parking area.

Base fire station doors from the access road and doors opening onto the flight line on the west side of the building.

Base Operations

Dock 22 along aircraft maintenance row.

B-1B on the taxiway, moving toward the south end of the taxiway for cross over to the runway for takeoff on a local training mission.

Surviving World War II Hangar near the south end of the taxiway that is no longer used for aircraft operations or maintenance.

Aviation fuel tank near the south end of the taxiway. A second tank is positioned nearby.

Red-white painted concrete building behind the engine blast defector is the ILS system for the runway.

Development Training for Airmen Center

Vehicle storage building at the south end of the taxiway/runway.

Defense Reutilization and Marketing Service

South end of the main runway, looking north from Kenney Road (service road) to the west side of the base.

Aircrew Flight Equipment Training and SERE

Samuel O. Turner PME (Professional Military Education) Center, former aircrew alert facility at the southwest end of the runway.

292

Former bomber/tanker alert parking apron, separately secured, lighted, and patrolled by armed security police to protect the bombers loaded with nuclear weapons on SIOP alert, referred to as a Christmas Tree (from the SAC period of nuclear alert).

Concrete K-rails which block off the former alert apron from the south end of the runway.

B-1B engine test cell facility.

Rushmore Center, base administration center.

The above looks at the majority of facilities on Ellsworth AFB in 2011, a big change from 1941, and now a modern ACC B-1B installation with worldwide commitments. As noted in a previous chapter, Ellsworth is under the concern/ program cleanup efforts of the U.S. Environmental Protection Agency, Region 8, Denver, Colorado. During over 70 years of operations, especially during the Cold War years, military operations deposited various levels of contamination of the base, with some leeching onto private land. The EPA added Ellsworth to its National Priorities List (NPL) on August 30, 1990. The Air Force, EPA, and State of South Dakota closely worked together to clean up contaminated areas on the base. Studies to identify hazardous substances were conducted in 12 general areas on the base, including landfills, a fire protection training area, spill sites, industrial areas, and an explosive ordnance disposal area. The hazardous substances found most often on the base were solvents and jet fuels, located in both soils and ground water. Unfortunately, some ground water contaminants have moved beyond Ellsworth's boundary to the east and south at low concentrations, but above Federal drinking water standards. Continued use of the contaminated ground water over long periods for household purposes, particularly as drinking water, could pose unacceptable health risks. The Air Force installed cleanup systems to address possible future health risks. Construction of clean up systems is complete at all contaminated areas. The clean up includes ground water pump-and-treat systems, bio-dechlorination, landfill covers, soil treatment systems, excavation activities, and natural attenuation. The systems are functioning properly.

Ellsworth obtains its drinking water from the Rapid City Municipal Distribution System. Ellsworth obtained its water supplies from five wells that had been drilled into deep bedrock aquifers below the base prior to hooking up to the Rapid City water system. These wells were taken out of service, and have been abandoned in accordance with the State of South Dakota's requirements. Shallow ground water in the area around Ellsworth is used for domestic water supplies and for livestock watering. Ground water contamination has impacted the drinking water wells of some homes to the east and south of the base. In 1999, Ellsworth completed a water supply line to provide treated water from the Rapid City Municipal Water Department's system to off base residents. Water was supplied to off base residents by Ellsworth until 2007, at which time a license for operation and maintenance of the water system line by the City of Box Elder became primary.

The Air Force capped landfills and has enforced institutional controls to prevent unauthorized access to those landfills, and to prevent the caps from being distributed. Contaminated ground water is pumped out of the ground and purified to drinking water standards. The treated water is then either discharged to a local drainage, to the Ellsworth water treatment plant, or re-injected into the aquifer. A contaminated ground water plume extends off site to the east.

293

However, the plume has been stopped at the site boundary and a gap in the plume is now evident. Natural attenuation of the remaining contamination will continue to be monitored. These ground water cleanup systems will be in operation for 20 to 30 years to complete the clean up. However, treatability studies are being implemented to enhance and possibly replace current pump-and-treat technologies. The relatively low levels of contamination in off base areas are expected to lessen within the same time frame. Clean up of the entire Ellsworth AFB, including 20 years of ground water treatment, is expected to cost approximately 30 million dollars. All cleanup activities are being performed by the Air Force. The EPA and State of South Dakota provide regular oversight.

The third five-year review of the remedial implementation at Ellsworth AFB has been completed by the Air Force, as required by Section 21, Comprehensive Environmental Response Compensation and Liability Act (CERCLA), commonly referred to as the Superfund. The review was conducted March through September 2010. The five year review determined that ground water remedies at Ellsworth AFB are in place and operating, and that they protect human health and the environment because contaminated ground water is contained at the base boundary, high concentration source areas have been identified and are being treated, and because land use controls and alternate water supplies prevent ground water use. All existing remedial systems require monitoring and occasional minor modifications. The Ellsworth AFB environmental flight staff continue to conduct these efforts and insure that the remedies remain protective of human health and the environment.

To elicit community involvement, in June 2010 16 individuals were interviewed and asked for their impressions of how the environmental cleanup work at Ellsworth was progressing. Each participant was asked a series of questions that were deployed in May 2010 by the Ellsworth AFB Resources Protection Management Group. The interviewees comprised a cross section of the local community, and included four county or municipal employees or elected officials, two local businessmen or developers, two on base residents, and eight off base residents. Interestingly, these interviewees said they were aware of the clean up and seemed satisfied with the progress being made; eleven said the effort is going well, with five making no comment or impressions.[2]

After years of heavy aircraft landings and takeoffs, Ellsworth's runway needs major upgrades and improvements. Howie Aubertin, 28th Civil Engineering Squadron Programs Flight Chief, noted:

Over six months beginning in May 2011, Ellsworth's main runway will be repaired at a current [January 2011] estimated cost of 5 to 7 1/2 million dollars. The repair work will consist of the center section of the runway, referred to as the "Keel." These are normal repairs that one would require of a 13,297-foot runway of this age. This section of the runway is where B-1Bs and other aircraft land on. A 3,400-foot section of the 40-foot wide Keel section of the south end of the runway, where B-1Bs land, has been deteriorating. The runway's last repair work was completed in 1998, when individual concrete slabs were removed and replaced. In the transition from B-52Hs to B-1Bs at Ellsworth, a 12-inch thick concrete overlay was poured in 1986 over the entire length of the runway with an expected life span of 20 years. The current concrete overlay is five years beyond its design life span. We've been planning and working on this for a few years. We have all the details of the timing and constraints during construction, so contractors who bid on the project know what we need. We're doing it as any large airport would. They want to get back to flying too, so we want to make sure we have a contractor that can fit the window. The main runway is an important piece that needs to be maintained with great care to ensure the safety of B-1B crews. Our mission is to put bombs on target, and we can't do that without the best pavement, so for us, it's the number one priority.

Alan Temple, 28th Civil Engineering Squadron Project Manager, stated:

During two of the four phases of construction, the flight line will continue to operate but in a reduced space for the B-1Bs. If funding comes through as planned from the Air Force, the project should be completed by mid-October.

In the summer of 2012, construction work was completed on the main runway, with new concrete sections in place.

Alan Temple went on to comment about the reconstruction of a section of the main runway:

The removal of existing concrete will be unlike standard road construction. Each giant slab will be cut out with concrete saws, lifted out of the runway onto a flatbed trailer, and removed without breaking up the concrete. This is possible because in 1998 "bond breaker" was laid between existing layers of previous concrete and asphalt.[3]

The contract proposal for runway repairs on Ellsworth reads as follows:

Federal Contract: 2011-01-14
Y-repair Keel section of runway, from 09-1015
Ellsworth AFB, South Dakota
Solicitation Number: W9128F11R0014
Agency: Department of the Army
Office: U.S. Army Corps of Engineers
Location: USACE District, Omaha
Notice Type: PRESOL
Solicitation Number: W9128F11R0014
Agency: Department of the Army
Contract Synopsis
1. On or about 28 January 2011, this office will issue a request for proposals for construction of the repair Keel section of runway, FXBM09-1015, Ellsworth AFB, South Dakota. The solicitation will close on or about 1 March 2011.
2. This solicitation is unrestricted and open to both large and small business participation.
3. Provisions in the solicitation documents will provide detailed site visit information (meeting place, date, time, points of contact, and access requirements).
Project Description
This is a full designed project. The work consists of repair to approximately 3,400 feet of deteriorated Keel section at the south end of runway 13-31. The repairs will consist of removal and replacement of the center twp slabs of the runway along with replacement of additional slabs as needed. Additionally, centerline joint resealing, crack sealing, and patch and repair work will be included with the runway repair.

RFP Evaluation Criteria
Firms participating in the solicitation will submit technical requirements in accordance with Section 0022 Proposal Submission Requirements Instructions and Evaluation. Submissions will be evaluated based on a best value approach considering technical elements for evaluation (i.e., Contractor Team Experience Evaluation Past Performance, Technical Approach/Key Personnel, Subcontractor's Qualification and Construction Experience, and Utilization of Small Business). An adjectival method of evaluation will be used to evaluate all technical factors. Prices will be evaluated separately for reasonableness and realism. A best value will be considered after reviewing price and technical factors. The estimated construction cost of this project is between $5,000,000 and $7,500.000.[4]

Centennial Heights housing area.

Changes continue to be made to the base area. One of the biggest is the transfer of the large Centennial Heights military housing area at the northeast side of the base, north of Rushmore Heights housing, which will transfer 828 military housing units to the City of Box Elder. The transfer was scheduled for August 1, 2011, but was delayed for one year.

The area needs to be fenced to secure the housing complex from the base, it needs road access to Box Elder, and utilities transferred from the base. The housing area is administered by Hunt Building, the development's owner, with renting available to the general public. Even though the housing area remained on base for another year, under private administration, no unescorted access was allowed until the transfer in 2012.[5]

295

The Wing operates under the 28th Bomb Wing as host unit. The Wing Commander's Staff consists of a Vice Commander, Executive Officer, Secretary, Director of Staff, Wing Inspector General, Command Chief-Master Sergeant, Historian, Information Management, Protocol, Public Affairs, Legal, Chapel, Military Equal Opportunity, Wing Plans, Treaty Compliance, Safety, Honor Guard, 28th Comptroller Contracting Squadron, and Sexual Assault Response Coordinator. The 28th Operations Group develops the best B-1B operations team to provide rapid, sustainable combat power anywhere, anytime in the world. Its mission is to provide combat-ready aircrew to project global power anytime in support of the Combatant Commander's objectives of maintaining combat ready aircrew and planning/executing training missions essential to attain versatile power projection and global reach. The Group's goals consist of the following:

Mission: Maximize combat-strike capability
People: Take care of the most valuable resource
Workplace: Improve our work environment

The Group's operating philosophy consists of the following:

Ethics and core values
Loyalty and leadership
War-fighting team

The Group consists of three squadrons:

28th Operations Support Squadron Ensures Wing combat readiness for worldwide operational taskings. Plans and supports combat operations and deployments for two tactical B-1B squadrons. Provides deployable planning and support teams for B-1B employment. Manages base airfield and radar approach control facility. Provides air traffic control, weather, airspace management, intelligence, life support, simulator, scheduling, and maintenance analysis.

34th Bomb Squadron Defeat America's enemies across the globe on a moment's notice.

37th Bomb Squadron Provides rapid, decisive, and sustainable combat airpower anytime, anywhere.[6]

The 28th Maintenance Group's mission is to enable the rapid global employment of air power by providing maintenance support to ensure combat ready B-1B Lance aircraft and associated support equipment. Provide mission capable aircraft, equipment, weapons, and trained personnel through aircraft maintenance, munitions, maintenance, and maintenance operations squadrons to sustain both training missions and wartime and contingency taskings. The Group is comprised of over 1,300 personnel distributed through four different squadrons:

28th Aircraft Maintenance Squadron Provides top-notch on-equipment maintenance to maintain to combat readiness for 29 B-1B Lancers, requiring 655 maintenance and support personnel. Provides the war fighter with combat mission-ready aircraft to conduct worldwide operations by employing air-to-ground precision guided munitions against any adversary, anytime and anyplace around the world on short notice.

28th Maintenance Squadron Provides on-and-off base equipment repair and maintenance support for 29 B-1B aircraft and 102 F-101-GE-102 engines in the 28th Bomb Wing and 366th Wing. Performs periodic inspections of weapon system and intermediate-level maintenance to support local training and worldwide conventional and contingency taskings as directed by Headquarters ACC and the National Command Authority.

28th Munitions Squadron Provides conventional munitions, armament systems, and trained munitions personnel in support of 29 combat ready B-1B aircraft to meet JCS operations plans taskings, including deployed theater operations and power projection operations from the United States. Maintains over 1,600 munitions line items, 30 conventional bomb modules, 36 conventional rotary launchers, and ACC's B-1B strategic launcher inventory.

28th Maintenance Operations Squadron Provides maintenance support of the B-1B to carry on a worldwide JCS contingency and exercise deployment taskings.[7]

28th Bomb Wing B-1B on Ellsworth's taxiway prior to moving to the south end of the runway for takeoff.

The following list of 28 B-1Bs was pulled from various sources which the Air Force or the 28th Bomb Wing does not confirm or deny. It is provided to give an indication of the range of production aircraft assigned to the Wing:

85-0060	85-0083	85-0086	86-0093	86-0096
86-0111	86-0116	86-0123	86-0129	86-0139
85-0079	85-0084	85-0087	86-0094	86-0099
86-0113	86-0118	86-0125	86-0130	
85-008	85-0085	85-0091	86-0095	86-0102
86-0114	86-0121	86-0127	86-0134	

The 28th Mission Support Group provides all essential base operating support services for Ellsworth AFB and combat support services for an Air Expeditionary Wing. A total of 1,600 military and civilian personnel in seven squadrons ensure civil engineering, communications, security, personnel support, contracting, logistical readiness, and base services to adequately support 16,000 active duty, dependent, DoD civilians, and military retirees living in the Rapid City area. The Group's squadrons manage more than 40 million dollars in budget authority, 3,833 facilities, and 5,000 acres of U.S. Air Force land. Its mission is to provide rapid, decisive, and sustainable combat/mission support anytime, anywhere. Its vision projects the dedication of professionals committed to excellence. Its squadron supports the 28th Bomb Wing.

28th Civil Engineering Squadron

28th Communications Squadron Provides reliable, efficient, and secure communications for rapid decisive and sustainable combat air power.

28th Contracting Squadron Serves as the business advisor to the Commander of the 28th Bomb Wing. Provides premier business advice to all organizations on Ellsworth AFB. Looks to the Small Business Community first when filling requirements for construction, services, and products.

28th Logistics Readiness Squadron

28th Force Support Squadron Composed of Military Personnel Flight, Civilian Personnel Flight, Family Support Center, Base Education & Training Services, Manpower, Airman Leadership School, First Term Airman Center, Professional Developmental Education, and Career Assistance Advisor. The Military and Civilian Personnel Flights and the Family Support Center are located in the Rushmore Center, providing personnel service to all active duty and retired military members and their families.[8]

The 28th Medical Group provides a fit and healthy force for global response. Its mission is to create an elite team providing a patient-centered medical home. The following are the 28th Medical Operations Squadrons:

Family Practice Clinic It is a multi-disciplinary clinic made up of four family health teams, each with two providers, a pediatrics team, and a Women's Health Team. It is dedicated to providing efficient, quality care to Tricare Prime beneficiaries from the ages of newborn to 65 years.

Optometry Clinic Provides primary eye care for active duty personnel and Tricare Prime beneficiaries. It contains routine eye examinations for glasses and gas mask inserts, as well as renewing contact lens prescriptions. It diagnoses, treats, and manages diseases and disorders of the eye.

Dental Clinic A full service dental clinic whose mission is to maintain active duty personnel in top dental health and ready to deploy at all times. Dentists and technicians provide a full range of dental services to include examinations, cleanings, general dentistry, prostrodontics, periodontics, endodontics, and oral surgery.

Flight Medicine Clinic Plans, coordinates, and provides oversight for base preventive, environmental, and aerospace medical operations. The Flight Surgeons provide medical care for aircrew and their families, and integrate key ORM analysis to minimize potential pilot fatality. The clinic also manages periodic physical/occupational exams for flying/special duty personnel.

Mental Health Clinic Consists of a number of helping programs aimed at maintaining mission readiness and enriching the lives of the base's military families.

Physical Therapy Clinic Specializes in the treatment and management of acute sprains/strains of lower and upper extremities, axial skeletal pain management, and post-operative orthopedic rehab. It offers an exercise area with bikes (upper and lower), elliptical, treadmill, UBE, BAPS (ankle/knee balance integrator), and many other rehabilitative tools.

Health and Wellness Center Enhances readiness through optimal health and total fitness by encouraging people to follow healthy lifestyles and avoiding harmful behaviors. The health programs provide people with the knowledge and skills to change lifestyle behaviors and move toward a balance of physical, emotional. intellectual, social, and spiritual health.

Public Health Clinic Consists of Community Health and Force Health Management Sections. Community Health encompasses Food Safety & Sanitation, Communicable Disease Prevention, Surveillance, and Entomology Programs. Force Health Management

297

covers Hearing Conservation, Occupational Health, Fetal Protection, Medical Standards, Deployment Medicine, Preventive Health Assessment Scheduling, and PIMR Admin Programs.

The 28th Medical Squadron

Laboratory Services Performs limited testing in house and sends all others to many different reference laboratories. Not all results are available to a patient's provider the same day.

Pharmacy Takes over the telephone refills on a 24-hour automated telephone refill system online through the Tricare online pharmacy refill option. Prescriptions are available for pick up after two days.

Radiology Services Its mission (Diagnostic, Imagery-Radiology Department) is to attain radiographs for routine X-ray examinations to assist Health Care Providers in their diagnostic treatment. MRIs and CT scans are also available on a limited basis and in cooperation with the Black Hills Veterans Administration's mobile unit.

Tricare Service Center Its western region contractor operates the center. Triwest, Hospital Tricare Operations, and Patient Administration Flight work together to assist patients' family members in accessing and understanding their health care benefits and responsibilities.[9]

The Air Force Financial Services Center provides customer support to over 389,000 active duty and reserve military and civilian customers throughout the world. It is responsible for processing temporary and permanent duty travel, as well as military pay transactions previously performed at base-level financial services offices prior to consolidation of Air Force Financial Services in September 2007. Its mission therefore is to deliver responsive, world class 24/7 financial services and to display skilled warriors supporting global operations. Its stated vision is to be customer focused while delivering financial services and to prepare warrior-ready financial professionals.[10]

The long line of Ellsworth AFB commanders continued on May 31, 2011, with a change of command ceremony conducted in the Pride Hangar. Col. Jeffrey Taliaferro passed command of the 28th BW to Col. Mark Weatherington. A large crowd of Ellsworth military, civilians, and local and state dignitaries attended.

CHAPTER FOURTEEN

SOUTH DAKOTA AIR AND SPACE MUSEUM

Display of Aircraft of Ellsworth Air Force Base

The B-1B as the central display at the South Dakota Air and Space Museum, outside the main gate of Ellsworth Air Force Base.
This and subsequent photographs: Lt. Col. George A. Larson, USAF (Ret.)

The B-1 program has been fraught with controversy almost from its inception. The aircraft, which was designed to replace the aging B-52 Stratofortress intercontinental bomber, proved too sophisticated and burdened with troubles to carry out the role. Many experts and military officials gave up on the aircraft. Despite the problems, the aircraft emerged in the Afghanistan "War on Terrorism" as a powerful strike bomber. In November 1969, the USAF issued a specification for a multi-role, long-range bomber capable of flying intercontinental missions without refueling. Once near the target, the aircraft was expected to penetrate and predict sophisticated enemy defenses. The aircraft would carry both conventional and nuclear weapons.

After careful evaluation of the proposed designs, the Air Force selected the Rockwell design in June 1970. Engineers at Rockwell solved the problem of having the aircraft perform at high speed and low altitude by designing a unique variable geometry wing, which could be swung out or in as flight characteristics for a given mission dictated. The aircraft was complex and revolutionary, but if if worked, it would provide the Air Force with an aircraft years ahead of any of its potential adversaries.

A contract for five test aircraft was issued and Rockwell went to work on construction. The first full-scale mock-up was finished in November 1971. The Air Force was impressed, but asked for 297 changes to the mock-up. A total of 22,000 hours of wind tunnel testing was conducted on the airframe even before the first full scale prototype was begun. With all the testing complete and proposed changes incorporated into the design, Rockwell began construction on the prototypes. The new aircraft was officially designated the B-1A.

The first aircraft flew in December 1974. It was sleek, fast, and sophisticated. The prototype's sing wing design allowed it to attain a speed of 1.25 Mach with its wings extended, and it could maintain stable flight at slower speeds for take-offs and landings. Four General Electric 010-GE-102 turbofan engines powered the aircraft. The engines allowed the aircraft to take-off with a maximum weight of 477,000 pounds.

The aircraft carried its offensive weapons in three separate weapons bays. It could carry a variety of different ordnance packages, including bombs and missiles. The aircraft could fly 7,455 miles without refueling, but with aerial refueling the aircraft could reach any point in the world.

Its four man crew consists of a pilot, copilot, weapon systems officer, and defensive systems officer. Initially, the crew was seated in an escape capsule that came off the aircraft as a complete unit in the event of trouble. This feature was later replaced with standard, single ejection seats.

The flight and bomb carrying characteristics of the aircraft were impressive. However, the heart of the new bomber was its sophisticated navigation, radars, and electronic countermeasures suite. The aircraft could fly "nap of the earth" missions; that is, it could fly very low over any terrain, up one side of a hill and down the other. The 147-foot long bomber could fly at 200-feet at speeds of 600 mph. It could jam enemy radars, confuse enemy radars, confuse enemy radar tracking missiles, and penetrate any air defenses on the planet. Flying at high speed and low altitude left no room for error. When the advanced electronics failed during several test flights, the aircrew on board was noticeably upset. The complex jamming arrays of electronic measures were so good that on occasion they jammed the electronics on the bomber itself. These and other problems led to extensive testing and modifications. Finally, in June 1977, President Jimmy Carter canceled the B-1 program in favor of the cruise missile program for the B-52 bomber.

Proponents of the aircraft were undaunted. They realized that cruise missiles could not carry the destructive power of a fully armed B-1 bomber. They knew that if they could resurrect the program and incorporate the modifications of the B-1 they would have a winner. On October 2, 1981, President Ronald Reagan authorized the Air Force to acquire 100 B-1Bs as part of the U.S. Strategic Modernization Program. The B-1B Lancer, as it was called, incorporated a stronger airframe, improved electronics, strengthened landing gear, upgraded and improved engine design, and many other new features. More importantly, the B-1B's electronic jamming equipment was improved. The infrared countermeasures, radar location, and warning systems were integrated to complement its low-radar cross section and form a sound defensive package for the aircraft. The upgraded version of the bomber was equipped with a new electronics suite that consisted of the AN/APQ-164 multi-mode offensive radar also used for navigation and terrain following. The radar also allowed targets to be located and engaged under the worst weather conditions.

The first B-1B took to the air in October 1985. Much of the development and testing program had been accomplished on the B-1A, so the test flight program proceeded rapidly. The aircraft was accepted by the Air Force, and the first production model was delivered in July 1985. Although it was officially known as the Lancer, crews nicknamed the aircraft the B-one or "Bone." The bomber was assigned to the nuclear deterrence mission up to the time of the Gulf War of 1991. Although it could have been used in attacking Iraqi targets in the Gulf, the Air Force decided to retain the aircraft's primary role, and it remained in the United States on nuclear alert to counter any possible nuclear threat and attack. Each of the bombers could carry up to (24) B-61 or B-83 gravity nuclear weapons, or the Lockheed AGM-69A Short Range Attack Missiles (SRAMs).

The fleet of B-1Bs is expensive to operate. After they entered service, the B-1B fleet never achieved its objective of meeting a 75 percent mission capable rate. In 1992 and 1993, the B-1B mission capable rate averaged 57 percent. According to the Air Force, a primary reason for the low-mission capable rate was the level of funding provided to support the B-1B logistics support system. Concerned about the low mission capable rate, the Air Force decided to leave several of the bombers as nuclear bombers while reconfiguring the others to carry conventional weapons. With proper funding and maintenance the availability rate increased to 84.3 percent. Although the availability rate varied, the B-1B was ready for combat.

On the night of December 17, 1998, two B-1Bs attacked and destroyed an Iraqi Republican Guard Barracks at Alkut, Iraq. The attack was part of Operation Desert Fox, designed to punish Saddam Hussein for failure to comply with United Nations directives. The attack was impressive, and both bombers returned safely to friendly airfields. Even as Air Force planners were evaluating the B-1B's combat debut, another crisis in Kosovo was drawing U.S. attention. Ethnic hatred between Serbs, Albanians, and Kosovars erupted into a bloody civil war. The Serbian Army moved to end the war by wiping out the Albanians and Kosovars. Despite United Nations demands to stop the massacres they continued, and the United Nations sent a military force to stop the Serbs. Part of the military component was six B-1Bs from the 77th BS and two from the 37th BS, EAFB. The Lancers bombed the oil refinery at Novi Sad and Ponkive Air Field in Yugoslavia. Both facilities were among the first to feel the fury of the B-1B. As the air campaign continued, other targets that required the Lancer's range, heavy bomb load, and accuracy were added to the mission list. The eight bombers attacked and destroyed dozens of other targets during their campaign.

300

The war in Afghanistan offered the aircrew of the B-1B a new opportunity to demonstrate the aircraft's conventional war capabilities. Enemy troops dug in on hillsides and in caves required massive bomb loads to either bring them to their knees or seal them inside their caves. The B-1B fulfilled the requirement when fully loaded. It can carry (84) MK-82 non-precision 500-pound gravity bombs. The Lancer can also carry up to 30 cluster bomb units (CBUs), 90 per mission for enhanced conventional capabilities against armor. Each CBU carries 72 anti-tank mines and 22 anti-personnel mines. In the event heavier bombs are needed, the Lancer can carry (24) 2,000-pound MK 84 mines. The B-1B can also carry (24) GBU-31 Joint Attack Munitions (JDAMs), which consists of a guided one-ton bomb. This bomb has a circular error probable (CEP) of under 30-feet. If the JDAM GPS system fails, the bomb's internal guidance system will guide the weapon to within 90-feet of that target.

The aircraft is also equipped with the CBU-103 Wind Corrected Munitions Dispenser (WCMD), which allows cluster bombs to be dropped in windy conditions with great accuracy; Joint Standoff Weapon (JSOW), able to be launched at a maximum range of 74 miles; and the AGM-158 Joint Air to Surface Standoff Missile (JASSM), a cruise missile that can be launched while the aircraft is up to a maximum range of 115 miles from the target.

Shortly after the start of Operation Enduring Freedom (the war against terrorism), the 28th Expeditionary Wing of the USAF was assigned B-1B bombers from the 7th (Dyess AFB, Texas), 28th (Ellsworth AFB, SD), and 366th (Mountain Home AFB, Idaho) BW. The aircraft were based on the British-controlled island of Diego Garcia in the Indian Ocean. Missions from Diego Garcia to Afghanistan lasted between 12 to 15 hours, covering a distance of over 5,500 miles. Initial attacks were conducted against Taliban anti-aircraft radars, guns, and command centers. These attacks were made using precision guided munitions. As the war progressed and sophisticated enemy targets were destroyed, B-1Bs dropped conventional bombs on front-line troop positions.

Guided to their targets by special forces troops using laser designators, the B-1Bs would come in at low altitude and destroy enemy positions with their 500-pound, 1000-pound, and 2,000-pound bombs. These strikes were so successful that the B-1B soon replaced the B-52 as the workhorse of the bombing campaign. The B-1B flew dozens of bombing missions over Afghanistan as well as over Iraq during the spring of 2003 war against Iraq. Its powerful bomb load, wide range of weapons, and ability to fly nap of the earth missions made it a valuable part of the USAF aircraft arsenal. During Operation Odyssey Dawn (March 2011), two 28th BW B-1Bs flew the first continental United States launched B-1B strike, flying non-stop from Ellsworth AFB to Libya. The two aircraft dropped 100 JDAMs on targets in Libya to support Libyan rebels fighting on the ground.

The B-1B on display at the museum was dedicated on February 4, 2005. It was retired from the Air Force and pulled from active duty status at Dyess AFB, Texas, in 2003. Eight B-1Bs were retired from operational status and offered to air museums in the United States. The B-1B at the museum was delivered to Dyess AFB in January 1988. It was named and carried nose art of the "Texas Raiders." The nose art came from a Commemorative Air Force Boeing B-17 Flying Fortress of the same name. The B-1B was painted gunship gray, removing the nose art. After it was flown to Ellsworth AFB, 28th BW maintenance personnel took two years to remove usable spare parts, given an estimated value of $400 million dollars. The parts removed from the bomber were transferred to B-1B spare stock inventory. While the B-1B was undergoing transformation into a museum static display aircraft, 28th BW support personnel prepared a display pad in front of the museum's building. A Rapid City salvage company removed a steel scale model replica of a B-2 Stealth bomber built by the Honda Corporation and donated to the museum as its entrance display. It took approximately 500 man hours to prepare the B-1B static display area. The repainted B-1B, after being stripped of usable equipment and electronics, was towed from Ellsworth's flight line to the museum, requiring light poles and road signs to be temporarily lowered to the ground to permit the bomber's wings to clear the vertical obstructions. Once at the museum, the bomber was backed onto the static display pad, with the wheel's secured to three hard points. It took another year for landscape work around the bomber to be completed.

Vought A-7D Corsair II.

Chief of Naval Operations looked at the Navy's future aircraft needs in 1963. The question was, could current Navy attack aircraft penetrate enemy air defenses and deliver a potent payload? At that time, the Navy's attack jets were one generation from the A-1 Skyraiders being phased out of service. These turbojets were the A-6 Intruders and A-4 Skyhawk. The A-6 had something going for it. It was an all-weather attack bomber with sophisticated electronics that was just then entering service. The A-4 Skyhawk was a small, single-engine fighter-bomber, designed to safely reach its target through its speed and agility, not to mention its small size. Because of this, however, it could not carry sophisticated electronics or a heavy payload. The Navy's 1963 study, called the "Sea Based Strike Study," determined that a new Navy attack aircraft would soon be needed.

Because of the growing war in Vietnam, the Navy decided not to design a new aircraft from scratch, but to modify a proven design. The Navy had four airframes to choose from: A-4 Skyhawk, A-6 Intruder, F-J1 Fury, and F-8 Crusader. Each aircraft manufacturer was asked to submit a proposal for an updated version of its aircraft. Whereas three manufacturers updated their existing airframe, Vought took a different and more radical approach. Vought elected to use the engine originally designed for the F-111. To make this engine fit, Vought's new fuselage was widened and shortened. The wings were strengthened to accommodate six hard weapons points, able to carry 15,000 pounds of ordnance. State of the art electronics were put into the fuselage: guidance systems, bombing system, and radar. These were placed in one central portion of the fuselage for easy maintenance. The design surpassed the Navy's proposal, with one exception. The aircraft had an empty weight of 15,036 pounds. The design was still the best of the competition and was selected. The new aircraft was named the A-7 Corsair II. The first squadron to be equipped with the A-7 was VA-147. The squadron embarked in the USS Ranger, flying its first combat mission over Vietnam in December 1967.

By 1967, the A-7 was becoming more than a carrier aircraft. In 1965, the USAF decided that it also needed a new attack aircraft. Because the USAF had prepared to fight a nuclear war against the Soviet Union, smaller and more delicate combat missions necessitated by the war in Vietnam caught it unprepared. The only USAF light fighter-bomber available was the North American F-100 Super Sabre and the British designed B-57 Canberra twin-engine bomber. In 1965, the USAF approached Vought to modify the A-7. The Air Force wanted a new engine,

electronics, and communications equipment, plus a 20mm cannon. Vought fitted the Air Force's A-7 with the TF41-A-1 Sprey engine. The fuselage was already wide enough to accommodate the new engine. The aircraft was armed with the M61 20mm, multi-barrel cannon below the cockpit, but only with 1,000 rounds of ammunition, which lasted only 10 seconds. An aerial refueling probe and advanced navigation/attack system were added. This allowed the aircraft to conduct low-level, terrain-following missions with regard to range restrictions. The Air Force's variant was called the A-7D.

On short range missions, without auxiliary fuel tanks, the A-7D could carry (18) 500-pound bombs. The aircraft could also deliver napalm, fire 2.75-inch rockets, and drop cluster bomb units. Even when carrying a full bomb load, pilots could outmaneuver North Vietnamese anti-aircraft gunners and deliver its ordnance with pin point accuracy. Using its sophisticated bombing computer and onboard navigation systems, the pilot could drop his bombs within 10 yards of the target. The A-7D was first assigned to the 354th Tactical Fighter Wing, first fighting in Vietnam in October 1972. By the time it reached Vietnam the ground war was winding down. The A-7 provided excellent support for helicopter rescues of downed pilots. The Navy took delivery of over 1,300 A-7s as the Air Force 460.

North American B-25J Mitchell.

When the North American B-25 Mitchell, twin-engine bomber is mentioned in conjunction with WWII events, the first American raid on Japan after Pearl Harbor comes to the surface. On April 18, 1942, USAAF Lt. Col. Jimmy Doolittle took-off from the rolling and pitching deck of the USS Hornet in a B-25, leading 15 other B-25s in a surprise, low-level bombing attack against the Japanese home islands (Tokyo and surrounding cities). The B-25 was a versatile aircraft with many different variants. A B-25J during WWII became a Very Important Person (VIP) transport, and is currently on display at the museum.

The B-25 airframe was eventually selected for modification into a VIP transport. North American Aircraft took out of service aircraft number 40-2165 to determine what modifications had to be made and the extent of these alternations. The B-25J was flown to North American's Kansas City production plant. North American engineers made the following modifications to the aircraft:

1. All non-essential military equipment was removed from the aircraft, including armor plate, guns, combat crew equipment, and bombing equipment.

2. Five over-stuffed passenger chairs were installed in the aft passenger compartment.

3. A drop-leaf desk was installed across the width, at the rear of the passenger compartment.

4. An intercom was installed, providing communications between the passenger compartment and the cockpit.

5. The nose was flared over.

6. Windows installed in the passenger compartment for natural light and view out the aircraft.

7. A flare unit installed behind the gunner's position to allow the aircraft to fly over a proposed landing area at night and release the flares to light the runway or landing area.

8. Two additional seats were installed ahead of the bomb bay, aft of the cockpit.

9. Bomb-bay converted to carry luggage.

10. One sleeping bunk installed in the crawl way between the cockpit and passenger compartment.

Aircraft number 40-4030, now on display at the museum, was modified for use by Allied Supreme Commander, U.S. Army General Dwight D. Eisenhower. The aircraft had additional modifications for General Eisenhower:

1. The top of the bomb bay was removed and lowered to create a more comfortable passenger compartment, as on standard U.S. airliners.

2. To accommodate bomb bay alterations, additional structural supports were added to the wing center section's carry through structure.

3. The area below the bomb bay was fitted with auxiliary fuel tanks, providing maximum range when carrying passengers and baggage.

4. Four leather passenger chairs were installed in the passenger compartment. Two chairs faced to the rear at the front of the compartment and two along the left side of the compartment.

5. Two metal arm chairs, with padded seats, were installed behind the cockpit, facing aft.

6. A fold-down, drop-leaf desk was fitted at the rear of the passenger compartment across the width of the fuselage.

7. A walnut storage cabinet installed above the desk.

8. Four windows installed on either side of the upper fuselage in the passenger compartment.

9. Luggage racks installed, airline style, above the passenger seats along the side of the fuselage.

10. To make room for the over-stuffed chairs along the left side of the passenger compartment, the tail gunner's entry hatch was sealed, and repositioned farther to the rear of the aircraft toward the tail.

11. The passenger compartment's walls covered with sound proofing to reduce noise from the aircraft's engines.

12. The insulation covered with airline type paneling.

13. Standard USAF interior olive drab paint color replaced by a lighter color, general's blue.

14. The rear gunner's position was removed to allow space for installation of a life raft, which could be jettisoned if an emergency ditching had to be made in water.

15. The bombardier's position covered with sheet metal.

The postwar history of General Eisenhower's B-25J aircraft is very interesting:

1. After WWII, the B-25J was flown back across the North Atlantic to Bolling field, outside Washington, D.C., and used as a cargo hauler under military designation CB-25J.

2. In 1948, the aircraft was again assigned the role of a VIP transport, under military designation VB-25J, assigned under operational control of the 110th Special Air Mission Group. The VB-25J was used to fly VIPS in the United States from one base to another as required.

3. January 1953, aircraft transferred to the 125th Air Transport Group, operating out of Washington National Airport.

4. August 1958, aircraft transferred to the 1001st Air Base Wing, Andrews Air Force Base, located outside Washington, D.C.

5. December 1958, the B-25J was flown to Davis-Monthan AFB, Arizona, to the Aircraft Maintenance and regeneration Center (AMARC) for open air storage after depot preparation for storage.

6. February 1959, the aircraft was stricken from active Air Force inventory and later sold to a commercial company for commercial operations.

303

7. In 1981, the USAF purchased back the B-25J, transporting the VIP aircraft to Ellsworth AFB. It was restored by museum volunteers from the museum back to its 1944 configuration and appearance when used by General Eisenhower.

B-26B Invader.

The story of the B-26 Invader, used in the air war over Korea, begins in the fall of 1940, when the USAAF requested a design for a new medium bomber to replace the existing twin-engine B-25 Mitchell, Douglas A-20 Havoc, and Martin B-26 Marauder. The apparent double designation of the B-26 ended after the Marauder was dropped from active inventory and the Invader replaced it in June 1948, with the A-26B becoming the B-26B and A-26C becoming the B-26C.

In 1939, the USAAC published a design requirement for a low-level medium bomber which was won by Douglas Aircraft Company with the A-20, referred to as the Havoc. By the time production ended on September 20, 1944, the Douglas Aircraft Company built 7,385. This was a successful design, meeting a requirement for a medium bomber to supplement the heavier B-24 and B-17 bombing operations by the 8th Air Force in Europe. The Douglas Aircraft Company modified its A-20 and developed a prototype with the designation XA-26 Invader. Design work began during January 1941, with advanced military capabilities built into the aircraft. The aircraft had a mid-mounted wing, with a laminar flow aerofoil fitted with electrically operated double-slotted flaps. It was powered by two 2,000 horsepower Pratt & Whitney R-2800-77 air-cooled radial engines.

The aircraft was designed to carry an impressive amount of armament: the internal bomb bay was able to hold a maximum of 4,000 pounds of bombs or two torpedoes in a side-by-side configuration and external under-wing hard-mounted ordnance racks. Two .50 caliber machine guns were mounted in remotely-controlled dorsal and ventral turrets. Both guns were operated by a gunner located behind the bomb bay. One version is a solid-nose with eight forward firing .50 caliber machine guns mounted in two parallel vertical rows of four each. This is the type of aircraft on display at the museum.

The prototype's first test flight was on July 10, 1942. Test flights indicated the aircraft's performance and handling exceeded design specifications. However, a series of changes were made on follow-on production aircraft. Engine cooling was improved by removing the propeller spinner from each engine. Two forward-firing .50 caliber machine guns were mounted on the left side of the aircraft's nose. The two twin .50 caliber machine gun turrets could be locked into the forward-firing position to increase the aircraft's fire power. Bomb bay capacity was reduced to 3,000 pounds, but was compensated for by the addition of two 500-pound bomb capacity racks under each wing.

The production variant consisted of two types: either a solid nose or transparent nose. The A-26B was the solid nose variant. The aircraft's designations became the B-26B and B-26C during the Korean War. The Korean War has been referred to as the "Forgotten War." Little is remembered about the night war fought by the Far East Air Force over South and North Korea. The primary night bombing and attack intruder aircraft used was the Douglas B-26 Intruder. The B-26 was pressed into combat service and used in an effort to slow the transport of military supplies and replacement troops by North Korea, and later by the Chinese Communist military units to front lines. Primary targets were Communist locomotives and railroad box cars and railroad tracks, including bridges over which Communist trains and trucks crossed on the way to the front. It was extremely dangerous for Communist troops to move supplies on the narrow and windy roads during the day, which were subject to the superiority of UN air power, which could rapidly be vectored to targets of opportunity. Communist troops quickly learned to dig into the hills, hiding supply dumps and troops from prowling U.N. aircraft. B-26 night attacks never completely stopped the flow of supplies to Communist front line troop positions. It was costly, losing 210 aircraft. A B-26 dropped the last bombs of the Korean War. It flew during the Vietnam War, and was eventually replaced by the OV-10 Bronco.

B-29 Superfortress.

The plane most responsible for ending WWII in the Pacific was the Boeing B-29 Superfortress. Research and development for the United States' best strategic bomber began in 1937. The first definitive design came about in 1940, and the first prototypes were ordered to be built by Boeing Aircraft in 1942. The first prototype flew on September 21, 1942. Fourteen pre-series models were built. These were known as YB-29s. The USAAF was so pleased with the prototypes that it ordered 1,500. This order exceeded Boeing's production capacity. Bombers were built by Bell Aircraft and the Glen L. Martin Company. Series production of the bomber began in 1943. The first B-29s arrived in India (the China-Burma-India Theater of Operations) to fight the Japanese in the spring of 1944. Bombing missions over Japan started in June 1944, continuing through the end of the war. The B-29 was, up to that time, by far the most expensive warplane built, costing $600,000, which was three times as much as the previous record holder, the B-17. In all, 3,970 B-29s were produced during the war. By far, the most famous of these B-29s were those assigned to the 509th Composite Group, North Field, Tinian Island: the "Enola Gay," which dropped an atomic bomb on Hiroshima, Japan, on August 6, 1945, and "Bockscar," which dropped an atomic bomb on Nagasaki on August 9, 1945.

During the Korean War, B-29s flew 1,076 days out of the 1,106-day war. In the face of many tactical obstacles of air operations in Korea, B-29 aircrew performed brilliantly, destroying industrial and military as well as strategic targets in North Korea with constant support for U.N. ground troops. FEAF Bomber Command lost 34 B-29s during the Korean War.

The Korean War ended the active combat duty of the majority of the B-29 fleet. During the Korean War, starting in 1950 and continuing after the war until 1957, SAC beefed up its combat wing strength, while FEAF Bomber Command maintained a B-29 combat strength of ninety-nine Superfortresses. SAC increased its strength from 19 to 51 bomb wings. By 1953, Sac's medium bomb wings had been equipped with the Boeing B-47 Stratojet, with B-29s flown to Davis-Monthan AFB for storage and disposal, starting in significant numbers during 1954.

The B-29 acquisition and restoration process illustrates how the museum preserves aviation history. In November 1985 a 12-member team departed Ellsworth, traveling to the Naval Weapons Test Center at China Lake, California, to retrieve the last available Superfortress from the desert. The team remained at China Lake for 30 days, preparing the B-29 for shipment to the museum. One former B-29 maintenance person led and directed the efforts of the volunteers; the team used heavy equipment on loan from the Navy. On November 21, 1985, members of the South Dakota National Guard loaded the B-29 parts on to seven flatbed trucks for the five-day trip to Ellsworth AFB. Soon after the convoy delivered the pieces to Ellsworth on December 6, 1985, a team of museum volunteers began the restoration process. Complete upper and lower gun turrets, still in their original packing crates, were located at the National Museum of the United States Air Force, Wright-Patterson AFB, Dayton, Ohio. A rear turret cover was purchased from an aviation parts company in Florida. Museum volunteers fabricated the rear gun turret assembly, since the tail of the display Superfortress had been badly smashed in the desert by a Navy front end loader. The entire vertical stabilizer had to be re-skinned, the propellers rebuilt, the entire external skin repaired, bomb bay doors repaired, and the bomber polished and remarked. A donation of a Norden bomb sight helped start the cockpit's restoration.

B-52D Stratofortress.

The B-52D, with its tall tail and distinctive Vietnam War camouflage paint scheme, is easy to identify. The B-52D came off the production line with aluminum finish on the upper surfaces and a white underside anti-radiation SAC paint scheme. The B-52D remained this color until the Vietnam War. In 1965, B-52Ds were painted black on the underside and tail, with the SEA camouflage color scheme (two shades of green and one of tan) on the top surfaces.

305

As with the entire B-52 fleet, the B-52D was the most extensively modified to increase the aircraft's effectiveness and to reduce structural damage from metal fatigue.

One of the hardest problems to fix on the B-52D was leaking fuel tanks. Beginning in September 1957, the Blue Band Program was implemented to repair weak sections on the bomber's interconnecting lines between wing fuel tanks. However, this did not completely fix the problem. The Hard Shell Program replaced the aluminum fuel line clamps with stainless steel clamps. This didn't correct the problem. The final fix, called Quick Clip, installed safety clips on the fuel lines, which finally stopped the wing leak problem.

The Big Four Modification was listed as the MOD 1000. The B-52D had to be able to penetrate targets at low level under Soviet SAM defenses and associated early detection and guidance radars. SAC B-52s, beginning in 1959, became vulnerable to Soviet SAMs. This was shown to be a credible concern when Soviet SA-2s downed a U.S. Lockheed U-2 spy plane on May 1, 1960. The U-2 was shot down near Sverdlovsk, deep inside Central Russia, while attempting an intelligence overflight for the CIA. To counter SA-2s, the B-52D was equipped with the AN/ALQ-27 ECM system.

The B-52D required extensive airframe maintenance to bring it back to acceptable flight (combat) status. This was called the High Stress Program. Beginning in 1960, Phase I was for Stratofortresses with 2,000 flying hours to have the fuselage bulkhead and aileron strengthened. The boost pump panels and wing splice plates were also reinforced. Phase II of the High Stress Program took place at 2,500 flying hours. This consisted of metal reinforcement of the upper wing splices on the inner engine pods, lower wing panels supporting inner and outer engine pods, the upper wing surface fuel probe access door, and the lower portion of the fuselage bulkhead. Phase III of the High Stress Program was referred to as IRAN (Inspect and Repair As Necessary). The program was designed to fix wing cracks, with the work completed at each B-52's home base.

Another program, called Malfunction Detection and Recording (MADREC), consisted of installing monitoring equipment in the bombardier and navigator compartment, along with auto pilot systems for the flight deck crew (pilot and copilot). This equipment was designed to alert the crew to malfunctions in critical aircraft flight systems.

The most famous upgrade for the B-52D was the 1965 Big Belly modification, which increased bomb load capacity, which was especially critical for the bombing support mission flown to assist U.S. ground forces fighting in Vietnam. The program increased the B-52D's bomb load from (27) 500-pound MK-82 bombs in the bomb bay to (84) MK-82s or (42) 750-pound MK-117s. These bombs were loaded into clips, then hoisted up into the bomb bay, decreasing munitions load time. An additional 24 (twelve on each external) under wing munitions pylons were carried. This provided 150 B-52Ds the capacity to carry and deliver a maximum conventional bomb load of 60,000 pounds.

B-52 bombing missions in SEA were referred to as ARC LIGHT strikes. A B-52D ARC LIGHT strike usually consisted of three aircraft, referred to as a cell. Three B-52Ds could deliver a huge amount of ordnance into a designated ground target box 3,000 feet wide and 9,000 feet long. Destruction on the ground was beyond description or comprehension.

In 1967, RIVET RAMBLER was implemented to give the B-52D improved electronic warfare capabilities to counter Soviet and Chinese produced radars and SAM equipment. This was called Phase V ECM upgrade, consisting of installation of one AN/ALR-18 receiving set, one AN/ALR-20 Panoramic receiving set, one AN/ALR-25 radar homing and warning system, four AN/ALT-68 or AN/ALT-22 continuous wave jamming transmitters, two AN/ALT-16 barrage jamming systems, one AN/ALT-32H or HIGH, and one AN/ALT-32 or LOW band jamming sets. The B-52D was also equipped with active defenses: six AN/ALE flare dispensers which held 96 flares and eight AN/ALE-24 chaff dispensers holding 125 chaff bundles.

The 1972 PLACER PLANK program, called ECP-151 and referred to as the Engineering Change Proposal, modified 80 B-52Ds to maintain a credible SAC conventional bombing force beyond the Vietnam War. The program consisted of the redesign and replacement of the lower wing skin with new alloy material to reduce metal fatigue. The center panel was redesigned and replaced. PLACER PLANK added new longerons, a partial replacement of the fuselage side skin, and installation of a new pressure bulkhead in the forward section of the fuselage. The program removed 3,000 parts from each B-52D, but in the process the aircraft gained 3,400 pounds. Even with the added weight the aircraft's aeronautical drag was reduced, which increased the B-52D's speed and range. The cost of the modification per aircraft reached $2,600,000.

After the Vietnam War the B-52D returned to its primary mission of nuclear deterrence. The 43rd SW, at Andersen AFB, was equipped with the B-52D, still wearing SEA camouflage. It was armed with four B28 nuclear free fall weapons. The four nuclear weapons were pre-loaded in the secure munitions area onto a clip and slowly moved on a munitions trailer to the alert area with the B-52D, then hoisted into the forward section of the bomb bay. There was sufficient room for another weapons clip, but this was not a standard munitions load configuration. The B28 could be dropped from high or low altitude. The B28 was pulled off B-52Ds when the aircraft was removed and retired from the USAF inventory in 1982, with the B28 remaining an operational weapon until 1990.

The B-52D and other alert systems were fitted with a nuclear command and control device called Coded Switch. Two Single Integrated Operations Plan (SIOP) officers – standard two-officer security policy for handling nuclear weapons – coded the switch in their intelligence facility, transported it to the alert aircraft, and installed it into the B-52 through a side fuselage access panel, then went inside the aircraft to the navigator's position, where the system was made safe (disengaged). Once airborne, launched under nuclear orders from the President or surviving national command authority, the switch was re-enabled after receipt of a valid nuclear launch order and verification by the onboard crew, allowing the airborne crew to proceed to its designated SIOP targets and drop their weapons.

B-52Ds became an important conventional weapons delivery platform in the defense of the Republic of South Korea. Using the full bomb load capacity of the B-52D, the aircraft's bombing accuracy was enhanced by the installation of electronic beacon offset aiming points. B-52Ds were equipped to receive location signals from the ground beacons while airborne to strike their targets with a high degree of accuracy. This system was used prior to the implementation and installation of precision guided munitions. These beacons were numbered and coded positions plotted on contingency charts as part of the 43rd SW's Contingency Combat Mission Folders. The CCMFs were complete, except for the addition of current code books and weather information, which were provided by wing intelligence personnel prior to aircraft launch. Today, ACC bombers use laptops or installed computers for instant airborne updates and target information using GPS and advanced INS. B-52 crews started this update process. In this process, a single B-52D or three in a cell could be vectored without time consuming pre-mission intelligence briefings and labor intensive (at that time, early 1980s) construction of basic CCMFs.

When B-52Ds entered South Korean air space from the south, they received a coded message as to what OAPs were to be loaded into the onboard computer to drop bombs on the designated target, in all weather conditions, night or day. In this manner, the bombing capability of the B-52D was enhanced to serve closely to that of a fighter-attack aircraft.

The B-52D was also used as a long range naval reconnaissance aircraft. B-52Ds, launched from Andersen, completed Busy Observer missions to the Indian Ocean and into the Persian Gulf, tracking Soviet and Communist Block freighters during the Iran-Iraq War. These 33-hour missions required multiple air-to-air refuelings to reach the patrol area and return to Andersen. The standard aircrew was augmented by two additional pilots and one navigator, making for crowded conditions inside the aircraft. The aircrew had motorized 35mm cameras with telephoto lenses and fast film to take photographs of Soviet naval ships. Aircrew were trained in Soviet ship recognition by wing intelligence personnel and on proper camera operations (before the introduction of digital camera systems). The flight profile used three aircraft: two primary and one spare. The spare flew to the first refueling point, returning to base if not required to replace one of the primary aircraft.

The B-52D's demise wasn't exclusively caused by age. There was a change in the Cold War's atmosphere between the U.S. and Soviet Union. By 1983 all B-52Ds were retired, with the majority destroyed to meet limitations under START. Twenty-five B-52Ds were flown to museums in the United States, South Korea, Guam, and the United Kingdom. They serve as memorials to those who flew the Stratofortress from 1956 to 1983. The museum's B-52D was moved on July 12, 1996, from a hillside west of the museum to inside the museum's fenced aircraft area.

EB-57B Canberra

After the Korean War began in 1950, the Air Force wanted to acquire a turbojet powered medium bomber to replace the piston-powered Douglas B-26 Invader. In 1951, the Air Force signed a production contract to build the British designed Canberra. By 1965 two EB-57B squadrons operated in SEA. The

Air Force's B-57 was a modified version of the RAF's electronic Canberra bomber built by the Glen L. Martin Company under a licensing agreement with English Electric. The first Martin Company produced B-57 flew on July 20, 1953. The Martin Company built 403 Canberras for the Air Force. The museum's EB-57B is an excellent example of a countermeasures aircraft.

BT-13 Valiant

With the need for more pilots, the USAAC wanted to increase its pilot training program in preparation for a war in Europe against the German Air Force. In 1938, the USAAC flew the Vultee Company's BC-3 basic combat trainer. The Vultee Company took this design, creating a USAAC trainer designated as the model-74. It had fixed landing gear, and a 450 hp engine that in 1939 gained a USAAC contract for 300 aircraft, redesignated as the B-13 Valiant. In 1939, this was the largest USAAC basic trainer contract at that time, the reason being war was looming on the horizon in Europe and the Pacific.

Although not an aerodynamically shaped trainer, the aircraft was rugged, and an ideal basic trainer. It was a low-wing, cantilever monoplane of all-metal construction, but with fabric covered controlled surfaces. To make it a suitable trainer able to sustain repeated take-offs and landings, the aircraft had oleo-pneumatic shock struts and was provided with a steerable tail wheel and hydraulic brakes.

The Valiant was the basic trainer most extensively used by the USAAF during WWII. It was used as a basic flight aircraft trainer. The aircraft was an advanced aircraft for the student pilot, equipped with a two-way radio for communications with ground controllers during flight operations around the training airfield. The student had to operate landing flaps to slow the aircraft for landing, as well as the two-position variable pitch propeller. By the end of WWII over 10,375 BT-13/BT-15 (R-97) engines had been built.

The student flew in the front (with the instructor behind) under a continuous transparent canopy, with dual controls and equipped for blind flying instructions and operations. By the end of 1944 over 11,000 of these rugged basic trainers had been produced, training thousands of military pilots.

C-45 Expeditor

The Beech C-45 was a light, twin-engine utility transport. The aircraft was developed from the Beechcraft Model 18 commercial transport. Between 1939 and 1945 Beech produced over 5,500 aircraft. The C-45 had two pilots, and could carry a maximum of six passengers or two litter-borne medical patients into and out of rough, short, unprepared landing strips. Although its designation started as C-45, it was changed to UC-45 in 1943. In 1942, the U.S. Army ordered 225 C-45s and 1,100 C-45F transports, with a total of 4,000 variants produced during the war. In 1951, the USAF ordered 900 of these to serve as a light utility aircraft. It flew during the Korean War, and was also used as a staff transport. Under designation C-45H, it entered Air Force service in 1950, serving through the 1960s and compromising a fleet of 432 rebuilt C-45s. SAC flew the aircraft for missile support until 1966. Beech produced the aircraft until 1970.

C-47 Skytrain

The twin-engine transport was developed from the Douglas OC-3 commercial passenger airliner of 1936. The USAAC ordered production under the designation C-47 for cargo transport. It was the

308

first U.S. military transport, with the first prototype flying on December 1935. It was designed to carry 28 paratroopers or 14 stretcher cases and three medical personnel, or 10,000 pounds of cargo, including oversized equipment, which could be loaded through a large cargo door positioned on the left side of the fuselage, aft of the training wing edge. It could tow troop-carrying gliders, WACO CG-4A. By the end of WWII production reached over 9,400. After the end of WWII, the USAF retained a large number of C-47s in service. Fortunately, they were available for transport duty during the Berlin Airlift and throughout the world. It was used during the Korean War in moving supplies, dropping paratroopers, evacuating wounded, and making headlines during the Marines' withdrawal from the Chosin Reservoir in North Korea after an attack by thousands of Communist Chinese troops in bitter cold temperatures. During the Vietnam War, the C-47 carried on its role as a supply transport and under modifications as an airborne gunship with the nickname "Puff the Magic Dragon," able to deliver devastating fire on Communist troops that were attacking U.S. or South Vietnamese Army troops. It was the workhorse of the post-Air Force military before being replaced by modern aircraft.

Douglas C-54 Skymaster

The Douglas C-54 was developed from the Douglas Aircraft Company's DC-4E passenger airliner, which began its development in 1933 for American Airlines, Eastern Airlines, Pan American Airways, Trans World Airways, and United Airlines. As with the C-47, the four-engine DC-4E was an ideal long-range military transport. The USAAC ordered the transport under the designation C-54A. Douglas engineers modified their design to accommodate additional fuel tanks to increase range to 3,000 miles, and it was able to carry a maximum of 33 passengers or paratroopers, or 9,000 pounds of cargo. Modifications to the aircraft included a large cargo door as on the C-47, beefed up cargo floor, and an interior cargo boom hoist. It was an ideal transport for use in the CBI to move supplies into Chinese western air bases from bases in eastern India. It was used during the Korean War and Berlin Airlift to

move supplies into the isolated city surrounded by Russian military forces. During WWII over 1,200 were built. The Air Force operated the C-54 until the early 1970s. The C-54G, the final variant, could carry 50 paratroops in seats or 32,500 pounds of cargo. It had an increased range of 4,000 miles, carrying 10,000 pounds of cargo.

C-131 Samaritan

The Convair C-131 military version of the Convair CV-240 was used by the Air Force as an air medical evacuation and VIP transport. It also carried the designation T-29 for training Air Force navigation. The C-131 began being removed from the Air Force inventory in the late 1970s, replaced by the C-9 Nightingale. The C-131A was fitted with large cargo doors for up to 27 stretchers, or for 37 sitting wounded. It could carry cargo under the designation VC-131H as a VIP aircraft. During Vietnam a few aircraft were modified as side firing gunships. Air National Guard aircrew flew the aircraft until the mid-1980s.

EC-135A Stratotanker

This aircraft on display at the museum was assigned to the 4th Airborne Command and Control Squadron. The EC-135A was a variant of the KC-135A Stratotanker. It was modified with internal stations for controllers and external antennas for increased air to ground communications. For long-range endurance, the aircraft was equipped for aerial refueling with a top fuselage mounted fuel receiver to take on fuel from a tanker's flying boom. Its maximum flight endurance was 72 hours, based on engine oil

309

consumption and fuel availability. EC-135As were still fitted with a flying boom. This gave the aircraft an interesting feature, as if for some reason it could not take on fuel through the receiver, its boom could be connected to a tanker with its receiver, pulling in fuel through the flying boom. The original J57 engines were replaced by JTBD/TF33 turbofan engines for an unrefueled range of 5,000 miles. From April 1, 1973, to September 1, 1991, the 4th ACCS was attached to the 28th BW, primarily to perform airborne launch functions for Minuteman ICBMs if they could not be launched from the underground launch centers. Only six KC-135As were converted to EC-135As.

Republic F-84 Thunderstreak

The Republic F-84F Thunderstreak evolved from the straight wing F-84. The first prototype flew on June 3, 1950, with the Air Force receiving the first operational aircraft for Tactical Air Command in 1954. The aircraft was assigned to perform in the role as a fighter-bomber. This was a large production run: Republic Aircraft produced 2,112 and General Motors another 599 for the Air Force. The U.S. offered NATO Air Forces the option to purchase 1,300 of this total production run. It was a subsonic aircraft and cheap to produce, with a 1954 cost averaging $800,000. The F-84F was replaced by the North American F-100. The F-84F was transferred to the Air National Guard. But during the Berlin Airlift, 200 were pulled back into active duty for forward deployment to West Germany in November 1961.

North American F-86A Sabre

The North American F-86A was the Air Force's first swept wing turbojet fighter, and was copied from the WWII German Air Force twin-turbojet fighter, the Me 262. The technology was also copied by the Soviet Air Force for its own turbojet, swept-wing fighter, the MiG-15. The F-86A first flew on October 1, 1947. Series production began on May 20, 1948. It was originally used as a high-altitude fighter during the Korean War, engaging Russian built MiG-15s. The Russians provided the fighter to the North Korean Air Force, as well as to the Communist Chinese Air Force in large numbers. Russian pilots also flew the MiG-15 against USAF F-86As, but the F-86A force was outnumbered during combat. During the Korean War, many pilots walked through an oriental style gate with a sign hanging below that said "MiG Alley 200 miles." MiG-15s were a great threat to the piston powered B-29s. The straight wing turbojet F-80s were not capable of operating against the MiG-15s on a one-to-one basis, but their pilots still were able to shoot down a few MiG-15s. F-86s were hard pressed to keep large numbers of MiG-15s from getting to the slow and vulnerable B-29s. B-29 aircrew fought on, taking heavy losses from attacking MiG-15s and forcing the FEAF to switch to night bombing operations. The air war between F-86s and MiG-15s was the first turbojet versus turbojet aerial combat.

North American F-101 Voodoo

The F-101 was conceived as an escort for long-range SAC bombers, such as the B-36 Peacemaker. But with the introduction of the B-52, escort fighters were no longer part of SAC's inventory, so the F-101 became a tactical air defense fighter. It took until May 1957 to become operational. Production ended in March 1961, with a total of 785 produced. In a reconnaissance variant, the RF-101C became the world's first supersonic photo-reconnaissance aircraft. It gained fame during the 1962 Cuban Missile Crisis and during high-speed, unarmed flights over heavily defended North Vietnam.

310

Republic F-105B Thunderchief

The Republic F-105 Thunderchief on display at the museum (F-105B-20-RE) was formerly assigned to the 4th Tactical Fighter Wing, Seymour Johnson AFB, North Carolina. The author's brother-in-law flew F-105s with the the 4th TFW, and was killed during an airborne training mission because of a missile malfunction. So the F-105 has great significance to the author. The Air Force began design work on a new fighter during the Korean War, and an advanced, long-range, supersonic, single-seat aircraft capable of delivering tactical nuclear weapons. Republic Aviation Corporation provided the Air Force with a heavy fighter. The first operational fighter was delivered to the 4th TFW on May 26, 1958. The first group of F-105s in Vietnam took part in Rolling Thunder on March 2, 1965. During the Vietnam War (1965-1971), F-105 combat losses were 321, with 61 lost to accidents. Total production run for the F-105 reached 833.

On October 17, 1980, the Air Force Logistic Center at Hill Air Force Base authorized nine AFRS/ANG F-105Bs for transfer to the National Museum of the United States Air Force for loan to U.S. aviation museums as static display aircraft. The museum received one of these aircraft. On October 22, 1980, an aircraft was scheduled for transfer and delivery by the 508th Tactical Fighter Group, Hill AFB, to the 28th BW. On December 11, 1980, the aircraft, with 3,839 flying hours, was set for transfer to the museum. The aircraft was placed on display in December 1982.

General Dynamics FB-111 Aardvark

The F-111 aircraft was designed by General Dynamics for SAC as a medium-range nuclear bomber with supersonic speed and terrain-following capabilities to fly under Soviet radars and missile batteries. As a swept-wing bomber it had a wide range of capabilities. It was modified from the F-111, with wings extended by 3 1/2 feet to 70 feet, and provided with a stronger landing gear, increased fuel capacity, updated electronics, and two Pratt & Whitney TF-30-P turbofan engines. The pilot and navigator-bombardier sat side-by-side. General Dynamics produced 76 FB-111s between 1968 and 1971. The bomber had an internal bomb bay that could hold (2) B43 or B61 or B83 nuclear weapons. The aircraft also had three under wing weapons pylons on each wing, capable of holding 31,500 pounds: (4) SRAMs or (24) 750 MK-117 bombs. The aircraft remained operational with SAC until 1990.

Bell H-13 Sioux helicopter

The H-13 was a light observation helicopter used as an observation aircraft early in the Vietnam War. It is classified as a three seat observation and training helicopter. It was also used as a medical evacuation helicopter. It is fitted with two blades for lift and powered by a single engine. With SAC it was used to support ICBM sites, beginning in 1967. At Ellsworth AFB, the helicopter was used to support missile sites around the base. It provided emergency medical evacuation for missile site support, crucial in all weather conditions.

Sentinel L-5

The L-5 Sentinel was the military production variant of the Stinson 105 Voyager. The U.S. Army

311

began ordering L-5s in 1942. Initially its military designation was O-62, then changed to the "L" for liaison designation in April 1942. During its war production over 3,600 were produced. The L-5 was able to conduct short and rough field operations. It became a light reconnaissance aircraft, a medical evacuation aircraft, and a supply aircraft that carried supplies into isolated locations. It could fly low and slow, unreeling communications wire from one military outpost to a command/artillery post. It could be used to spot enemy locations. It could land military personnel for support to outlying military installations, even landing on country roads near EAFB missile sites. The aircraft remained in service with the Air Force until 1955.

Cessna O-2A Forward Air Control aircraft

The O-2 was the military variant of the Cessna Model 337 Super Skymaster. It had twin tail booms with a pusher-puller set of engines. In 1966, the Air Force ordered production to perform forward air controller (FAC) duties in SEA. It was a survival aircraft with its two engine power plant configuration, taking Communist ground fire and still making it safely back to base. The pilot marked enemy targets with smoke rockets, communicating directly with orbiting aircraft to coordinate ground attack missions.

Lockheed Shooting Star

The Lockheed T-33A Shooting Star was a two-place turbojet designed for training pilots who passed or completed basic propeller driven aircraft flight training. It was a modification from Lockheed's single-seat P-80 fighter by the lengthening of the fuselage by three feet for the second pilot seat for the instructor. The aircraft first flew in 1948, with production continuing until 1954, and over 5,790 produced. Besides the Air Force, over twenty nations operated the T-33.

Northrop T-38 Talon

The Northrop T-38 Talon is a twin-engine, high-altitude, supersonic turbojet trainer used in a variety of roles because of its design, economy of operations, ease of maintenance, high performance, and exceptional safety record. The Talon first flew in 1959, and more than 1,100 were delivered to the Air Force between 1961 and 1972, when production ended. As the T-38 fleet aged, specific airframe, engine, and system components were modified and replaced. PACER CLASSIC is the name given to the aircraft's sustainment program, which integrates essential modifications and includes major structural replacement into one process. ATC began receiving T-38C models in 2001 as part of the Avionics Upgrade Program. T-38C models underwent a propulsion modernization program, replacing major or engine reliability and maintainability, and an engine inlet/injector modification to increase available take-off thrust. These upgrades and modifications with the PACER CLASSIC program should extend the service life of T-38s to 2020.

The T-38 has swept wings, a streamlined fuselage, and tricycle landing gear with a steerable nose wheel. Two independent hydraulic systems power the ailerons, rudder, and flight control surfaces, along with critical aircraft components that are waist high and can be easily reached by maintenance crews. The T-38C incorporates a "glass cockpit" with integrated avionics displays, heads-up display, and an electronic "no drop bomb" scoring system. The AT-38B has a gun sight and practice bomb dispenser.

312

The T-38 needs as little as 2,300 feet to take-off and can climb to 30,000-feet in 60 seconds. The T-38 modernization program updates the propulsion system, adding 19 percent more thrust and reducing take-off distance to 2,100-feet. The instructor and student pilot sit in tandem on rocket-powered ejection seats in a pressurized, air-conditioned cockpit.

Cessna U-3 Blue Canoe

The aircraft was the military variant of the Cessna 310 twin piston-engined light transport the Air Force used as a light administration liaison, cargo, and utility transport. The Air Force purchased 160 upgraded to the U-3B, with all-weather capabilities fitted and more powerful engines, providing the aircraft with a longer nose, swept vertical fin, and additional cabin windows for VIP transport.

Beechcraft U-8 Seminole/RU-8 Seminole

The aircraft was the military variant of the Beechcraft L-23 Seminole. The twin-engine light transport and liaison aircraft entered U.S. military service in 1953, and was operational for nearly 40 years until 1992. It served primarily as a U.S. Army aircraft, with only one delivered to the Air Force. The aircraft was used during the Korean and Vietnam wars.

Bell UH-1P Iroquois helicopter

The UH-1 evolved from a 1955 U.S. Army competition for a new utility helicopter. The U.S. Army employed it in various roles, including as an armed escort or attack gunship. The Air Force extensively used the helicopter, which was ideal for supporting SAC's widely separated ICBM silos and launch control facilities. These helicopters started operational duty with the Air Force in 1963. It continues to be used, but has been replaced by the Blackhawk helicopter in support and liaison operations.

F-100D on display at the museum was flown by the South Dakota Air National Guard.

The Air Force wanted a high-performance fighter to replace the Korean War era F-86 and to counter follow-on, next generation Russian fighters (MiG-17 and MiG-21). The F-100A was the world's first production fighter able to break the speed of sound in level flight prior to the deployment of the Russian Air Force MiG-21. The first XF-100A flew on May 25, 1953, at Edwards Flight Test Center, Edwards AFB, California. The F-101A went into series production and immediately suffered problems. It had an accident rate higher than other USAF fighters, which North American engineers determined was due to a short tail fin creating roll problems. Some pilots with as many as 100 flight hours in the aircraft had problems flying the F-100. Engineers increased the tail fin height and lengthened the wings to create the F-100C. The F-100D was redesigned to incorporate broad, slotted landing flats and a taller vertical tail that also incorporated a flaring for dumping fuel, as well as installation of a radar warning receiver antenna. The F-100A was a day fighter. The USAF Thunderbirds were equipped with the F-100C in September 1955.

Once the 200 F-100As had their tail fins lengthened the aircraft became popular with its pilots. The F-100A was a pure fighter. The Air Force retired its F-100s in 1972, but the Air Force National Guard flew them until 1980. Over 2,300 of the aircraft

313

were produced for the U.S. Air Force and foreign air forces. From 1964 to 1974, F-100Ds were heavily used during the Vietnam War as a ground attack aircraft, beginning on June 9, 1964, and operating out of Da Nang Air Base, attacking targets in the Plaines des Jarres in Laos. The F-100D flew 360,290 strike sorties, losing 186 aircraft to enemy ground fire and mechanical problems.

Nike-Ajax surface-to-air missile.

In the 1950s, U.S. National Intelligence agencies believed the Soviet Long-Range Air (LRA) Force was developing a four-engine turbojet bomber in the class of SAC's B-52 to carry atomic weapons to reach targets in the Continental United States. The Nike Ajax SAM system was designed to be used as a weapon of last resort for air defense around important U.S. military installations, primarily SAC bomber bases, such as Ellsworth AFB. Ellsworth AFB was defended by four Nike Ajax SAM sites; later three were inactivated and one was converted to configuration as a Nike Hercules SAM site.

Minuteman II LGM-30F

The 44th Missile Wing operated the Minuteman I and II ICBMs. The MMII was distributed throughout three missile squadrons, five flights per squadron, with 10 missiles per flight. This gave EAFB a total of 150 ICBMS. Beginning in 1971, EAFB transitioned from the MMI to the MMII. The MMII (LGM-30F) was an improved ICBM. It could lift the heavier MK II W-56 thermonuclear warhead with a yield of two

megatons at a range of 7,000 miles. ICBMs remained on alert at EAFB until an announcement by the Secretary of Defense in February 1991 to remove the missiles, with the MMIII assuming the nation's ground based nuclear alert at other bases. All 150 missiles and their 15 Launch Control Facilities were deactivated by April 1994. The 44th MW was deactivated on July 4, 1994.

Hound Dog GAM-77/AGM-28 air to surface missile.

The latest addition to the museum (as of 2012) is a restored Hound Dog ASM, dedicated on May 7, 2011. It was carried on EAFB assigned B-52Gs and B-52Hs on alert. It was classified as an air-to-surface missile in SAC's inventory from 1961 to 1976. The B-52 could use the missile's engine for auxiliary power on take-off or to improve airborne performance. Because of a direct fuel connection to the B-52's fuel tanks, it could be refueled (internal tanks topped off in flight) prior to launch. The missile was equipped with an INS, which was updated prior to launch from the B-52. With terrain following capabilities and active anti-radar technology, it was a capable missile, flying either high or low altitude penetrations to target. The missile carried a one megaton, W28 thermonuclear warhead.

This small winged aircraft is the McDonnell GAM-72/ADM-20 Quail. It was developed as a decoy missile to be dropped from penetrating SAC B-52s to confuse Soviet radars as to which was a real or decoy penetrating bomber. Each B-52 could carry two Quails in their aft bomb bay area, due to its wings being folded, without reducing nuclear weapons payload. The decoy first flew in 1958. SAC B-52s were equipped with the Quail from 1960 until 1978. Its modest range of 265 miles made it a close to target penetration aid. The quail was designed to allow the B-52 to reach heavily defended targets inside the Soviet Union.

314

Diorama of a Predator that EAFB will control from a dedicated facility on base. At this time, no Predators are to be physically flown from the base, but instead will be flown from forward, overseas operating locations.

Stratobowl flight exhibit

B-1B F-101 engine.

Southwest Asia exhibit

B-1B flight simulator.

One of many aviation artworks on display at the museum.

MMII missile procedures trainer with emblems assigned to the 44th MW.

The museum's gift shop offers a wide variety of aviation memorabilia, books, clothes, aviation art, posters, and models.

315

APPENDIX I: ELLSWORTH COMMANDERS

Ellsworth AFB Commander's Name	Date of Command on the Installation
Lt. Col. William H. Crom	1 February 1940 to 12 February 1940
Major Lotha A. Smith	12 February 1940 to 1 September 1940
Major William O. Eareckson	1 September 1940 to 21 October 1940
Captain (Major) Donald E. Titus	21 October 1940 to 27 May 1941
Major Norman D. Sillin	7 November 1941 to 1 January 1942
Major (Lt. Col.) William O. Eareckson	1 January 1942 to 23 January 1943
Col. Earl H. Deford	23 January 1943 to 19 March 1943
Major Robert Corth	19 March 1943 to 27 March 1943
Lt. Col. Jack N. Donohew	27 March 1943 to 18 April 1943
Lt. Col. Ralph W. Rodieck	18 April 1943 to 27 October 1943
Lt. Col. John W. Massion	27 October 1943 to 4 January 1944
Lt. Col. Alexander W. Bryant	4 January 1944 to 1 April 1944
Col. Robert H. Herman	1 April 1944 to 24 July 1944
Lt. Col. John W. Massion	24 July 1944 to 26 August 1944
Col. Robert H. Herman	26 August 1944 to 26 May 1945
Lt. Col. John C. Larson	26 May 1945 to 27 June 1945
Col. Robert H. Herman	27 June 1945 to 21 July 1945
Lt. Col. Walter L. Wheeler	21 July 1945 to 27 September 1945
Lt. Col. John C. Larson	27 September 1945 to 20 October 1945
Col. Richard M. Montgomery	4 August 1946 to 15 August 1946
Col. Thomas J, Gent, Jr.	15 August 1946 to 17 September 1947
Col. Raymond L. Curtice	17 September 1947 to 28 September 1947
Lt. Col. Everett W. Holstrom	28 September 1947 to 9 October 1947
Col. Raymond L. Curtice	9 October 1947 to 15 October 1947
Col. Thomas J. Gent, Jr.	15 October 1947 to 10 July 1948
Col. William W. Jones	10 July 1948 to 19 August 1948
Col. Albert T. Wilson	19 August 1948 to 15 November 1950
Brig. Gen. Richard E. Ellsworth	15 November 1950 to 18 March 1953
Col. Howard C. Moore	18 March 1953 to 30 April 1953
Col. Areil W. Neilson	30 April 1953 to 15 June 1955
Col. Neil D. Van Sickle	15 June 1955 to 20 April 1957
Col. Loran D. Briggs	20 April 1957 to 6 July 1959
Col. Francis W. Nye	6 July 1959 to June 1962
Col. Allen B. Rowlett	June 1962 to 1 July 1962
Col. Leo C. Lewis	1 July 1962 to 10 July 1965
Col. Albert H. Schneider	10 July 1965 to 27 December 1966
Col. Edgar S. Harris, Jr.	27 December 1966 to 4 May 1967
Col. Albert H. Schneider	4 May 1967 to 2 June 1967
Col. Edgar Harris, Jr.	2 June 1967 to 24 July 1968
Col. James Wagner	24 July 1968 to 1 February 1970
Col. Richard L. Lawson	1 February 1970 to 18 June 1971
Col. Junior B. Reed	18 June 1971 to 30 May 1973
Col. Harold B. Coffee	30 May 1973 to 8 April 1974
Col. James E. Light, Jr.	8 April 1974 to 21 April 1976
Col. Harold J.M. Williams	21 April 1976 to 11 May 1976
Col. Clinton H. Winne, Jr.	11 May 1976 to 24 April 1978
Col. Clarence R. Autrey	24 April 1978 to 5 March 1979
Col. Alan B. Renshaw	5 March 1979 to 18 August 1980
Col. Wayne W. Lambert	18 August 1980 to 26 October 1981
Col. Robert F. Durkin	26 October 1981 to 20 May 1983
Col. Carl B. Krueger	20 May 1983 to 28 August 1984
Col. Steven W. Nielsen	28 August 1984 to 21 March 1986
Col. Harold B. Adams	21 March 1986 to 15 April 1988
Col. Thomas A. Twomey	15 April 1988 to 4 April 1990
Col. Charles R. Henderson	4 April 1990 to 25 November 1991
Col. Edgar A. Ott	25 November 1991 to 22 January 1992
Brig. Gen. J.C. Wilson	22 January 1992 to 30 November 1994
Brig. Gen. Robert C. Hinson	30 November 1994 to 22 August 1995
Col. Leroy Barnidge, Jr.	22 August 1995 to 19 February 1997
Col. William M. Frazer III	19 February 1997 to 27 May 1998
Col. Anthony F. Przybyslawski	27 May 1998 to May 2000
Col. Edward Rice	May 2000 to April 2002
Col. James Kowalski	April 2002 to February 2004
Col. Joseph Brown IV	February 2004 to June 2005
Col. Jeffrey Smith	June 2005 to June 2007
Col. Scott Vander Hamm	June 2007 to 5 June 2009
Col. Jeffrey Taliaferro	5 June 2009 to 13 May 2011
Col. Mark Weatherington	13 May 2011------------------[11]

END NOTES

Chapter 1

1. "Setting the Stage, 1941-1947, Site Selection and Constriction," Ellsworth Air Force Base, South Dakota: 28th Bomb Wing, Historian.

2. "Completion Report, Rapid City Army Air Base, Rapid City, South Dakota, Job No. Rapid City A-2," completed September 30, 1942. Rapid City, South Dakota: U.S. Engineer Office, October 30, 1942.

3. "South Dakota World War II Army Airfields," Pierre, South Dakota: South Dakota Air National Guard Museum.

4. "Completion Report."

5. "Rapid City, S.D., Names Site $8,500,000 Army Air Base; Bomb Range near Badlands," Rapid City, S.D., The Rapid City Daily Journal, Wednesday, December 10, 1941.

6. "Air Base Planes Heard by Cosmos," Rapid City, South Dakota: The Rapid City Daily Journal, January 12, 1942.

7. "Plans Progress for Air Base," Rapid City, South Dakota: The Rapid City Daily Journal, January 5, 1942.

8. "Inland to Use Rapid City Port," Rapid City, South Dakota: The Rapid City Daily Journal, February 20, 1942.

9. "Commissioners Told Governor the City Will Take Over Rapid City Army Air Base," Rapid City, South Dakota: The Rapid City Daily Journal, February 25, 1942, 1-2.

10. U.S. Army Corps of Engineers, Defense Environmental Restoration Program for Formerly Used Defense Sites, Ordnance and Explosive, Archives Search Report, Armstrong County Air-to-Air Gunnery range, Dewey and Sully Counties, SD. Huntsville, Alabama: Huntsville Engineering Support Center.

11. "Air Force Joins Tribe in Bombing Range Cleanup," Rapid City, South Dakota: Rapid City Journal, June 2006, A1 & A8.

12. "Commissioners Told Governor the City Will Take Over Rapid City Army Air Base."

13. "Completion Report"

14. "Air Base Work Increased here," Rapid City, South Dakota: The Rapid City Daily Journal, February 5, 1942, page 2.

15. "Rail Spur Fence Bid Opening Set-Well is now down 3,000 Feet." Rapid City, South Dakota: The Rapid City Daily Journal, March 4, 1942, 1.

16. "Contracts of Local Airbase Announced," Rapid City, South Dakota: The Rapid City Daily Journal, March 30, 1942.

17. "Local Airbase Still Unnamed-A Special Board has that Job," Rapid City, South Dakota: The Rapid City Daily Journal, April 3, 1942, 2.

18. "Airbase Radio Station Named," Rapid City, South Dakota: The Rapid City Daily Journal, May 7, 1942, 2.

19. "Veteran Air Officer, Once Buck Private in Army, Taking Command of Rapid City Airbase," Rapid City, South Dakota: The Rapid City Daily Journal, June 2, 1942, 2.

20. "Paving Underway as Rapid City Air Base Receives Official Title," Rapid City, South Dakota: The Rapid City Daily Journal, June 5, 1942.

21. John F. Welch, RB-36 Days at Rapid City or Rapid City Free Style, The People, The Airplanes and the Times at Rapid City/Ellsworth Air Force Base, Facilities (Rapid City, SD.: Silver Wings Aviation, Inc., 1944), 27-28.

Chapter 2

1. "Airbase Vanguard Arriving in City," Rapid City, South Dakota: The Rapid City Daily Journal, July 2, 1942, 2.

2. Thomas Manning, History of Air Education and Training Command, 1942-2002 (San Antonio, Texas: Office of History and Research, Headquarters Air Education Training Command, 2005.

3. Captain George A. Larson, USAF, B-17 Flying Fortress, World War II Aerial Tactics," academic paper for United States Air Force Air Command & Staff College, non-resident program, Maxwell Air Force Base, Alabama, June 1973.

4. "Boeing B-17F and Boeing B-17G," Dayton, Ohio: Wright Patterson Air Force Base, National Museum of the United States Air Force.

5. 303rd Bombardment Group (H), "8th Air Force Bomber Crew Composition."

6. 303rd Bombardment Group (H), "Duties and responsibilities of the Airplane Commander."

7. Wesley Frank Craven and James Lea Cate, Ed., The Army Air Forces in World War II, Volume 6, Men and Planes, Chapter 18, "Combat Crew and Unit Training," (Washington, DC: Office of Air Force History, 1983), 606-607.

8. 303rd Bombardment Group (H).

9. Craven

10. 303rd Bombardment Group (H).

11. 303rd Bombardment Group (H), "Requirements for Navigators on or before a mission," Headquarters, England, October 15, 1944.

12. 303rd Bombardment Group (H), "Duties and Responsibilities of the Navigator," Headquarters, England, October 15, 1944.

13. Airmen 1st Class Alessandria N. Hurley, "Historian Opens Door to Ellsworth Past, Future," Ellsworth Air Force Base, 28th Bomb Wing Public Affairs, February 16, 2011.

14. The following notation from Paul Marcello's Master thesis at the University of Wisconsin on the World War II participation of his grandfather, Alfred Marcello. Grant A. Fuller to Paul J. Marcello, October 20, 2004.

15. The following notation from Paul Marcello's Master thesis at the University of Wisconsin on the World War II participation of his grandfather, Alfred Marcello. Arthur H. Juhlin to Paul J. Marcello, September 15, 2004.

16. The following notation from Paul Marcello's Master thesis at the University of Wisconsin on the World War II participation of his grandfather, Alfred Marcello. Grant A Fuller to Paul J. Marcello, December 10, 1990.

17. Lt. Col. George A. Larson, USAF (Ret.), "The Shuttle Bombers, B-17s and B-24s in Russia," Combat Aircraft, Volume 4, Number 1, April-May 2002, 80-86.

18. "Mission Diary of Art Juhlin, Navigator, lead crew of Donald A. Jones", http://www.100thbg.com

19. Larson

20. "Mission Diary."

21. 303rd Bombardment Group (H), "Oxygen and equipment," Headquarters, England, October 15, 1944.

22. 303rd Bombardment Group (H), "Clothing, parachutes, flak suits & life vests," Headquarters, England, October 15, 1944.

23. The Army Air Forces in World War II, Volume 6, Men and Planes, "Bombardier Training," (Washington, DC: United States Army Air Forces Historical Division, 1955.

24. 303rd Bombardment Group (H), "Duties and Responsibilities of the Bombardier," Headquarters, England, October 15, 1944.

25. 303rd Bombardment Group (H), "Operational duties of Bombardier," Headquarters, England, October 15, 1944.

26. 303rd Bombardment Group (H), "Duties and Responsibilities of the Engineer," Headquarters, England, October 15, 1944.

27. 303rd Bombardment Group (H), "Duties and Responsibilities of the Radio Operator," Headquarters, England, October 15, 1944.

28. 303rd Bombardment Group Heavy (H), "B-17 Crew Requirements and Standard Operating Procedures, Duties and Responsibilities of the Airplane Commander,

Duties and Responsibilities of the Gunner," *B-17 Pilot Training Manual (1943)*, Headquarters, England, October 15, 1944.

29. 303rd Bombardment Group (H), "B-17 Crew Requirements and Standard Operating Procedures, Duties and Responsibilities of the Airplane Commander, Combat Gunners," Headquarters, England, October 15, 1944.

30. 303rd Bombardment Group (H), "B-17 Crew Requirements and Standard Operating Procedures, Duties and Responsibilities of the Airplane Commander, Aerial Observers," Headquarters, England, October 15, 1944.

31. "Getting Ready, July 1942-April 1943, 96th Bombardment Group (H)," http://96bg.org/96hist.html

32. Mauer Mauer, Ed. *Air Force Combat Units of World War II*, "88th Bombardment Group," (Washington, DC, Office of Air Force History, 1983), 154.

33. Mauer Mauer, Ed. *Air Force Combat Units of World War II*, "383rd Bombardment Group," (Washington, DC, Office of Air Force History, 1983), 270.

34. Mauer Mauer, Ed. *Air Force Combat Units of World War II*, "95th Bombardment Group," (Washington, DC, Office of Air Force History, 1983), 163.

35. Mauer Mauer, Ed. *Air Force Combat Units of World War II*, "452nd Bombardment Group," (Washington, DC, Office of Air Force History, 1983), 326.

36. Mauer Mauer, Ed. *Air Force Combat Units of World War II*, "398th Bombardment Group," (Washington, DC, Office of Air Force History, 1983), 284.

37. "B-17F crash, 398th Bombardment Group (H), 603rd Bombardment Squadron, Rapid City Army Air Field, January 17, 1944," Pierre, SD: South Dakota State Historical Society.

38. Mauer Mauer, Ed. *Air Force Combat Units of World War II*, "447th Bombardment Group," (Washington, DC, Office of Air Force History, 1983), 321.

39. Mauer Mauer, Ed. *Air Force Combat Units of World War II*, "457th Bombardment Group," (Washington, DC, Office of Air Force History, 1983), 332.

40. "World War II Airmen return to their Training Site," Ellsworth Air Force Base, 28th Bomb Wing Public Affairs.

41. Mauer Mauer, Ed. *Air Force Combat Units of World War II*, "463rd Bombardment Group," (Washington, DC, Office of Air Force History, 1983), 338.

42. "463rd Bombardment Group (H) Heavy," http://www.463rd.org/history.htm.

Chapter 3

1. "58th Reconnaissance Squadron (Weather)," Ellsworth Air Force Base, 28th Bomb Wing Historian.

2. "Ellsworth Air Force Base," Ellsworth Air Force Base, 28th Bomb Wing Public Affairs.

3. Stephen L. McFarland, *A Concise History of the U.S. Air Force* (Air Force Fiftieth Anniversary, Commemorative Edition) (Washington, DC: Air Force History and Museum's Program, 1997), 40-41.

4. "Brief History of the 28th Bomb Wing," Ellsworth Air Force Base, 28th Bomb Wing Historian.

5. J.C. Hopkins and Sheldon A. Goldberg, *The Development of the Strategic Air Command, 1946-1986* (The Fortieth Anniversary History), Offutt Air Force Base, Nebraska: Office of the Historian, Headquarters Strategic Air Command, September 1, 1986, v.

6. Lt. Col. George A. Larson, USAF (Ret.), *The Superfortress and Its Final Glory, The Korean Air War, The Cold War's First Aerial Combat* (Bloomington, IN: Xlibris Corporation, 2010), 43-44.

7. "Brief History of the 28th Bomb Wing."

8. Larson

9. Hopkins, 1-4.

10. Lt. Col. George A. Larson, USAF (Ret.), Strategic Air Command Intelligence Officer, 1971-1981 and 1989-1992.

11. George A. Larson, "South Dakota Air and Space Museum Features Air Power from Cold War," *Aviation Heritage*, January 1994, 60.

12. Lt. Col. George A. Larson, USAF (Ret.), "The B-29 Atomic Bomb Option During the Berlin Airlift," *Friends Journal*, Volume 26, November 4, Winter 2006/2007, 37-42.

13. "Brief History of 28th Bomb Wing."

14. "South Dakota Air and Space Museum Features Air Power from Cold War."

15. "Brief History of 28th Bomb Wing."

Chapter 4

1. "Guide to Air Force Installations Worldwide, Major Active Duty Installations, Ellsworth Air Force Base, *"Air Force Magazine*, "2010 USAF Almanac," Arlington, VA.: Air Force Association, May 2010, 114.

2. George A. Larson, "Six Turning, Four Burning," *United States Air Force Year Book*, RAF Fairford, United Kingdom: The Royal Air Force Benevolent Association Fund Enterprises, Douglas Badar House, 2000), 31-34.

3. Lt. Col. George A. Larson, USAF (Ret.), "B-36 Peacemaker, America's Nuclear Big Stick," *Combat Aircraft*, June 2000, 748-754.

4. John F. Welch, *RB-36 Days at Rapid City or Rapid City Free Style, The People, The Airplanes and the Times at Rapid City/Ellsworth Air Force Base, Facilities* (Rapid City, SD.: Silver Wings Aviation, Inc., 1944), 28-31.

5. "A Brief History of Ellsworth Air Force Base."

6. Welch, 31-36.

7. "History of the 717th Bombardment Squadron," Ellsworth Air Force Base, SD.: 28th Bomb Wing Historian.

8. Welch, 149-166.

9. Ibid, 87, 209-217.

10. "How Ellsworth Air Force Base got its Name," Ellsworth Air Force Base, 28th Bomb Wing Historian.

11. "RB-36H 51-13721, Newfoundland, Canada, March 18, 1953." Synopsis of Air Force Accident Report (added material as of July 17, 2003). Montgomery, Alabama: United States Air Force Historical Research Center, Maxwell Air Force Base.

12. Raleigh H. Watson, Berryvill, Virginia. Interview, B-36 Reunion, Atwater, California: Castel Air Museum, October 13-15, 1995.

13. "RB-36H 51-13721."

14. Richard Stoker, "Burgoyne's Cove RB-36H Crash Site," *The Atlantic Canada Aviation Museum Newsletter*, Nova Scotia, Newfoundland, Summer 2005, 2-5.

15. Richard J. Stoker, St. John's Newfoundland, to Lt. Col. George A. Larson, USAF (Ret.), November 10, 2010.

16. Welch, 222-223.

17. "History of the 717th Bombardment Squadron."

18. "History of Ellsworth Air Force Base."

Chapter 5

1. Interview of members of the 3081st Aviation Group, Ellsworth Air Force Base, South Dakota, Lt. Col. George A. Larson, USAF (Ret.), August 14, 1995.

2. Information extracted from "Status of Work Items as of January 7, 1955," prepared by the office in charge, Midwest Engineering District and Construction Office, U.S. Army Corps of Engineers, Kansas City, Missouri.

3. Lamb Associates, Inc., "Site Summary, Ellsworth Air Force Base (Rushmore Air Force Station), South Dakota," Albuquerque, New Mexico: Sandia National Laboratories, 1994, 2-6 & 2-8.

4. "National Atomic Museum." Today it is named, The National Museum of Nuclear Science and Technology. Albuquerque, New Mexico (Moved outside the southeast gate to Kirkland AFB to provide more access to visitors to the museum and larger interior and exterior display areas), National Atomic Museum Foundation, 21.

5. Bill Yenne, *The History of the US Air Force* (New York: Bison Books, 1984), 157-158.

6. "National Atomic Museum," 24.

7. 3081st Aviation Group News Release, August 10, 1995," printed in the *Rapid City Journal*, Rapid City, South Dakota.

8. Environmental Protection Agency, National Oil and Hazardous Substances Pollution Contingency Plan; National Priorities List; "Notice of intent for partial deletion of the Ellsworth Air Force Base Site from the National Priorities List," Washington, DC: June 8, 2006.

9. Chuck Hanson, *U.S. Nuclear Weapons, The Secret Story* (Arlington, Texas: Aerofax, 1988), 147.

10. Interview with members of the 3081st Aviation Group.

11. Airman 1st Class Ashley New, "Airmen Compete in Load Competition," *The Patriot*, Ellsworth Air Force Base, 28th Bomb Wing Public Affairs, Volume 6, Issue 11, March 18, 2011, 1.

12. "Ellsworth Air Force Base Fact Sheet," South Dakota: Ellsworth Air Force Base, 28th Bomb Wing Public Affairs.

13. Lamb Associates, Inc., 2-5.

14. Kevin Cameron, "Taking Apart the Bomb," *Popular Science*, April 1933, 102-103.

Chapter 6

1. "Nike Missile Sites," Carlisle Barracks, Pennsylvania: United States Institute for Military History, United States Army War College.

2. "Nike Missile Sites at Ellsworth Air Force Base," Ellsworth Air Force Base, 28th Bomb Wing Historian.

End Notes

3.	http://www.nikeordnance.com/RapidCity/htm
4. "Nike Missile Sites at Ellsworth Air Force Base."
5. United States Corps of Engineers, Chicago District, "Frontline of Defense, Nike Missile Sites in Illinois," Denver, Colorado: Department of Interior, National Park Service, Rocky Mountain Support Office, 1996.

Chapter 7
1. "History of Ellsworth Air Force Base," Ellsworth Air Force Base, 28th Bomb Wing Historian.
2. George Larson, "Last of the Titans," *United States Air Force Yearbook 1998*, England, The Royal Air Force Benevolent Fund Enterprises. Bader House, RAF Fairford, 58.
3. "Titan Is at Ellsworth Air Force Base," Ellsworth Air Force Base, 28th Bomb Wing Historian.
4. "Titan I ICBM Site Construction," Denver, Colorado: Wings Over the Rockies Air & Space Museum.
5. Larson
6. "Titan II," Green Valley, Arizona: Titan II Museum.
7. Larson
8. "History of Ellsworth Air Force Base."
9. http://www.airforce/nuc.acc.htm
10. Larson
11. SAC.com

Chapter 8
1. Department of Interior, National Park Service and Department of Defense, United States Air Force, Legacy Management Program, "Minuteman Missile Sites, Management Alternatives Environmental Assessment," Denver, Colorado: National Parks Service, Rocky Mountain Region, 1995.
2. "Minuteman Missile Technology," SAC.com
3. Department of Interior.
4. "The 44th Missile Wing," Ellsworth Air Force Base, 28th Bomb Wing Historian.
5. "66th , 67th and 68th SMS," Ellsworth Air Force Base, 28th Bomb Wing Historian.
6. Major Aida E. Roig-Compton (USAFR) and MSgt Paul Gremse, "A Day in the Life of a Missileer," Ellsworth Air Force Base, 28th Bomb Wing Historian, November 14, 1997.
7. "The 44th Missile Wing."
8. "Minuteman Missile Technology."
9. "The 44th Missile Wing."
10. "History of Ellsworth Air Force Base," Ellsworth Air Force Base, 28th Bomb Wing Historian.
11. :History of Minuteman Sites," Washington, DC: National Park Service, United States Department of Interior, 2003.
12. Lt. Col. George A. Larson, USAF (Ret.), "Dedication of the Minuteman Missile National Historic Site, September 27, 2002.
13. Lt. Col. Petrov Stanislaus, retired former Soviet officer (assigned to an ICBM early warming command center), interview at the Minuteman Missile National Historic Site, D-09 and D-01, Lt. Col. George A. Larson, USAF (Ret.), May 7, 2007.

14. Association of Air Force Missileers (www.afmissilers.org).
15. Lt. Col. Stanislaus.
16. Lt. Col. George A. Larson, USAF (Ret.), "Legacy of the Cold War, Contaminated Defense Sites."

Chapter 9
1. "The 4th Airborne Command and Control Squadron (4th ACCS)," Ellsworth Air Force Base, 28th Bomb Wing Historian.
2. *The Development of Strategic Air Command (The Fortieth Anniversary History)*, "1965", Offutt Air Force Base, Nebraska: Headquarters Strategic Air Command, Command Historian, September 1, 1986, 134.
3. "28th Bombardment Wing," SAC.com
4. "EC-135A," Offutt Air Force Base, SAC Command Historian.
5. "4th ACCS."
6. "28th Bombardment Wing."
7. "EC-135A."
8. "The History of the Looking Glass," 2nd Airborne Command & Control Squadron.
9. *The Development of Strategic Air Command*.

Chapter 10
1. "A Brief History of the 77th Bomb Squadron," Ellsworth Air Force Base, 28th Bomb Wing Historian.
2. "B-52Ds, Ellsworth Air Force Base," Offutt Air Force Base, Headquarters Strategic Air Command, Command Historian Office.
3. "Accident Report of KC-135 Crash, September 10, 1962," Offutt Air Force Base, Nebraska, Strategic Air Command Safety Office, Command Historian Office.
4. "A Brief History of the 77th Bomb Squadron."
5. Lt. Col. George A. Larson, USAF (Ret.), "Linebacker II, Strategic Air Command's B-52s over North Vietnam, *Combat Aircraft, The International Journal of Military Aviation*, Volume 3, Number 6, November 2001, 565-566.
6. *The Development of the Strategic Air Command, 1946-1986 (The Fortieth Anniversary History)*, Offutt Air Force Base, Nebraska: Office of the Historian, Strategic Air Command, September 1, 1986, 180.
7. "A Brief History of the 77th Bomb Squadron."
8. "A Brief History of the 28th Bomb Wing," Ellsworth Air Force Base, 28th Bomb Wing Historian.
9. "KC-135R Stratotanker," *US Air Force Fact Sheet*, Grissom Air Force Reserve Base, 434th Air Refueling Wing Public Affairs, November 2009.
10. "Alert Facilities," Offutt Air Force Base, Headquarters Strategic Air Command, Command Historian Office, June 1990.
11. Terry Vanden-Heuvel, AMARC Public Affairs Office, "B-52 Treaty Reduction Program," Davis-Monthan Air Force Base, Phoenix, Arizona. April 4, 1997.
12. Terry Vanden-Heuvel, AMARC Public Affairs Office, "Preservation vs Storage," Davis-Monthan Air Force Base, Phoenix, Arizona. December 16, 1996.

13. Lt. Col. George A. Larson, visit to AMARC, March 12, 1997.
14. "A Brief History of the 28th Bomb Wing."

Chapter 11
1. Gordon B. Lease to Lt. Col. George A. Larson, USAF (Ret.), February 15, 2001.
2.	http://Rapidcitylibrary.org/lib_info/1972/flood
3. Lease
4. "B-1B Lancer," *U.S. Air Force Fact Sheet*, Langley Air Force Base, Virginia: Air Combat Command, Public Affairs office, 2005.
5. John Williams, DCSI Transport Maintenance, 28th Bomb Wing, Ellsworth Air Force Base, to Lt. Col. George A. Larson, USAF (Ret.), December 12, 2010.
6. "The 99th Wing," Ellsworth Air Force Base, 28th Bomb Wing Historian.
7. "Weapon System Officer," San Antonio, Texas: Randolph Air Force Base, 563rd Flying Training Squadron, Public Affairs Office.
8. "B-1B and B-52H CAT/CWD FTU Training," Hampton, Virginia: Langley Air Force Base, Air Command Combat.
9. "History of the B-1B Strategic Bomber," Bellevue, Nebraska: Offutt Air Force Base, Strategic Air Command, Command Historian Office.
10. "A Brief History of Ellsworth Air Force Base," Ellsworth Air Force Base, 28th Bomb Wing Historian.
11. "Operation deployments of the B-1B Lancer," Ellsworth Air Force Base, 28th Bomb Wing Historian.
12. "A Brief History of Ellsworth Air Force Base."
13. "Departure of the 77th BS," Ellsworth Air Force Base, 28th Bomb Wing Historian.
14 "Quick Strike mines," Newport, Rhode Island: United States Naval War College, Maritime History Department.
15. Air Force Airmen First Class Allen A. Marshall, "They're Back, START Team Arrives Again," *The Patriot*, June 16, 1990, 1.
16. "A Brief History of Ellsworth Air Force Base."
17. "The Kosovo Campaign: Airpower Made it Work, "*Air Force Association*, http://www.afa.orr/media/reports/april.asp
18. US Army General Henry H. Shelton, Chairman JCS, *ABC's This Week*, April 11, 1999.
19. Joe Lockhart, White House Spokesman, Washington, DC, White House news briefing, April 9, 1999.
20. "The Kosovo Campaign," Ellsworth Air Force Base, 28th Bomb Wing Historian.
21. "A Brief History of Ellsworth Air Force Base."
22. Lt. Col. George A. Larson, USAF (Ret.), "Mercy Mission in the Balkans, Army Air Force Exchange Service."
23. Andy Louder, Rapid City, South Dakota, interview Lt. Col. George A. Larson, USAF (Ret.), June 22, 2010.
24. "Operation Provide Promise," American Forces Press Service, news release, February 7, 1966.

25. Letter to AAFES personnel by US Army General Dennis J. Reiner, Chief of Staff, U.S. Army. Courtesy Andy Louder.

26. "A Brief History of Ellsworth Air Force Base."

27. "Operation Enduring Freedom," Ellsworth Air Force Base, 28th Bomb Wing Historian.

28. "John Thune Campaign Statements," Rapid City, South Dakota: John Thune U.S. Senator election headquarters, October October 17, 2004.

29. "Base Realignment and Closure," Washington, DC: U.S. Department of Defense, the BRAC Commission, August 27, 2005.

30. "B-1B Lancer inventory, AMARC," Tucson, Arizona: 309th Aerospace Maintenance and Regeneration Group, e-mail to Lt. Col. George A. Larson, USAF (Ret.)

31. "Base Realignment and Closure."

32. Master Sgt. Mona Ferrell, "Ellsworth B-1 Lancer Takes to Russian Sky," U.S. Air Forces in Europe News Service, August 18, 2005.

33. "B-1B Capabilities," Joint Base Langley, VA.: Air Combat Command.

34. Air Force Senior Airmen James Pyeatt, 28th Munitions Squadron, 28th BW, Ellsworth Air Force Base, interview Lt. Col. George A. Larson, USAF (Ret.), May 15, 2008.

35. Air Force Staff Sgt. Ryan Walker, 34th Aircraft Maintenance Squadron, 28th BW, Ellsworth Air Force Base, interview, Lt. Col. George A. Larson, USAF (Ret.), May 15, 2008.

36. Air Force Major Brian Witkowsky, 34th BS, 28th BW, Ellsworth Air Force Base, interview, Lt. Col. George A. Larson, USAF (Ret.), May 15, 2008.

37. Air Force Technical Sgt. Russell Wicke, "B-1B Performs as Never Envisioned after 20 Years," The Patriot, Ellsworth Air Force Base, Air Combat Command (ACC), Office of Public Affairs, April 23, 2008, Volume 3, Issue 18.

38. Witkowsky

39. Air Force Major Joseph Kleeman, 34th BS, 28th BW, interview, Ellsworth Air Force Base, interview, Lt. Col. George A. Larson, USAF (Ret.)

40. Witkowsky

41. Kleeman

42. Witkowsky

43. Kleeman

44. Witkowsky

45. Major Bob Liebman, Electronic Warfare Officer (B-1B Lancer), 28th BW, Ellsworth Air Force Base. " B-1Bs in Southwest Asia," presentation to Black Hills Veterans Writer's Group, Western Dakota Technical Institute, Saturday, December 11, 2010.

46. "Combat mission log," Major Bob Liebman to Lt. Col. George A. Larson, USAF (Ret.), December 18, 2010.

47. "Sniper Pod for B-1B," Ellsworth Air Force Base, 28th Bomb Wing Public Affairs, June 11, 2008.

48. Air Force Lt. Col. Lucien Case, Chief of Plans, 28th Bomb Wing Operations Group, Ellsworth Air Force Base, interview, Lt. Col. George A. Larson, USAF (Ret.), May 15, 2008.

49. Air Force Major Eric Upton, "A is for Airpower," The Patriot, Ellsworth Air Force Base, 28th BW Public Affairs, May 30, 2008.

50. Robert Crawford, "The Air Force Song."

51. Airman First Class Alessandra N. Hurley, "Ellsworth Satisfies ORI," Ellsworth Air Force Base, 28th Bomb Wing Public Affairs, November 8, 2010.

52. "Whiteman, Ellsworth win APA Mission," AirForcMagazine.com, March 31-April 2, 2011.

53. "Ellsworth, Whiteman to host UAV Units," Air Force Times, May 6, 2010.

54. Senior Airman Jared A. Denton, "Ellsworth activates detachment," Ellsworth Air Force Base, 28th Bomb Wing Public Affairs, April 8, 2011.

55. "Reaper UAV," Factsheet, United States Air Force.

56. Air Force Airman 1st Class Alessandra N. Hurley, "Ellsworth Airmen return home from Southwest Asia," Ellsworth Air Force Base, 28th Bomb Wing Public Affairs, July 28, 2010.

57. Airmen 1st Class Alessandra N. Hurley, "Ellsworth Practices Life-Saving Skills for upcoming ORI," The Patriot, Ellsworth Air Force Base, 28th Bomb Wing Public Affairs, August 30, 2010.

58. Airmen 1st Class Jared Denton, "Ellsworth Conducts Final Exercise before ORI," The Patriot, Ellsworth Air Force Base, 28th Bomb Wing Public Affairs, September 24, 2010.

59. "B-1B Accidents," Ellsworth Air Force Base, 28th Bomb Wing Historian.

60. "Ellsworth Remembers Lost Airmen," Ellsworth Air Force Base, 28th Bomb Wing Historian.

61. Christie Anderson, Air Combat Command Public Affairs, "ACC Commander discusses safety stand-down day," Plainsman, Ellsworth Air Force Base, 28th Bomb Wing Public Affairs, September 26, 1997.

62. "Ellsworth Remembers Lost Airmen."

63. Captain Bob Russell, 28th BW, Chief B-1B Weapons Systems Trainer, "Silent 'Aircraft' Trains Ellsworth Crews," Plainsman, Ellsworth Air force Base, 28th Bomb Wing Public Affairs, September 12, 1997.

64. Senior Airman Jennifer Blake, "Ellsworth Remembers Fallen Crew, Not Forgotten," Plainsman, Ellsworth Air Force Base, 28th Bomb Wing Public Affairs, September 26, 1997.

65. Senior Airman Jennifer Blake, "Operational Error Blamed for Sept. 19 B-1B Crash," Plainsman, Ellsworth Air Force Base, 28th Bomb Wing Public Affairs, December 12, 1997.

66. "B-1B accident, September 19, 1997, Ellsworth Air Force Base," Air Combat Command, Accident Investigation Board, December 1997.

67. "B-1 Accidents," Ellsworth Air Force Base, 28th Bomb Wing Historian.

68. Senior Airmen Jared A. Denton, "Wings of Freedom," Ellsworth Air Force Base, 28th Bomb Wing Public Affairs, January 24, 2011.

69. "Federal Budget Reduces B-1Bs," The Patriot, Ellsworth Air Force Base, 28th Bomb Wing Public Affairs, February 25, 2011.

70. Senior Airmen Jarad A. Denton, "Ellsworth Airmen Return from Libya Mission," The Patriot, Ellsworth Air Force Base, 28th Bomb Wing Public Affairs, April 8, 2011.

Chapter 13

1. "Defense Biometric Identification System," Ellsworth Air Force Base, Patriot, February 2011.

2. "Superfund Program, Ellsworth Air Force Base," Denver, Colorado, United States Environmental Protection Agency, December 8, 2010.

3. "Ellsworth Air Force Base Runway to be rebuilt," Ellsworth Air Force Base, 28th Bomb Wing, Civil Engineering Squadron, January 2011.

4. "Y-Repair Keel Section of Runway, FXBM-09-1015, Ellsworth AFB, South Dakota," http://implu.com/federal_contracts/listing/W9128F11R0014

5. "Centennial Housing on Base Transfer Delayed until 2012," Ellsworth Air Force Base, 28th Bomb Wing Public Affairs, May 11, 2011.

6. "28th Operations Group," U.S. Air Force Fact Sheet.

7. "28th Maintenance Group," U.S. Air Force Fact Sheet.

8. "28th Mission Support Group," U.S. Air Force Fact Sheet.

9. "28th Medical Group," U.S. Air Force Fact Sheet.

10. "Air Force Financial Services Center," U.S. Air Force Fact Sheet.

Appendix I

1. "Thunder Across the Plains, A Pamphlet Dedicated to the Heritage of the 28th Bomb Wing," Ellsworth Air Force Base, 28th Bomb Wing Historian.

End Notes